The 100 Inventions

Inventions

A ranking of the most important inventions in human history

Fakrudeen Ali Ahmed

Copyright © 2021 by Fakrudeen Ali Ahmed
Photographs and illustrations by Norah Fakrudeen

All rights reserved

Dedicated to my father.

Contents

Introduction — 9

PART 1 — 19

1. Cloth — 21
2. Fire — 24
3. Wheel — 27
4. Agriculture — 30
5. Writing — 33
6. Numbers — 36
7. Paper — 39
8. Book — 42
9. Clock — 45
10. Calendar — 48
11. Money — 51
12. Ruler — 54
13. Map — 57
14. Ship — 60
15. Iron metallurgy — 63
16. Knife — 66
17. Pot — 69
18. Furniture — 72
19. Toilet — 75
20. Religion — 78
21. Pen — 81
22. Letter — 84
23. Shoe — 87
24. Stove — 90
25. Road — 93

PART 2 — 97

26. Thermometer — 99
27. Steam Engine — 102
28. Railways — 105
29. Internal combustion engine — 108
30. Automobile — 111
31. Electricity — 114
32. Light bulb — 117
33. Fertilizer — 120

34. Sewing machine	123
35. Vaccine	126
36. Antibiotic	129
37. Bleach	132
38. Radio	135
39. Telegraph	138
40. Telephone	141
41. Flying shuttle	144
42. Spinning Jenny	147
43. Cotton gin	150
44. Newspaper	153
45. Glass	156
46. Anesthetic	159
47. Computer	162
48. Transistor	165
49. Internet	168
50. Compass	171

PART 3   175

51. Firearm	177
52. Nuclear weapon	180
53. Eye glasses	183
54. Microscope	186
55. Petroleum	189
56. Plastic	192
57. Rubber	195
58. Lock	198
59. Battery	201
60. Electric motor	204
61. Photography	207
62. Pasteurization	210
63. Television	213
64. Refrigerator	216
65. Airplane	219
66. Paint	222
67. Tea and Coffee	225
68. Cinema	228
69. Phonograph	231
70. Rocket	234
71. Satellite	237
72. Mobile phone	240
73. Soap	243
74. Pain killer	246
75. Telescope	249

PART 4	253
76. Alcohol	255
77. Tobacco	258
78. Bridge	261
79. Dam	264
80. Canal	267
81. Bicycle	270
82. X-ray	273
83. Cement	276
84. Elevator	279
85. Skyscraper	282
86. Dynamite	285
87. Pesticide	288
88. Safety match	291
89. Kerosene lamp	294
90. Vacuum Tube	297
91. Washing machine	300
92. GPS	303
93. Radar	306
94. RFID	309
95. Laser	312
96. Traffic light	315
97. Barcode	318
98. ATM	321
99. Tank	324
100. Zipper	327

Ten principles of human innovation	331
Table 1: When were they invented?	343
Table 2: They were invented in which countries?	344
Bibliography	347
Index	431

Introduction

When thinking about inventions I am reminded of the quote, "We wanted flying cars, instead we got 140 characters". This book aims to list 100 greatest hits of humanity in terms of inventions to introduce and motivate future generations of inventors to the true range of human inventions. The hope is they will go out and invent their own inventions of the flying cars variety instead of yet another meaningless invention like a social media or photo sharing site or inventing yet faster means of executing transactions a few milliseconds faster.

Manhattan and *Silicon Valley* is sucking up human talent to produce at best useless inventions like sharing your comment without thinking it through in 140 characters and at worst inventions like *credit default swaps* which destroys all of the economy on its way down like in the *2008 financial crisis* or dissing out conspiracy theory of your choice in an endless echo chamber threatening democracy itself in the process. This is an attempt to motivate future generations of inventors to see the full potential of humanity.

One of the books which made an impact on me as a child was *The 100* by Princeton professor Michael Hart. This book introduced the most important people in history to me and inspired me to follow in their paths and motivated me to strive to be one among them. This book takes a similar style and I hope will motivate a few future inventors. This is a book which ranks all the innovations we people invented and played a role in shaping humanity itself. This book ranks innovation based on impact to human cultural evolution irrespective of whether it was purely positive. I have tried to give the rankings rational justification as much as possible, particularly by comparing an invention with its closest competitors and why it is ranked in a particular place relative to them.

Think about the collective loss to humanity if Jonas Salk went to work on financial innovations for Goldman Sachs or advertising optimization for Google in the 1950s. How many of us know Robert Cochrane's work in India on leprosy (chapter on *Antibiotic*) or Maurice

Hilleman's work who invented 40 vaccines including *MMR* (chapter on *Vaccine*) while we know all about the umpteenth billionaire selling a battery driven car or yet another useless form of social media. If humanity doesn't get our priorities right, the innovation that powered human productivity can slow down and human talent will be wasted away in serving advertisements a few more milliseconds faster or serving celebrities' thoughts in the toilet in one sentence bites or trading stocks using sophisticated deep learning.

My guiding principles in ranking the inventions:

1. It needs to be a thing instead of just an idea or concept. This rules out the idea of the "assembly line" by itself as an invention.
2. It needs to be a thing instead of a scientific theory. This rules out the *theory of gravitation* or the *theory of evolution*.
3. It needs to be an invention instead of discovery. This rules out the discovery of Oxygen or the discovery of the Americas.
4. It needs to be an invention instead of an institution. This rules out innovative institutions such as schools, courts, governments, stock market or federal reserve. A possible exception to this rule: religion in my thinking can be classified both as an invention and an institution and gets a ranking in the book. All religions are not institutional religions anyway and it is much more personal than courts and laws, for instance.
5. It needs to be an invention by the human mind instead of a gift from evolution. This rules out *language* as it is mostly innate.
6. I group a set of related inventions together into one invention. Few examples will clarify this easily. X-ray, MRI and ultrasound are ranked together. Pen and pencil are ranked together. Bicycles and Motorbikes are ranked together. RFID cards and magnetic strip cards are ranked together.
7. Interestingly some Inventions have the rare honor of being included in multiple sections. One of the greatest inventions, *the Jacquard loom* for instance features in Flying shuttle grouped in looms and in computers grouped in automation technology.
8. I have excluded the inventions which are promising but haven't had the time to make an impact in history just yet. CRISPR falls

into this category, for instance. This also explains the lack of any inventions from the 21st century and not because we have stopped being innovative. (I hope; However you may want to reread *paragraph four*).
9. For the same reason, I have tried to stay away from the latest fashions and fads, some promising and some pure hype, taking the *long run* view. This excludes self driving cars, robots, quantum computing, artificial intelligence and bitcoins from the list. That is, this book still should make some sense a 100 years from now, similar to many books from the earlier centuries I have myself referenced; Classics such as Jean Baptiste Du Halde's *General History of China* from 1736, Marco Polo's travelogue from around 1300, Roger Bacon's *Opus Majus* from 1267 and Christiaan Huygens's *Horologium Oscillatorium* of 1673.

Ultimately in an exercise of this kind there is human judgement involved and as the author of the book I reserve judgement in ranking the inventions and I tried to do this with as much *objectivity* as possible. However, when readers point out important omissions or justification for a change in ranking, I hope to adjust the ranking in future revisions of this work. Some of the factors considered in the ranking:

1. The number of people in history affected by the invention. This ranks cloth or wheel higher than electricity or steam engines.
2. The number of people currently living in the world affected by the invention. This ranks pen higher than eye glasses.
3. Age of the invention. The more ancient the invention the more time it had to impact human history and society. This ranks electricity higher than computers.
4. The frequency of the use of the invention in our lives. This ranks pot higher than x-ray.
5. The impact of living without an invention. This ranks plastics higher than television.
6. How the invention affects human beings (*Homo sapiens*) as a species. Nuclear weapons rank higher than petroleum mainly because it is an existential threat to our species and has the power to send us back to the middle ages. It is rarely used (and I hope

will never be used) making it trade-off the existential factor with the frequency of use factor for an invention like electricity.
7. How the invention affects our basic necessities: *food, cloth and shelter*. This makes the fertilizer rank higher than the telegraph.
8. How the invention of our life and health as it is a prerequisite for everything else. This ranks vaccines higher than batteries.
9. How the invention impacts human productivity. As the day will also be fixed to *24 hours*, increasing human productivity is the best way for us to improve society and most of the important inventions in history have been productivity inventions, flying shuttle, Jacquard loom, steam engines and cotton gin.
10. The *productivity of the invention*, how the invention powers other inventions. This makes electricity, as the parent of so many other inventions in the list, rank higher than anesthetic. It also makes the wheel rank higher than agriculture due to the former's descendants in the list such as automobiles and bicycles.

This is a book mostly should read for the fun and the exhilaration of it instead of being a scholarly reference. However I have tried my best to be as accurate as possible in all the statements of facts. There is a rich list of references in the *bibliography* for people looking for evidence of the facts and further readings. I have excluded *Wikipedia* from the list of references as it is common to all the chapters in the book and is an obvious starting point for the research on many inventions. However, I have independently confirmed any claims in Wikipedia using another source instead of taking it at face value, sometimes from the references in the wikipedia article themselves. All the monetary values in the book are denominated in the current US dollars (2021) unless otherwise specified. I have preferred the SI system over imperial units in most places as it is the more widespread system by a *country mile*, nay a *kilometer*, around the world (see the chapter on *Ruler*). I have restricted each invention to exactly three pages to be fair to all the inventions and make the book readable given that it has hundred chapters, in many cases by resisting the temptation to write more on some inventions or omitting some partially relevant details.

Human inventions fall into a few periods in our history:

1. Our foundational inventions from the *start of our species*. This covers the period from around 200,000 years ago to around 5,000 years ago. This category includes clothes, fire, agriculture and so on.
2. Our foundational inventions at the *start of history*. This covers the period from the start of history around 5,000 years ago to the beginning of the middle ages around 1,500 years ago, 500 CE. This includes writing (the very definition of the start of history), wheel, numbers and so on.
3. Our inventions in *the middle ages*. This covers the period 500 CE -1500 CE. This includes the book (printing press), eye glasses, thermometer, telescope and so on.
4. Our inventions during *the industrial revolutions*. This covers the period 1500 CE-1925 CE. This includes spinning Jenny, cotton gin, steam engines and so on.
5. Our inventions in *modern history*. This covers the last 100 years from 1925 onwards to now. This includes computers, vaccines, lasers and so on.

This typically means the inventions from an earlier age rank higher than the one from the later period simply because the former had more time to have an impact on our history and it has the *"classic movies"* effect; The movies from 30s and 40s that survive to our time have done so mainly because they are very good and withstood the competition from so many others. That's the reason most of the old movies we watch look good but the new movies are "hit or miss". Time filters out all the bad movies as well as all the bad inventions. It holds true as a first approximation, however there are exceptions in movies, *Saving Private Ryan*, the great 1998 movie in the AFI top 100 list and in inventions, *GPS*, from around the same time in the inventions list.

I have learned many things myself in the process of writing this book and tried to share it with the readers as lucidly as I can.

- James Watt's invention is not the steam engine but the *condenser* in the more efficient and practical steam engine.

- How exactly weaving works and *warp* and *weft* threads and the importance of the *shed*, a beautiful process and important part of our shared human history we should know.
- I learned, somewhat embarrassingly for a computer engineer, what a *7nm process* exactly means. It brought home the realization that the high tech computer industry hyper-specializes, mainly because of narrow and perverse incentives, for our own serious detriment in the long term.
- I also learned the shocking difference between countries in the world; In Chad, with the same population as the Netherlands, only around one in nine people, 12%, had access to electricity. A fact - people who live inside their smartphones and talk of bitcoin replacing paper currency or betting on battery cars taking over the world anytime now and ridiculously overvaluing some companies at 3% of the US GDP in the process - would do well to remember.
- I learned about some of humanity's priceless treasures for the first time: *Leningrad codex, Sanaa palimpsest* and, the confusingly named, *Dresden codex*.
- Another lesson is the deepening of the perspective on the importance of some inventions; Telegraph, although obsolete after India stopped the last service in 2013, is one of the most important inventions ever in human history; It for the first time enabled us to send pure information to distant places instead of sending it through a physical carrier such as the *pony express* or trains. This meant we could send information *lightning* fast, literally as both use electricity, a billion kilometers per hour instead of hundreds of kilometers per hour with physical carriers. This was the biggest "jump" in communications for humanity. Everything else on top of this such as telephone or email is just a minor change, relatively speaking.

My own knowledge advantage as a subject matter expert in some chapters made me present them in a different way, a more profound and interesting way, I hope. For example, the computer chapter emphasizes the *Turing machine*, the notion of computing itself and The *Jacquard loom* instead of developments such as mainframes, microcomputers or operating systems and hard disk drives.

When I started the book, I expected that the US and the UK would top the list of inventions but I didn't realize till I finished how big the gaps are. The US dominates the list so heavily that almost 40% of the top inventions are by the American inventors. The second ranking country, England accounts for another 25% of the invention mainly because it was the cradle of the industrial revolution. Germany comes in third more along the expected lines, given the country is long known for inventions from the printing press to the painkiller; The mechanical inventions such as clocks (*Nuremberg*), knives (*Solingen*) and cars (*Stuttgart*). If anything, I had expected a bit more from Germany when I started out. France comes in as the surprising fourth; I didn't know that France is this innovative. I would have guessed the vaccine and pasteurization but I was pleasantly surprised by her role in computers (*Jacquard loom*) and internal combustion engine.

I have provided some reasons for the dominance of the US and the UK in the inventions: *immigration* and *the tradition of strong intellectual property (IP) protection*, respecting the work of people who do intellectual work. I have come to the realization through the process of writing my own book with its countless hours of hard work for months that IP protection is very important for people who depend on it for their livelihood whether in books, movies, drugs or software. It is one of the main factors in inventions coming out of the US instead of India or Nigeria. The story of Eli Whitney and his *Cotton Gin* stresses the same point; The trouble he had with protecting his inventions as compared to another American great inventor who was fortunate enough to live in a time of better IP protection in the same country, Thomas Alva Edison. This is further explained in the chapter on *Cotton Gin* and in the *principles* at the end. When I say "fortunate enough", it is not just the fortune of Edison but all the rest of us who benefited from his inventions whether it is electric light or cinema. It would have been a great loss for all of humanity, if he ran around trying to protect his microphone patent from thieves instead of spending his time productively, inventing the electric light, cinema and phonograph.

I hope reading the book provides a fun, entertaining but also a productive experience where you are surprised by one or two surprises of your own and learn something useful similar to my own experience writing the book, whether it is knowing the *warp* and *weft* threads, the difference between all the "confusing" inventions of the industrial revolution, spinning Jenny, flying shuttle etc., why Nitrogen molecules are common in all the explosives or what is the difference between soap and detergent; Or something else altogether that you start thinking about because of the book and makes your day.

Fakrudeen Ali Ahmed
Sunnyvale, California, USA
2 June 2021

PART 1

Cloth

"But of the fruit of the tree, which is in the midst of the garden, God hath said, Ye shal not eate of it, neither shall ye touch it, lest ye die.

For God doeth know, that in the day ye eate thereof, then your eyes shalbee opened: and ye shall bee as Gods, knowing good and euill.

And when the woman saw, that the tree was good for food, and that it was pleasant to the eyes, and a tree to be desired to make one wise, she tooke of the fruit thereof, and did eate, and gaue also vnto her husband with her, and hee did eate.

And **the eyes of them both were opened, & they knew that they were naked**, and they sewed figge leaues together, and made themselues aprons.

And they heard the voyce of the LORD God, walking in the garden in the coole of the day: and Adam and his wife hid themselues from the presence of the LORD God, amongst the trees of the garden.

And the LORD God called vnto Adam, and said vnto him, Where art thou? And he said, I heard thy voice in the garden: and I was afraid, because I was naked, and I hid my selfe.

And he said, **Who told thee, that thou wast naked? Hast thou eaten of the tree, whereof I commanded thee, that thou shouldest not eate?**

...Unto the woman he said, I will greatly multiply thy sorowe and thy conception. In sorow thou shalt bring forth children: and thy desire shall be to thy husband, and hee shall rule ouer thee.

And vnto Adam he said, ... In the sweate of thy face shalt thou eate bread, till thou returne vnto the ground: for out of it wast thou taken, for dust thou art, and vnto dust shalt thou returne."

Almighty God was onto something when he said knowing that you are naked is part of the important wisdom from the *tree of knowledge*. Cloth is the innovation which started to differentiate people from other animals, our cousins from the evolution on earth, and kick started the human cultural revolution. *Tarkhan dress,* named for the town 30 miles from Cairo in Egypt where this dress was discovered, is considered the oldest known piece of dress; It currently resides in University College London *Petrie Museum*; It was discovered in 1913 and is radiocarbon dated to between 5500-5100 years ago. However clothes predate this 5000 years old cloth by tens of thousands of years if not by hundreds of thousands of years. Interestingly, a study on lice dates our clothing to around 170,000 years ago in Africa.

Important advance in clothing occurred with the invention of the sewing needle. Oldest known sewing needle is from 61,000 years ago, found in *Sibudu Cave* near Durban in South Africa. Another well known specimen with an age of 50,000 years is from *Denisovan cave* near *Novosibirsk*, the third most populous city in Russia after Moscow and St.Petersburg, in Siberian district of Russia near Kazakhstan border and Xinjiang region of China. There is some dispute over whether people (*Homo sapiens*) or Neandarthals made this needle though. Both needles were made of bones as iron and steel were far in the future. In any case, this invention also very likely predates these specimens by thousands of years. Initial clothing was likely from animal hides and once the needle was invented multiple fur pieces were stitched together and worn.

Another major advance was the invention of fibres which helped make clothing without animal skins. Earliest known flax fibres are from *Dzudzuana cave* near capital Tbilisi in Georgia, and date back to 36,000 years. The fact these fibres were dyed shows that this date is an underestimate for the invention of fibres.

Cloth is the single most important innovation of human beings. It freed *homo sapiens*, an animal, from the tyranny of elements including midnight sun and cold at night for the first time, the start of controlling nature through our intelligence and inventions as in the promise, *"ye shall bee as Gods"*. It helped us spread all over the world including parts of Africa with its intense heat and Europe with its intense cold. Only innovation on the same scale of importance as clothing is fire. However, cloth protects the person all the time while fire is required only for part of human lives. The wheel was invented much later and some transports via river or sea don't even require them. Agriculture was also a much later invention and human societies have survived without it. Writing or numbers are more modern inventions with only a few thousand years of history; It was also restricted to tiny elites for much of our history unlike clothes which were always universal. There were some rare abominations like the *Breast Tax*, a tax imposed on the lower caste women by the *Kingdom of Travancore* (Kerala, India) if they wanted to cover their breasts in public, until 1924. Another instance was the sumptuary laws on clothing of medieval Europe to prevent commoners from wearing clothes of nobles. On balance, clothing was used by vast majorities of people who ever lived unlike writing or numbers.

Among our basic necessities of food, shelter and cloth, almost all animals need and use food and shelter, a cave or a nest for example, with no known animals using clothes. This marks it out as a true human invention coming from our culture rather than biology. This may also be the invention which survived the longest in our history and may yet survive all our other inventions in this book going obsolete. Clothing deserves its rank as the top human invention which impacted the most people over humanity's history. This is one invention used by all the 7.8 Billion of us almost all the time, more than wheel, pot, pen, electricity or computers or anything else we can think of in the future, richly deserving its top place.

Fire

Human beings evolved to eat cooked food. It is literally possible to starve to death even while filling one's stomach with raw food. In the wild, people typically survive only a few months without cooking, even if they can obtain meat. Wrangham cites evidence that urban raw-foodists, ... are often underweight. Of course, they may consider this desirable, but Wrangham considers it alarming that in one study half the women were malnourished to the point they stopped menstruating. ... for the same amount of calories ingested, the body gets roughly 30 percent more energy from cooked oat, wheat or potato starch as compared to raw, and as much as 78 percent from the protein in an egg.

...In essence, cooking—including not only heat but also mechanical processes such as chopping and grinding—outsources some of the body's work of digestion so that more energy is extracted from food and less expended in processing it. Cooking breaks down collagen, the connective tissue in meat, and softens the cell walls of plants to release their stores of starch and fat. The calories to fuel the bigger brains of successive species of hominids came at the expense of the energy-intensive tissue in the gut, which was shrinking at the same time... Cooking freed up time, as well; the great apes spend four to seven hours a day just chewing, not an activity that prioritizes the intellect.

The trade-off between the gut and the brain is the key insight of the "expensive tissue hypothesis," proposed by Leslie Aiello and Peter Wheeler in 1995. Wrangham credits this with inspiring his own thinking—except that Aiello and Wheeler identified meat-eating as the driver of human evolution, while Wrangham emphasizes cooking. "What could be more human," he asks, "than the use of fire?"

The article, *Why Fire makes us human?*, in the Smithsonian magazine of June 2013 describes the theory that fire, through cooking, made us human by building our big brains which are notoriously energy expensive; Our brain, with only 2.5% of our body weight, takes up 20% of our energy consumption. This theory by Richard Wrangham in the book, *Catching Fire, How cooking made us Human*, proposes that we traded off our gut for our brain with the invention of fire and cooking. It makes fire the reason for our brain size and intelligence, a prerequisite for all the inventions in this book. *Wonderwerk cave in So*uth Africa is cited as the earliest human control of fire dated to more than 1M years ago. However it is dated too early to be credited to modern humans; We, *homo sapiens*, are only a 200,000 years old species. The dispersal of us, modern humans from Africa happened only around 50,000 years ago. If we knew how to make fire before we left Africa, the earliest fire control sites should be in Africa. On this basis, the *Kalambo Falls* site in Zambia, dated to 110000-60000 years ago near *Lake Tanganyika*, one of the African great lakes and the second-largest freshwater lake in the world by volume and depth, seems a far more promising candidate.

Fire is one of the greatest inventions people ever invented and it changed the trajectory of human civilization. It expanded the kind of foods we can eat and made food more nutritious by softening them and releasing more nutrients in human digestible form in that critical process called cooking. Many scholars claim that cooking was instrumental in the evolution of our brain which is after all the most important organ for human inventions. Based on this claim, fire via cooking set us on a path to differentiate ourselves from all other animals with the explosion of our brain by providing more efficient calories. This may have even led to the language itself. People could have only reached and survived colder latitudes like Europe with the help of fire.

It enabled us to hunt animals far bigger than us, again providing far more calories for the exploding brain. It provided protection for early people against animal predators. It would have allowed people to come down from trees safely even with other animal predators. This step is crucial for *bipedalism* and freeing up of our hands for using tools and other inventions. It also greatly increased human productivity, as the day will always be twenty four hours long, by providing light in the dark.

Fire was so important in many early human societies it was worshipped as a God, *Hephaestus* of Greeks, *Agni* of Indians and *Logi* of Norse being examples. Importance of controlling fire to modern society in the form of cooking, heating, iron and other metal works, external and internal combustion engines and electricity generation plants is very clear. It plays a critical role in the military in the form of firearms and bombs. In the form of rockets it has been critical in our exploration of space and satellites as well.

It clearly rates as a human invention as our closest relatives in the animal kingdom, chimpanzees, can't make or control fire. Darwin himself says, "This discovery of fire, probably the greatest ever made by man, excepting language, dates from before the dawn of history." Fire is ranked high as one of the greatest inventions because Fire was important for large scale food production at the heart of modern civilizations and it was important in metallurgy including the emergence of iron tools. Modern industrial revolutions wouldn't have been possible without the ability to control Fire. Only inventions which are even on the same scale are clothing, wheel, writing and numbers. However, writing and numbers came much later in our history and didn't play as big a role in human cultural evolution. Widespread use of wheels required roads which only came in much later. Unlike Darwin, I don't consider Language as a human [cultural] invention but more of an evolutionary byproduct of an exploding brain, a *language organ with its innate grammar*, following modern linguistic views of eminent scientists particularly of Noam Chomsky, Steven Pinker et.al. So language doesn't get a place in this list of purely cultural human inventions.

Wheel

"a wheeled vehicle required not just wheels but also an axle to hold the vehicle. The wheel, axle, and vehicle together made a complicated combination of load-bearing moving parts. The earliest wagons were planed and chiseled entirely from wood, and the moving parts had to fit precisely. In a wagon with a fixed axle and revolving wheels (apparently the earliest type), the axle arms (the ends of the axle that passed through the centre of the wheel) had to fit snugly, but not too snugly, in the hole through the nave, or hub. If the fit was too loose, the wheels would wobble as they turned. If it was too tight, there would be excessive drag on the revolving wheel."

David Anthony, in his book *The Horse, the Wheel, and Language*, explains the main complexity behind inventing the wheel. This chapter, with equal justice, can be called the *Axle*. The *Bronocice Pot*, a decorated clay mug of the *Trichterbecker* culture, discovered in *Małopolskie* in southern Poland near Kraków, depicts one of the earliest known wheeled vehicles, dated to around 5500 years ago; It shows a wheeled vehicle with a shaft for drawn animals and four wheels. First wheels, the actual physical ones, were discovered, not very far from this

pot, from the graves of *Yamnaya* culture at Balki *Kurgan* (earthen grave mound serving similar function to a *gravestone*) in Ukraine north of the Black sea dated to around 5330-4900 years ago. King Tutankhanun's tomb, one of the few tombs of *pharaohs* not looted by thieves and discovered in pristine condition in 1922, had multiple chariots showing that wheel was a very mature technology by his time, around 3300 years ago. Wheel is a more recent invention in our history; It is only around 5500 years old as compared to the history of hundreds of thousands of years for clothing and fire. It is surprising that wheels were invented after boats, musical instruments, metal works and many other complex inventions despite our use of the term "reinventing the wheel" to indicate its antiquity.

Wheel is a quintessential invention of humanity. All modern land transportation depends on wheels and underlies modern innovations such as automobiles, railways and airplanes. It was also instrumental in the creation of pots with potters' wheels. It is also used in water mills, gears and electric turbines. However it is for the first reason, that of transportation of people and goods, the wheel is ranked at this place.

Most complex part of the invention is not so much the wheel as the axle. Wheel has to fit snugly so that it goes forward without wobbling and it can't be too tight as to make the wheels unmovable. Wheel and Axle went through multiple evolutions with axles rotating with wheels at first to stable axle with rotating wheels with some way of reducing the friction between the fast rotating wheel and the axle. Stable axle made it possible for the system to bear much bigger loads. Nowadays wheel bearings made of hardened steel balls make it much easier to both reduce friction and ensure a snug fit. *Ball bearings*, such as the ones in the automobile wheels, reduce friction by changing normal, static friction to rolling friction exactly mimicking the main advantage wheels have. Wheels change normal friction to rolling friction and thereby produce much farther movement for the same amount of power applied. We feel this difference when our shopping cart wheel gets stuck and we have to drag it part of the way.

Wheels require good roads for them to be useful. It also required draught animals before the invention of internal combustion engines. Lack of good draught animals is considered as one of the main reasons why native americans didn't invent the wheels as they didn't have any good draught animals like oxen or horses. This invention is apparently hard enough that it was invented so late and many civilizations including native americans and australians didn't invent it at all, other than as toys. Main reason for this may be that in biology there are no examples of wheels other than microscopic bacterial flagellar motors which were not known until well after the wheel itself was invented.

Wheel ranks lower than clothing or fire due to the amount of time, hundreds of thousands of years, they had to impact human civilization. Wheel ranks higher than writing and other foundational inventions from around the same time as it impacted more people over history, particularly before the invention of Gutenberg's printing press and books. I counted a full one-fifth of the inventions in this list related to the wheel, demonstrating its importance. It is also the first major invention which acted as a force multiplier enabling us to do much more than what purely the muscle strength of people and animals could do, acting as a pioneer for other inventions using *non-animal mechanical power* such as steam engines, internal combustion engines for trains, motor vehicles and airplanes, in addition to actually providing the wheels for them to move in the first place. This marked the first transition for us from being another animal species using its muscle strength to a species using its superior mental abilities for its world conquering advantage. Metallurgy and ships were invented before the wheel, but not by a whole lot more; The wheel ranks higher because of this force multiplier effect on humanity in addition to its productivity for generating further high ranking inventions. Agriculture was also invented before the wheel but doesn't have either of these effects, force multiplier or as the model for other high ranking inventions. People had alternative ways of obtaining food including hunting and fishing; Wheel doesn't have a real alternative on land where we spend more time as a land based species and other animals like donkeys could only carry a fraction of the load of that of wheels. It is this lack of alternatives, for the purpose it is used, that ranks the wheel above agriculture.

Agriculture

"*A small step for a man, a giant leap for mankind*: that's the phrase grandly applied to the first moon landing. But it suits even better an event whose outlines are emerging ever more clearly from the mists of prehistory: the discovery that wild grasses could be deliberately planted, cultivated and harvested as cereals. That step opened the way to an unprecedented expansion of food supplies and of human population that, in turn, made cities and civilization possible.

...The earliest archeological sites with evidence of domesticated grain lie in the southern Levant, at the north end of the Dead Sea, and date to about 10,000 years ago. Long before that, people in the region had gathered and eaten wild grain. Scientists now believe they have pinpointed the transition from wild to domesticated cereals to an area within 50 to 100 miles of the Dead Sea, just before 10,000 years ago.

...Their study focused on people belonging to the Natufian culture, named after an archeological site in the Judean hills known as Wadi-al-Natuf.

The Natufians just preceded the time span at which domesticated plants begin to appear in the archeological record, and the Yale archeologists believe they were the probable inventors of agriculture in the Near East."

Natufian culture in current day Jericho in Palestine near Jerusalem is credited with the first origin of agriculture according to some scientific theories presented in this 1991 article from *The New York Times*. In any case, the first records of agriculture are from the greater *Levant*, modern day Syria and neighboring areas including *Wadi-al-Natuf* ("Natuf valley"), around 12,000 years ago. It was an easier step for the people of *Levant* as wild grains were already abundant in Levant and people came to rely on wild grains for food. According to the somewhat controversial theory mentioned above, sudden climate change forced them to turn to artificial planting of the wild grains giving birth to agriculture. In any case, It definitely wasn't as hard a leap as the invention of wheel and axle. At least 11 places are considered as independent origins of agriculture. However the *Fertile Crescent*, of which Levant is a part, played the most important role followed relatively later by China [Rice, 6200 BCE] and India [Cattle, 8500 BCE]. Earliest cereals were *Einkorn wheat* (diploid) and *Emmer wheat* (tetraploid), precursors of the modern wheat, a hexaploid (six sets of chromosoms).

Animal domestication (*animal husbandry*) is considered a part of agriculture. Cattle were domesticated from wild aurochs around 10,000-11,000 years ago in Turkey and India. Pig and sheep were domesticated in Mesopotamia around the same time. Horse and Camel were domesticated much later. Horses were domesticated around 5500 years ago in *Akmola* province of Kazakhstan bordering Russia based on analysis of *Botai settlements*. Camels were domesticated in Arabia around 5000 years ago based on the analysis of camel bones and their possible use as pack animals. Chicken domestication is mired in controversy with *Cishan* in China cited as one of the earliest sites around 5000 years ago. Dogs and cats were likely domesticated around 12,000 and 10,000 years ago respectively in Levant.

Some scholars consider Agriculture as a mixed blessing. Jared Diamond considers it the cause of deep class divisions with a tiny elite over-lording over the vast majority of people. In a way it is similar to recent technological progress with a tiny elite in *big tech* and other industries becoming exorbitantly rich while most of the people are running in place or becoming poorer in the US. This applies to even technology workers themselves, as current CEO to worker salaries in the US are at least 100:1 and close to 300:1 in some companies; It was a more tolerable 30:1 in 1980. Most of the benefits of agriculture were sucked up by a tiny elite while people became much worse including malnutrition due to montonic diet, starvation and reduced life expectancy. Jared Diamond says:

"Skeletons from Greece and Turkey show that the average height of hunter-gatherers toward the end of the ice ages was a generous 5'9" for men, 5'5" for women. With the adoption of agriculture, height crashed, and by 3000 B. C. had reached a low of only 5'3" for men, 5' for women."

On balance, agriculture for the first time supported huge populations living a sedentary lifestyle which was a prerequisite for the rise of civilization, cities and all the inventions. People turning from nomadic lifestyle to agriculture started modern human civilizations. As Jared Diamond explains in *Guns, Germs and Steel*, people with agriculture and hence complex technology conquered people all over the world, from the Americas to Australia and Africa, without agriculture. *Corn*, cereal grain in British English, is the quantity which provides value to everything else in the modern economy including gold, silver and money, according to Adam Smith in *Wealth of Nations*. Agriculture acts as the foundation for the modern economy and all other jobs. It ranks below clothing and fire as it has been only with us for 10,000 years instead of hundreds of thousands of years for the former. Wheel ranks above agriculture for being an invention with no precedent in biology (other than at the microscopic level of ATP synthesis and flagellar motor that won't be known until much later). Agriculture was an easier leap from eating wild grains with a clear predecessor in nature. It is an easier invention as can be seen from the fact many isolated civilizations in North America and Papua New Guinea developed agriculture but not the wheel.

Writing

"§195. If a son strike his father, they shall cut off his fingers.

§196. If a man destroy the eye of another man, they shall destroy his eye.

§197. If one break a man's bone, they shall break his bone.

§198. If one destroy the eye of a freeman or break the bone of a freeman, he shall pay one mana of silver .

§199. If one destroy the eye of a man's slave or break a bone of a man's slave he shall pay one - half his price.

§200. If a man knock out a tooth of a man of his own rank, they shall knock out his tooth ."

This is from the *code of Hammurabi*, the king of Babylon around 3700 years ago; It is a legal text prescribing the code of justice, laws and punishments in the Babylonian society, inscribed on a diorite (a kind of igneous rock) *stele* in Mesopotamian cuneiform script now at the *Louvre Museum* in Paris. We know about this because, fortunately for us, the

Babylonians inscribed it on the stele but more importantly, writing had been invented. Before the invention of writing, knowledge was very much localized by space and time. You could only learn the information from the person who is *near you* at the *same time*. Human memory is also notably fallible which means this information may not be fully accurate. Writing laid the foundations for modern civilization with books and record keeping. For the first time records could be transferred across generations and centuries which lead to the exponential explosion of ideas. Writing enabled the steady accumulation of knowledge which meant every generation could start from a higher base of knowledge using standard forms of instruction such as in schools. This laid the foundation for information technologies such as books, newspapers, computers and the invention of money, expansion of trading and to the industrial and scientific revolutions including the steam engine and electricity.

Earliest known writing systems include Sumerian *Cuneiform*, used in the code of Hammurabi above, and Egyptian *Hieroglyphs*. According to the British Library, writing systems were invented independently in four places, excluding Indus valley that is not conclusive. First in Mesopotamia around 5400-5300 years ago and shortly afterwards in Egypt around 5200 years ago. Around 3300 years ago, there is evidence of a writing system in China. Mesoamerica invented the writing system around 2900-2300 years ago. Of those, egyptian invention is the most important as the *alphabet* and almost all the world's writing systems arose from this invention via *Phoenician* (Canaanites of Levant around Lebanon) traders who traded with Egypt and Greece.

In 2015, ISIS terrorists destroyed the beautiful *Temple of Bel* in *Palmyra* and killed the 82 year old syrian archaeologist Khaled al-Asaad. The Temple of Bel (or Baal) was the most important heritage site in Palmyra with the temple standing for 2000 years. (This is the same Baal who is mentioned in Moses' story in the Bible). Baal himself was one of the gods of Phoenicians who invented the world's first alphabet, called *Phoenician* or *Proto-Canaanite*, from Hieroglyphs via the *Rebus principle* (using pictures to represent sounds instead of the object, the picture of the tree to represent the sound "t", for instance). The alphabet itself was invented only once as far as we know. First archaeological evidence is from *Wadi el hol* ("Terrible Valley") near Luxor in Egypt, dated to around 3800 years ago. In fact modern alphabets in many languages are surprisingly similar to this original. *Alep, Bet* of Phoenician alphabet is very similar to *Aliph, Ba* in Arabic, *Alef, Bet* in Hebrew or A, B in English. Like Hebrew and Arabic, Phoenician alphabet was written from right to left. The switch from right to left (RTL) to left to right (LTR) apparently happened along with the switch from stone tablets to writing with ink, on *papyrus* for instance. The apparent reason is most of the people in history were right handed and this wouldn't smudge the ink when writing.

Charles Wheelan in his book, *Naked Economics,* claims India is one of the poorest countries in the world mainly because of illiteracy with 25% of the population being illiterate, down from 50% in the early 1990s. For the same reason countries topping the inventions list, the US, England, Germany, the Netherlands etc., were universally literate a century or two ago. It is also reflected in India's per capita GDP (*purchasing power parity*) in 2019 of 6,996 dollars as compared to the US which has a per capita GDP of 65,297 dollars, an order of magnitude difference. Writing in the form of literacy forms the core of the *human capital* which is far more important for a country's economy than physical assets like oil.

Writing is one of the inventions which formed the foundation for all other inventions. Writing is important in history for the simple reason that writing marks the start of written history, *by definition*. Writing ranks below clothing and fire as their timelines are an order of magnitude different. Wheel is from around the same age as writing; Throughout human history most of the human population would have been illiterate, till formal schools were developed recently, but very likely would have used some form of wheeled vehicles. However it is fair to say human history was determined by the few people who had access to this invention as readers or writers, particularly in its modern reincarnation as books with Gutenberg's printing press; *Elements, Thirukkural, Principia Mathematica, Wealth of Nations, Origin of the species* being a few examples. It is very likely that without the Bible and the Qur'an in their written form these great religions, for better or worse, may not have survived these many centuries and may owe their existence to writing.

Numbers

"31. Rule for addition of affirmative and negative quantities and cipher: § 19. The sum of two affirmative quantities is affirmative; of two negative is negative; of an affirmative and a negative is their difference; or, if they be equal, nought. The sum of cipher and negative is negative; of affirmative and nought is positive; of two ciphers is cipher.

35-36. Rule for division : §23—24. Positive, divided by positive, or negative by negative, is affirmative. **Cipher, divided by cipher, is nought**. Positive, divided by negative, is negative. Negative, divided by affirmative, is negative. Positive, or negative, divided by cipher, is a fraction with that for denominator: or cipher divided by negative or affirmative."

Indian mathematician Brahma Gupta, in his book BRAHMA-SPHUTA SIDD'HÁNTA written around 628 CE and translated by Henry Thomas Colebrooke of England and British India in 1817, explains the rules for zero in Arithmetic; Emphasis is added on one of the rules where Brahma Gupta, one of the greatest minds of his time, gets it wrong, 0/0 = 0, showing that the concept was still new and original at the time. The *Bakhshali manuscript,* a birch bark mathematical manuscript found in 1881 at Bakhshali near Peshawar in British India, contains heavy use of decimal symbols including zero that look strikingly close to modern decimal numerals. However its age (from carbon dating by Oxford University) is given as multiple ranges, 224-383 CE, 680-779 CE, and 885-993 CE, making it inconclusive as the oldest use of the decimal system. Numbers, in particular decimal number systems, built the modern world. True positional number system with a zero to signify *nothing* in a particular position of a number was a truly revolutionary invention. It is vastly superior to old systems such as the *Roman numerals*. It simplified and accelerated mathematics, science, economy among other fields. We can't imagine modern science, money and computers without it. It spread to the arabic world and Europe in the 9th century, particularly with the works of Persian mathematician *Al Khwarizmi* (Muḥammad ibn Mūsā al-Khwārizmī) at Baghdad in Iraq in the early 9th century. It was used for business transactions in Italy, particularly Pisa in the 15th century and spread to all of Europe in the 16th century.

Tally marks, a simple counting (*unary*) system with a grouping mechanism, was one of the earliest number systems to be invented. Earliest evidence of this system is a 25,000 year old *Ishango bone* from Congo in Africa near Lake Edward, the smallest of the African great lakes, and Uganda border. Additive systems like Roman numerals evolved from this system as unary systems take too long to read and write. (In computer parlance, number N requires N digits in unary while positional systems only require *logN* digits making them *exponentially* more compact and easy to grasp).

Babylonian numerals, from around 4000-5000 years ago, used a base 60 system but with only two symbols, one to represent a one ͐ and the second for ten ⟨. Numbers between 1-9 were represented by combining the symbol for one vertically in various ways. Multiples of ten between 10 and 50 were represented by combining the symbol for ten horizontally. It had no zero which meant 19 and 109 were written the same except possibly for an extra bit of space. This meant it was not a true positional system. However vestiges of the system are still with us today in the form of minutes and seconds and 360 degrees of a circle. A clay tablet in the Yale University babylonian collection ("YBC 7289") in the US shows these numbers beautifully with an approximation to square root of 2, without a radix point to separate the fraction; The ancient student using the tablet was computing the diagonal of a square with sides 30 units long using this approximation (The tablet is easy to decode and worth a look to appreciate it). Babylonian system was the most advanced ancient system as it came closest to a proper positional system. Egyptians invented another early system which was more primitive. It contains one symbol for every power of 10, a symbol for 1, a symbol for 10 and a symbol for 100 and so on. It is depicted in *Stele: princess Nefertiabet and her food* in Louvre Museum in Paris, a funeral art for princess Nefertiabet from 4000 years ago. This was an additive system similar to Roman numerals; This system doesn't have a zero and doesn't need it as it is not a positional system. Another interesting system was the *Mayan numerals*. It is depicted in the *Dresden Codex*, pre-book from Mesoamerica, from the 11-12th century Americas. It uses a base-20 system with vertically written positional notation and it has a zero. It built 20 symbols required for a base 20 system from ., - (reminiscent of Morse code) with a special symbol, a shell for 0. Another well known system surviving to this day despite its deficiencies is the Roman numerals. *The Colosseum* in Rome built in the first century has its entrances numbered with these numerals. It is a very inefficient system in not being positional, no zero, both additive and subtractive and many extraneous symbols for 5, 10, 100, 500, 1000 and so on.

As the foundation for all of modern science, *numbers* rank much higher than most of the inventions driven by science such as thermometer, electricity or telegraph. Its importance may be inferred from the stunting of European science and mathematics in the middle ages (*"dark ages"*) and their taking off just after the introduction of Indo-Arabic numerals in the 16th century. Indian and Islamic science was in an advanced state during this same period and the number system played at least a partial role in their success. It can also be inferred from the advanced state of Babylonian Astronomy with their base-60 system. Binary (*base-2*) systems simplified the design of modern computers. (Some early computers such as ENIAC used decimal numbers). It ranks very close to writing in its importance and is more important than the printing press. Based on the history of Europe before and after the introduction of the decimal system, numbers in the form of Indo-Arabic numerals (base-10 system with a zero) should be called the mother of all Inventions or at the least "co-mother" along with the printing press. In the end analysis, it is "only" an improvement in efficiency, of reading and writing numbers, just like the printing press is only an improvement in efficiency over the manual copying of the books by scribes. However these efficiency improvements made a world of difference for humanity as further explained in the chapter on paper.

Paper

"At length , in the Year 95 of the Chriſtian Æra, under the *Tong han*, that is under the *Han* who had removed their Court into a more Easterly Province than the *Han* their Predecessors, a great Mandarin of the Palace, whose Name was *Tſai lun*, invented a better sort of Paper under the Reign of *Hoti*, which was called *Tſai beou tchi*, Paper of the Lord *Tſai*.

This Mandarin made use of the Bark of different Trees, and of old worn-out pieces of Silk and Hempen Cloth, by constant boiling of which Matter he brought it to a liquid consistence, and reduced it to a form of thin Paste, of which he made different sorts of Paper ; he also made some of the Knots of Silk , which they called Flaxen Paper : Soon after the Industry of the Chinese brought these Discoveries to perfection , and found out the Secret of polishing the Paper, and giving it a Lustre."

The French historian, Jean Baptiste Du Halde, a well-known expert on China in his 1736 book *The General History of China* describes the origin of the paper, *Tchi (*纸*)* in chinese. *Cai Lun (Ts'ai Lun)*, an imperial official who lived in the first century is traditionally credited with inventing the paper. However it likely predates him and paper was invented in China around the first century. Paper documents from this period, one of them likely a map, were discovered in Northern China near Mongolia. A factory for making paper from cotton was established at Samarkand, on the old *silk road* now in Uzbekistan, around the year 706 CE in the islamic world. In the 11th century it spread from the islamic world to Europe. Paper, with the printing press, kindled the fire of the industrial revolution in Europe. Paper made writing inexpensive, storing large amounts of information cheaper and made widespread libraries possible. Before the invention of the paper, there were very few well known libraries in the ancient world, among them the *library of Alexandria*, a port city in mediterranean in northern Egypt with its *papyrus* scrolls and the *library of Pergamum*, a city in mediterranean, western Turkey with its *parchment* manuscripts, both from the 3rd century BCE.

Before paper, people used parchments (or *vellum,* a fine quality parchment) which is made up of animal skin. A "page" of this parchment was very expensive. Modern parchments cost on the order of a few hundred dollars per "sheet". This means that this book would cost around 15,000-30,000 dollars for parchments alone [assuming parchments weren't more expensive in ancient times which makes this an underestimate]. Another alternative was *Papyrus* from the papyrus plant, *Cyperus papyrus*, of Egypt; Papyrus consists of two layers with strips of papyrus stem laid out side by side to make a layer. Strips of each layer are at right angles to the other layer much like *warp* and *weft* threads in weaving (see the chapter on *flying shuttle*) but with all the "weft" strips above "warp" strips unlike in weaving. It had many disadvantages over paper. It didn't last as long as paper as it broke apart along the strips and didn't last for more than a decade or two particularly outside the climate of Egypt. Papyrus is relatively cheaper as compared to parchments but it is a trade off in durability. Papyrus is available for around 4 dollars a page which makes the cost of this book around 1,200-2,500 dollars. However the Papyrus book wouldn't last as long as the paper or parchment book. Both of them would be much bulkier than a paper book.

Paper documents and books are easily accessible and transportable due to their reduced weight and volume. Typical paper thickness is around *0.1 mm* while both papyrus and parchment thickness is around *0.5 mm*, around five times thicker. There was a medium which was in use before papyrus or parchment that was much heavier and thicker, *stone or clay tablets*. A clay tablet in British museum weighs 210g or 2 pages of a clay book weighing a pound. This means a 300 page book would weigh 30 kilograms (60 *lb*). With a width of *30 mm* per tablet "page", it would be 300 times thicker than the paper book.

Paper is ranked very high mainly for the knowledge revolution it enabled via books. Paper accelerated the spread of knowledge and information by its accurate recording and reproduction to a wider populace, affordably. It is ranked just after the foundational inventions such as clothing, fire and writing. Although paper is *only* an incremental improvement from tablet, papyrus or parchment this reduction in size and cost enabled the possibility of mass producing the books using a printing press. It would have been unthinkable to use block or *movable type* printing to print on top of a clay tablet. (Although it was done on *vellum* by Gutenberg for his *42-line bible*, see the chapter on *book*). This trajectory is analogous to miniaturized transistors eventually enabling smart phones which would have been unthinkable with the vacuum tube technology. Inventions which are "only" an incremental improvement over existing technology can play a breakthrough role in human development. Another similar example is James Watt's *steam condenser* which is "only" an improvement in energy efficiency over *Newcomen's steam engine* which made all the difference in making the steam engines more useful, affordable and mainstream (see the chapter on *steam engine*). Newcomen's engine couldn't have brought in the industrial revolution nor did papyrus or parchment bring in the knowledge revolution, only paper did. *Incremental inventions matter, a lot.*

Book

"The art of printing which has been discovered in Mayence is the *art of arts, the science of sciences*, by means of which it will be possible to place in the hands of all men treasures of literature and of knowledge which have heretofore been out of their reach.

An infinite number of works which very few students could have consulted in Paris, or Athens or in the libraries of other great university towns, are now translated into all languages and scattered abroad among all the nations of the world."

The *Carthusian* (an order of Roman Catholic church) monk at Cologne in Germany, Werner Rolewinck, wrote about the printing press, in *Fasciculus Temporum* ("Encyclopedia of history"); It is the best selling book of the 15th century by a living author (that is, just after the Bible) and one of the best selling books of all time with the 1475 edition still available at the *National Library of the Netherlands* (and digitized by

Google books). Johannes Gutenberg, a goldsmith in Germany invented the printing press in the 1440s. Gutenberg established the first printing press in Mainz near Frankfurt in western Germany near Belgium. One of the first printed books was the beautiful *42-line bible* ("Gutenberg bible") printed in the early 1450s and now available in many libraries such as the *British Library* in London and the *Library of Congress* in Washington, DC in the US. Books and in particular the printing press of Gutenberg played a crucial role in catapulting humanity to a higher level of civilization as the good monk observed, a mere 25 years after the invention. Although books existed before the printing press they were very expensive as book production was a very manual, labor intensive process having to be copied by scribes one book at a time.

Codex, made up of *parchments* or *papyrus* pages (see the chapter on *paper*), written by scribes existed before the printing press. This in itself was an important invention in the history of the books evolving from ancient scrolls (Latin, *"volumen"*). (In computer parlance, it is the difference between tape and disk drive with codex providing quick access to a page without having to unroll the unwieldy scroll). One of the well known codices is the beautiful *codex sinaiticus* ("Sinai Bible") from the 4th century discovered at Saint Catherine's Monastery near Mount Sinai in Egypt bordering Israel, currently at the British library; This is one of the oldest known bibles in the world, written manually on parchment pages. However, codices were rare and expensive, both because of the expense of the parchment and the manual effort required to reproduce each copy, making codices out of reach of most people at the time. There were some precursors to Gutenberg's press including *Jikji* ("Anthology of Great Buddhist Priests' Zen Teachings") in Korea printed in 1377 with movable types now at *Bibliothèque nationale de France* in Paris (available online).

A BBC program on the printing press says that every 8 pages of parchment "requires a sheep or a goat". This would have made codices not only very expensive but very cruel for animals leaving aside the cost. It is a form of human progress that we now hesitate to print on paper to save the environment when modern paper and books themselves were a vast improvement over parchments. Interestingly, Gutenberg himself printed some of his original bibles in *vellum* (expensive, high quality parchment) as expensive *premium editions.*

For the printing press to happen multiple technological factors had to come together which happened by the 15th century in Europe:

- Concept of a *pre-book* consisting of multiple pages in the form of *Codex* as described above
- Mature *paper technology* which made paper widely available and cheap
- *Movable type* or metal blocks representing alphabet and numbers, an innovation over full page wood blocks used for illustrations in codices
- A *mechanical press* which was used to press agricultural commodities such as grapes and olives in agriculture
- In addition Gutenberg himself invented an important factor, an *oil based ink* ("varnish") with lead and copper pigments, now a signature of the master's work, instead of typical water based ink used in fountain pens

Printing press, *the art of arts, the science of sciences,* ranks very high as the "mother" of all inventions as books spread the knowledge required to create all the other inventions. It created important inventions such as the newspapers, *"the press",* a foundational pillar of modern democracy. Printing press is used to print the paper currency of the modern economy. (Interestingly *Koenig & Bauer*, founded in 1817 at Würzburg near Mainz in Germany provides printers for modern government mints printing money). The printing press lifted Europe out of the dark ages and into the *Renaissance* and the industrial revolution. It created modern democracy, economy, science and the very enlightenment itself. It provides a way for us mortals to become immortal by communicating to people far into the future and in a far away place, more than any religion can hope to provide. I read Adam Smith's *Wealth of Nations,* written in 1776 before American independence, *245 years later* in 2021 in California, a *distance of 8000 km* from San Francisco to Edinburgh, a real example of books and human ideas truly transcending time and space.

Clock

"It is now generally conceded that the production of a portable timekeeper was accomplished by Peter Henlein or Hele, a clock-maker of Nuremberg, who was born in 1480 and died in 1542. He, shortly after 1500, used a long ribbon of steel tightly coiled round a central spindle to maintain the motion of the mechanism. The invention has been ascribed to Habrecht and others, at a much later date, but Johannes Coccleus, who was born in 1470, in his commentary dated 1511, accurately describes a striking watch and distinctly credits its introduction to Henlein.

...The earliest watches are scarcely to be distinguished from small table-clocks. The case was a cylindrical box, generally of metal, chased and gilt, with a hinged lid on one side to enclose the dial, the lid being engraved and usually pierced with an aperture over each hour, through which the position of the hand might be seen. Most of the watches were provided with a bell, on which in some cases the hours were sounded in regular progression; in other instances the bell was merely utilized for an alarum. The cases were usually worked à jour to emit the sound.

At the South Kensington Museum is a circular table clock, about three inches in diameter, in an engraved brass case having a perforated dome

surmounted by a small horizontal dial as shown in Fig. 64. On the inside of the bottom cover is inscribed , " P. H. Nor . . 1505." Mr. Mitchell, of the South Kensington Museum, suggests that " Nor " stands for Norimbergue, " at Nuremberg," and accepting this explanation one may reasonably conjecture that the original parts of this clock are the handiwork of Hele. The plates of the movement are of steel , but the balance and its accessories are comparatively modern. "

The book, *Old Clocks and Watches and Their Makers*, published in 1899 describes the very first (portable) watch of 1505 by Peter Henlein of Nuremberg, now known as the *Watch 1505* and valued at US $50-80 million. We need some periodic activity which serves as the smallest unit of time we can measure. Incidentally, this also means *spacetime* must be *discrete* [quantized]. If we need some periodic activity to measure time, the period of that activity which is some finite quantity must be the *smallest quantum of time* we can measure. Currently in the *SI* system, this is defined as the transition period between two *Cesium* energy levels and the *second* itself is defined as (exactly) *9 192 631 770* times this smallest quantum. That is we can't measure or distinguish a time smaller than 1/9 192 631 770 of a second. According to Einstein's *relativity* (and finiteness of *speed of light*) time and space are equivalent which means we can't measure a distance smaller than the space quantum corresponding to this time quantum. (In fact now *metre* is defined in terms of second and the *speed of light* in the SI system). This makes all of spacetime discrete and quantized as we always need this periodic activity to measure time; Continuous, infinitely sliceable, spacetime of *real numbers* and *calculus* is just a figment of our imagination. We can't measure any time period less than around one in 10 billionth of a second at present.

Early people realized many phenomena around them are periodic, day and night, waxing and waning of the moon, seasons and so on. They used these periodic activities as the *quantum of time* to measure time. However a day is too coarse to be a useful unit of time. Sundials (shadow clocks) were invented to split a day into fine grained units called "hours". Oldest known sundial is from the *Valley of the Kings* near Luxor in Egypt from around 3500 years ago. Interestingly this device has

12 hours which is where our 12-24 hour clocks must have originated. It also has some dots between hour lines likely to measure parts of the hour. From then on the precision of the technology with water clocks, hourglasses and few mechanical clocks didn't improve much till the middle ages. In Europe in the middle ages, church towers had the only clock in the town and indicated every hour with a bell (*Glocke* in German). A big breakthrough came when Galileo had the idea to use a pendulum as the periodic activity (quantum of time) for clocks. In 1657, Christiaan Huygens at the Hague in the Netherlands built the first pendulum clock which was able to measure a second. They were the most accurate clocks for 300 years. The invention of the *Chronometer*, a very accurate clock unaffected by the motion of ships, by John Harrison in England to accurately measure time in ships was a critical maritime and cartographic technology (see the chapter on *map*) in addition to a big step in time keeping. First quartz crystal clock was invented by Canadian Warren Morrison at Bell labs in New York in 1927. *Seiko*, founded by Kintarō Hattori in 1881 at Tokyo in Japan, released the first quartz watch in 1969. Quartz watches typically use a frequency of 32768 (2^{15}) Hz. The first atomic clock with better accuracy than a quartz clock, based on *caesium-133*, was built by Louis Essen and Jack Perry in 1955 at the *National Physical Laboratory* in London in the UK.

Clock ranks as one of the most important inventions of all *time* (pun intended). Modern science and technology, Newton's laws, Maxwell's equations, computers and mission to the moon, is not possible without a way to accurately measure time. Modern life is tied to the clock in terms of our schedules and appointments at work and leisure. Modern computers with their clock cycles and doing one unit of work per clock cycle would have been meaningless without a clock to produce this constant time quantum. It is hard to rank among *numbers*, the *printing press* and *time keeping* given their critical importance in modern human history spanning the last 500 years. Numbers and books played a more widespread role in modern science and mathematics. It is particularly true in areas such as biology, as in Darwinian evolution or Mendelian genetics, where [fine grained] time keeping didn't play much of a role. Accurate time keeping, starting with the pendulum clocks, is more recent than both numbers and books. This leads us to rank the clock very high and just below numbers and the printing press.

Calendar

Calendar is very important for human history for the simplest of reasons. Without a calendar we can't even say that the world war started on *Sept 1, 1939*. Technically we can still describe it in terms of elapsed days by saying it started 29,753 days ago. But it obviously has the problem that we need to say when this particular statement was made, Feb 15, 2021 in this case, as the number of days will increase as the time passes. We can see why it is particularly important for political or economic contracts. In short, we need a reference day identified by a special event like the birth of the Christ or *Hijra*, Prophet Muhammad's flight from Mecca to Medina which counts as the day *one* and from which we start counting. This technically solves the fundamental problem associated with the calendar, that of identifying a reference date all of us can agree to. We can count months by dividing with a certain number of days, say 30 days, roughly the time between two full moons. We can count 12 months to get a lunar calendar or skip the months altogether and divide by 365 days to count the number of years once we know that earth's revolution around the sun takes that many days. A year as 365

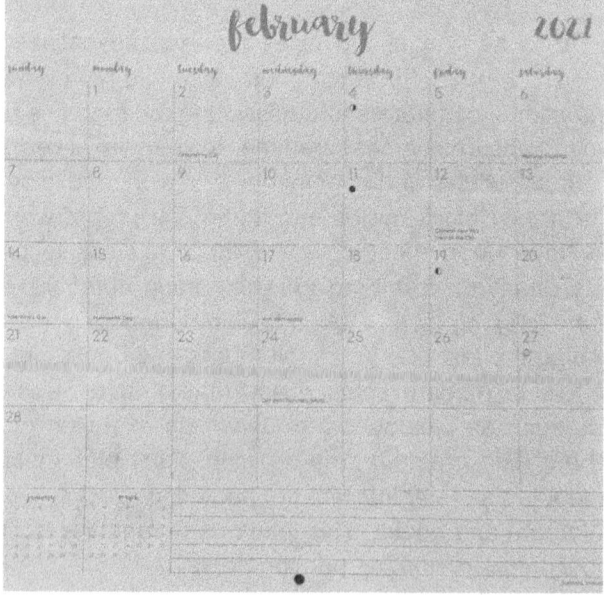

days would have been defined by carefully observing the duration of day and night and in particular *equinoxes* - days of equal day and night, 20 March and 23 September in the *Gregorian calendar*. The time period between two *vernal* [summer] equinoxes is [roughly] 365 days or a *tropical year*. Incidentally, the incommensurability between earth and moon's periods of rotation, year and month respectively, causes all the confusion between lunar and solar calendars.

First calendar was invented by Sumerians around 2300 years ago and was (roughly) a *lunar* calendar. However it was inaccurate with ad-hoc extra months added whenever seasons went out of sync with the calendar. This formed the basis for the well known and advanced *Babylonian Calendar*. These calendars used both the moon and the sun and resolved the incompatibility between their periods using some ad hoc schemes such as inserting *intercalary* days or months. First proper calendar is from Egypt that used a 365 days calendar for the first time with 12 months and 5 extra days. There is a beautiful papyrus legal document from Egypt in the British museum showing a date of *"month of Mesore, year 4 of Ptolemy III Euergetes I"* (September 18-October 17th 243 BCE). This calendar lost a day every 4 years as it didn't include the extra one-fourth of a day in a *solar year*. In 46 BCE, Julius Caesar introduced the *Julian calendar* with modern calendar months and days but without the leap year. Although it was the most accurate calendar for more than a millenium even this went noticeably out of sync with the seasons. This difference was important for religious reasons such as the day of *Easter*. Hence Pope Gregory XIII introduced the modern calendar called the *Gregorian Calendar* correcting for leap days in 1582 and Roman Catholic nations adopted it immediately. Britain and her colonies including America adopted the Gregorian calendar in 1752; Greece adopted it only in 1923. Aloysius Lilius (Luigi Lilio) at Verona near Milan in Italy provided the fundamental mechanism behind this calendar.

There is no particular physical reason the period of earth's orbit, the *year*, should be an integer multiple of earth's rotation period, the *day*. That is there is no reason earth should complete a full rotation when it reaches a *reference point* in its orbit after completing a full orbital motion. (They are two independent parameters of the system in modern parlance). This scientific understanding gave rise to the Julian calendar. Using the same argument, there is no physical reason that earth should complete a nice fraction of a full rotation such as ¼ of the full rotation Julian Calendar assumes when it completes a full orbital motion at a reference point. This led to the Gregorian Calendar that improved a year from 365.25 days to 365.2425 days. To add to further complication there are two kinds of *year*.

- The *tropical year* is the year we measure from one equinox to the next equinox [of the same kind, two successive summer equinoxes for instance]. That is, it is the time for the sun to return to the same position (with respect to the reference frame of far away stars, a constant *reference frame* for practical purposes) *as seen from the earth*. Tropical year is what matters for plants and animals on earth as beings dependent on *photosynthesis* as it controls the seasons. This is the typical year we refer to when we say *year* without further adjectives.
- The *sidereal year* is the time earth takes to complete one full orbit around the sun with respect to far away stars, which is longer than the *tropical year* by about 20 minutes due to the *precession of the equinoxes*.

Hipparchus at Nicaea near Athens in Greece was the first to describe the difference between these years after observing the precession of equinoxes, slow rotation of earth's axis in a cone once every 26000 years - approximately, minutes in a year(525600) / 20 minutes.

Modern society can't function without a calendar for planning and scheduling. Its importance is just below writing, numbers or time keeping and ranks higher than most of the inventions in history. Modern trade and economy can't exist without a calendar with its dates and contracts; Hence money as an instrument of modern economic exchange ranks lower. As mentioned in the beginning it is very important for recording human history accurately. Calendars are older than the printing press. However, unlike the calendar, books played a crucial role in the explosion of knowledge, a prerequisite for all modern inventions.

Money

Charles Wheelan, in the book *Naked Money* explains the three purposes of an *ideal* money whether it is gold, currency or jail cigarettes. (It is part of his excellent *Naked* series that explains very hard concepts to non-expert readers in an intuitive and funny way, *Naked Economics* for instance.):

1. Money serves as a *unit of account*. It can be used to place monetary value on things in a uniform way so that everyone has the same understanding of value. That is "The house costs 250,000 dollars" is more understandable than "The house costs 10,000 bags of rice" or even "The house costs 250 bitcoins" (as bitcoin fluctuates too wildly for us to have an intuitive understanding of the worth of one bitcoin).
2. Money is a *store of value*. It gives us a way of accepting payment for something now and using this purchasing power later. That is, it is part of our wealth or asset just like a bag of rice.

3. Money is used as a *medium of exchange*. It can be used to conduct transactions with relative ease. Jail cigarettes serve the same purpose in jail even if (or particularly if) you are a non-smoker.

In a primitive state with no division of labour there is no need for money as everyone builds their own house, grows their own food and so on. Once the complex society developed and division of labour started in full force, money instead of complex barter exchange would have been the next logical step in the evolution of the economy. It is impossible to organize a complex economy without a universal medium of exchange. The site *themoneyquestion.org* explains the crux of the issue with barter exchanges [emphasis added]:

"According to Smith's story, **money emerged with increasing productivity and the division of labour,** as individuals found themselves without many of the necessities they required but at the same time an excess of their own products. Without a common means of payment individuals had to resort to barter in order to trade, which was problematic **as both sides of the deal had to have something the other person wanted (the "double coincidence of wants").**"

Money is very different from wealth as *"the father of capitalism"*, Adam Smith makes very clear. Wealth or *capital* is something which is productive such as house, factory, stocks and more importantly nowadays, based on American economist Gary Becker's theory, *human capital* which explains the difference between Singapore being wealthy without any natural resources and Nigeria being poor despite huge oil wealth. In fact, as most modern investors know, keeping one's assets in money is actually one of the most unproductive forms of capital. However all assets should be converted to this universal medium of exchange for us to be able to buy anything in the economy.

A sedentary society with agriculture and division of labour is necessary for the idea of money to form. Cattle and *cowry shells*, of organisms literally called *Monetaria moneta* ("money cowry"), were used as the first forms of currencies. Then rare and precious metals like gold and silver started to get used as money. First standardized coins made of *electrum*, a natural alloy of gold and silver, were invented in the *kingdom of Lydia,* at Mediterranean sea in Western Turkey overlooking Greece, around 2700 years ago. First paper money called *Jiaozi* originated in *Song dynasty* China around 900 years ago in the 11th century. This evolved from merchant receipts, used by merchants as a shortcut to carrying sacks of heavy coins. (For comparison, in currencies backed by gold, currency was a shortcut to carrying sacks of gold). Marco Polo's book *The Travels of Marco Polo* has a chapter titled *"How the Great Kaan Causeth the Bark of Trees, made into something like Paper, to pass for Money over all his Country."* This shows it must have been a very novel idea in 13th century Europe. Most of these paper currencies were backed by a real asset like gold (*gold standard*) till a century back. However modern forms of Money like the US dollar is literally *fiat money* and is not backed by any commodity such as gold, for good reasons having to do with the depressions in the early 20th century particularly in the US and the UK. British *Pound Sterling* is now the oldest living currency in the world with more than 1200 years in continuous use. The *Euro*, European common currency, is used by 19 countries in Europe.

As a foundation for the modern economy, money is ranked just after foundational inventions of humanity and among the top inventions of modern history. The printing press and books are ranked higher as they had impact in multiple areas of humanity including politics, religion, economy and science. Efficient division of labour and hence complex civilization required for other inventions would have been impossible without money. It also greatly expanded trade as carrying money is much more convenient than any commodity. Trade networks served to spread ideas far and wide which was critical for spreading of modern foundational inventions like the printing press quickly. Money and division of labour made it possible for inventors such as Thomas Edison or Henry Ford to specialize in their fields and earn a living instead of having to farm their own field for food or make their own shoe. It [along with patent laws] made it easier to sell modern inventions efficiently, providing incentives for *risk taking* required for inventions.

Ruler

"WASHINGTON, Oct. 30. - The president has officially promulgated, through the state department, the convention which has been ratified between the United States and the nations of Europe (except Great Britain.) together with most of the South American Republics, providing for the establishment and maintenance, at Paris, of an International Bureau of Weights and Measures. The duties of this bureau will embrace the custody of the international prototypes of the metre and kilogramme, and periodical comparisons of them with the national standards and with the fundamental standards of the non-metric weights used in the different countries, for scientific purposes. The bureau will also, among its other operations, make comparisons and verifications of scales of precision &c. at the request of scientific societies, constructors, or men of science. The treaty was concluded in May 1875, but the ratifications thereof have only recently been exchanged."

This is an article titled, "THE WEIGHTS AND MEASURES TREATY" on Oct 31, 1878 from *The New York Times*. The treaty it talks about is the metre convention on 20 May 1875 that created the International Bureau of Weights and Measures (BIPM), the standards body that manages the International System of Units (SI). Now, the United States, Myanmar and Liberia are the only countries in the world still clinging to miles, pounds and gallons and don't use metric systems (with expensive consequences as we will see later in the chapter). In

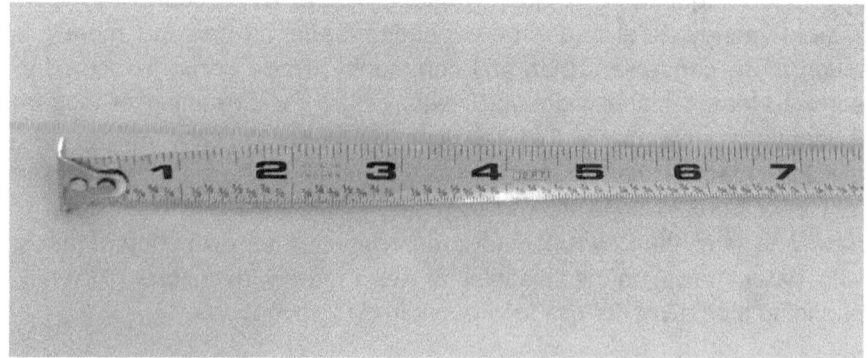

1790, French Parliament asked the French Academy of Sciences to propose a system of measurements that would be suitable for adoption by the entire world. France adopted the work of the academy as the *metric system* in 1795. It defined (among others):

- *Meter*, the measure of length equal to the ten-millionth part of the arc of the terrestrial meridian between the north pole and the equator.
- *Gram*, the absolute weight of a volume of pure water equal to the cube of one hundredth part of the meter at the temperature of melting ice.
- *Liter*, the measure of capacity, both for liquids and for dry matter, the capacity of which will be that of the cube of tenth part of the meter.

Many countries in Europe, Central America and South America started adopting the system in the mid 19th century. In 1875, 17 nations including France and the US met in Paris in the Metre convention mentioned above, making the metric system a worldwide system that would eventually lead to the SI units. The SI (from the French *Le Système International d'Unités*) was established in 1960 by the 11th General Conference on Weights and Measures (CGPM, *Conférence Générale des Poids et Mesures*). In 1965, the United Kingdom, birthplace of imperial units, adopted the SI units followed by other nations in the commonwealth such as Canada (1975).

Now units are being redefined using eternal fundamental constants of nature instead of man-made objects such as the *platinum-iridium standard weight* in Paris. In the 1967 CGPM conference, *second* was defined as the duration of 9 192 631 770 periods of the radiation corresponding to the transition between the two hyperfine levels of the ground state of the *Cesium-133* atom. This implies we can't measure a duration smaller than 1/9.2 nano seconds (see the chapter on *clock*). In the 1983 Conference, *metre* was defined using the speed of light c; One metre is the length of the path travelled by light in vacuum during a time interval with the duration of 1/299 792 458 of a second. In the 2018 conference, *kilogram* was defined by taking the value of the Planck constant *h* to be (exactly) 6.626 070 15 x 10^{-34} kg m^2 s^{-1} and working out the unit for kilogram from the units of metre and second.

Origin of the system of measurements is unknown as each culture locally adopted systems based on human anatomical features such as hand, foot or fingers. The *mile* comes from *milia passuum*, Latin for a thousand paces, *pace* referring to normal walking steps; The Romans were the first to use the mile. The *pound* is from Latin *libra* including abbreviations **lb.**, **£** that reflect the medieval custom of keeping accounts in Latin. The *gallon* originated from wine measurements in England, with French *jale* probably referring to a bowl. Confusion between measurement systems can be very costly as NASA found out the hard way. *The New York Times* of Oct 1, 1999 says:

"Simple confusion over whether measurements were metric or not led to the loss of a $125 million spacecraft last week as it approached Mars, the National Aeronautics and Space Administration said today. An internal review team at NASA's Jet Propulsion Laboratory said in a preliminary conclusion that engineers at the Lockheed Martin Corporation, which had built the spacecraft, specified certain measurements about the spacecraft's thrust in pounds, an English unit, but that NASA scientists thought the information was in the metric measurement of newtons.

It is not known with certainty what happened to the Mars Climate Orbiter. At first there was speculation that it crashed on Mars or burned up in the atmosphere. But the review team now tends to think that the spacecraft might have left Mars's orbit and is now orbiting the sun. Under this theory, the spacecraft approached too close to Mars and got too hot, causing the engine that was supposed to bring the craft into orbit to stop functioning, so the Orbiter went back into space."

A system of measurements is a critical invention for the modern economy and the overall society. The incident above illustrates the importance of a uniform measurement system in human endeavors and cooperation. Its importance is comparable to Money's role as the *unit of account* in the modern economy and ranks just after Money.

Map

Three Japanese tourists came unstuck on their planned Australian holiday on Thursday when they abandoned their hire car in Moreton Bay after they tried to "drive" to North Stradbroke Island. The low tide and a GPS navigation system lured them into the bay at Oyster Point at Cleveland.

Their planned adventure to Straddie ended at 11am and the incoming tide soon forced them to seek help and abandon the vehicle. By 3pm the car was stranded in two metres of water and the subject of much amusement from onlookers on the shore and passing boat and ferry traffic.

The Tokyo students had wanted to take a day trip to Straddie and believed their GPS unit would be able to guide them there. The GPS forgot to mention the 15 kilometres of water and mud between the mainland and the island.

Yuzu Noda, 21, said she was listening to the GPS and "it told us we could drive down there". "It kept saying it would navigate us to a road. We got stuck . . . there's lots of mud."

In 2012, three tourists from Japan tried to drive a car to *North Stradbroke Island*, an island near Brisbane in Australian east coast by literally following maps in their GPS unit through the sea from the mainland. This is an extreme case of how much we have come to rely on maps [and GPS] in our society. *Imago Mundi* at the British Museum is one of the oldest maps in the world. It is from the babylonian period around 2600 years ago, discovered near Baghdad in Iraq. Most of the old maps from this age were inaccurate and assumed a flat earth. Ptolemy's Greek text *Geography* written in the 2nd century CE at Alexandria in Egypt assumes a spherical earth and introduces the *latitudes* ("breadth") and *longitudes* ("length") for the first time. However no map actually survived from his text and maps bearing his name were later reconstructed in the middle ages from his text. Columbus used these *Ptolemy maps* which didn't know about the Americas and also used longer Arabian miles, leading him to think that Asia was much closer, a map accident but with a better ending.

Maps allowed people to navigate the world and equally importantly, get back home. Maps helped Columbus and other explorers find new worlds such as North America and South America. In 1507, German Martin Waldseemüller at Basel in Switzerland produced the first world map with the Americas and the name *"AMERICA"* for the new continents. (The *Museo Galileo* and the *Library of Congress* have digitized this map in stunning detail). Belgian Gerardus Mercator at Duisburg near Dusseldorf in Germany introduced the *Mercator projection* in his world map of 1569, the first modern map with all the modern conventions. This shows the familiar north south alignment of all the continents we see in modern maps [including its distortion of size near the poles]. This was a great navigational aid, as dead reckoning in ships would become straight lines in this projection albeit at the cost of distortion near the poles by enlarging land masses. For instance, Greeland looks to be the same size as the whole continent of Africa; In reality, Africa is bigger than India, China and the US combined with room left over for more countries as shown in a popular map called *"True size of Africa"*. In Mercator map lines of constant bearing, called *rhumb lines*, on spherical earth are the straight lines, not great circle segments; In particular, longitudes are the vertical, north-south, straight lines and latitudes are the horizontal straight lines.

Timekeeping devices are required to correctly compute longitudes based on local time [from the sun, when directly overhead with no shadows at noon] and the time at a reference location like Greenwich. This means every hour corresponds to 15 degrees of longitude, 360 degrees of the circle/24 hours in a day. In fact the invention of the *Chronometer* by John Harrison in England to accurately measure time in ships can be considered as one of the important breakthroughs in *cartography* ("map making") as it helped measure longitudes accurately. This invention was driven by a prize offered by the British government after more than 1400 sailors lost their lives in the *Scilly naval disaster* of 1707. It literally became a matter of life and death to accurately know the position, latitude and longitude, of the ship on the map. The ability to calculate the position accurately gave the british empire a powerful maritime advantage for military and trade.

Map as an invention has both the modern impact in our daily lives and the historical impact in human society. Maps are very important for our day to day navigation in their modern technological forms such as Google maps with GPS. It is required in all forms of travel - ships, automobiles and airplanes. Maps rank very high among human inventions in recorded history just behind our foundational inventions of pre-history such as fire and wheel. It is ranked after writing, numbers and *ruler* as maps require them as a prerequisite. Printing press had a much bigger impact in expanding knowledge and producing future inventions including making the very process of cartography easier and affordable. Map ranks above other modern inventions, particularly the ones from the industrial revolution. For a concrete comparison, between electricity and maps humanity has been living with electricity for around 200 years and with maps for around 2000 years. Given the time, maps played a bigger role in human history than electricity and deserves its rank above electricity.

Ship

"Be it so. This burning of widows is your custom; prepare the funeral pile. But my nation has also a custom. When men burn women alive we hang them, and confiscate all their property. My carpenters shall therefore erect gibbets on which to hang all concerned when the widow is consumed. Let us all act according to national customs."

71% of our planet is covered with water, 96.5% of it as oceans. Then it is no wonder that one of the most important human inventions was to tame the high seas. Egyptians invented the boats around 5,500 years ago based on the drawings on egyptian jars at the Brooklyn museum in New York dated to 3450-3350 BCE. Egyptians very likely used the boats to sail in the *Niles* river which was their lifeline and main transport mechanism. Austronesians from Taiwan island spread to Indonesia and all the Austronesian islands such as Guinea, Tonga, Hawaii and surprisingly Madagascar to the west starting 5000-3000 years ago in some form of outrigger boats such as dugout canoes with outriggers. Given the distances involved, around 10,000 km (6,000 miles) from East Asia to Madagascar, they definitely had very good boats. Ships played a big role in the discovery of new worlds such as the Americas and Oceania and the integration of the whole world into a single, global village for the first time. It also played a big role in Eurasian's dominance of almost all parts of the world and elimination of indigenous populations such as the native Americans. It played an

important role in the colonisation of countries such as India, changing their history completely. India went from the *Mughal empire* with myriads of kings and princes to an exploitative corporate rule by the *East India company* to a colony of the British empire with recurrent famines and poverty with Indians as second class citizens in their own country to the biggest modern democracy it is today, in 500 years from the arrival of Portugese explorer Vasco de Gama to *Kozhikode* (Calicut) port near Bengaluru in India on 20 May 1498. Without ships, India's historical trajectory would have been very different as there would have been no *East India company* and very likely no British empire spanning the globe. It would have remained an amalgam of princely states at war with each other or likely evolved to be an union of multiple independent countries with different languages similar to the European union instead of a unified modern democracy (or three if we include Pakistan and Bangladesh, parts of former British India). India also wouldn't have seen railways, telegraph and other inventions of the industrial revolution and enlightenment as early as she did. Commander in chief of India in 1848, Charles James Napier's quote on *Sati* above shows the positive impact on India of this collision thanks to the ships.

Spain's position as a global power in the 16th and 17the centuries was due to a large extent to her global exploration using ships and trade with her colonies. Dutch *Golden Age* (1581-1672) was defined by their maritime trade, with dutch accounting for more than half of all the European trade during this period. Later on, Britain became dominant due to her naval power. As recently as 1982, British were able to defeat Argentina in the Falklands war due to the superiority of her navy. Much earlier, vikings with their longships became rulers of the seas in the 9-13th century. In every century in recent history, the country with the best navy and ships became the dominant global power including the US in the 21st century with her aircraft carriers and nuclear submarines. In 2019, ships carried 11 Billion tons of goods in a year while only 62 million tons (0.5%) of goods were shipped by air, demonstrating their importance to modern trade. The March 2021 blockade of the Suez Canal by *Ever Given* ship held up 10 Billion dollars in trade every day.

The 19th century saw the replacement of sail ships of the past five millennia with steam powered ships combining two high ranking human inventions. American inventor Robert Fulton at Bethlehem near Philadelphia built the first commercially successful steamboat, *North River Steamboat* (*Clermont*), in 1807. (He had also built the first submarine in history for France, *Nautilus*, in 1800). This boat carried passengers between New York City and Albany in northern New York in the US. *SS Savannah* was the first steam ship to cross the Atlantic in 1819 but it was a hybrid sail/steamship and the steam engine was used only for part of the journey. By the 1830s steam ships became common across the Atlantic. *HMS Argus* of the British navy was the first *aircraft carrier*, ship carrying military aircrafts, and was launched in 1918. The 20th century saw the steam engines replaced with diesel engines, gas turbine engines and interestingly, nuclear powered steam engines. Nuclear powered steam turbine engines are now used in military aircraft carriers, submarines and icebreaker ships.

Ships rank very high as an invention which played a significant role in human society. It was critical in the discovery of new worlds recently by Columbus and other explorers and by Austronesians around 3,000-5,000 years ago. It changed the course of history in the Americas, India and Africa among other places due to colonialism. Slave trade changed the lives of many Africans who were transported like commodities in large numbers and still suffer its effects in places like the United States. There were 12 million people transported as slaves, when the world population in 1600 was only around 500-600 million people, working out to 3% of the people in the world at the time (4%, if we include the few million people who died during the voyages). British empire mainly owed its strength to its excellent navy and ships; English language owes its status as the *lingua franca* of the world to ships. Ships like the *Mayflower* are very important in American history and culture. Without pearl harbor attack on American navy ships by Japanese ships America may not have entered the second world war, at least not as early as 1941. Money and maps play a more important role in the daily lives of most people and ranked higher. More than airplanes or the internet, ships made the world a *global village*, particularly when there were no passports till a century back. For this reason, ships rank very high in the list of inventions.

(Image credit: Deutsche Welle)

Iron metallurgy

The *Petrie Museum* in University college London, England contains small iron beads, once part of a necklace, dated to 5200 years ago. This is considered the earliest known use of iron by people. However this iron didn't come from the earth; It literally fell from the sky, as a meteorite. Although an iron dagger from *Alaca Höyük* near Capital Ankara in Turkey was initially claimed to be made of *smelted iron,* that is extracted by human effort from ore, later studies dispute this and it very likely has *meteorite iron* as well. Iron age in Asia and Europe started around 3200 years ago with iron metallurgy and tools widespread by around 3000 years ago. This ancient smelting is called *bloomery smelting* in which iron doesn't become liquid as it is not hot enough and produces a bloom which is worked on directly. In modern blast furnaces with a temperature of around 2000°C iron melts completely and forms *Pig iron*, raw material for steel and other iron alloys. In 1856, Henry Bessemer at London invented a process for mass scale steel production by mixing cold air into the molten pig iron in a converter called *Bessemer converter*. This started the modern industrial scale steel age including trains, skyscrapers and industrial equipment. Steel is so ubiquitous now, almost 2 Billion tonnes of steel is produced every year in the world. For comparison, cereal production is around 2.7 Billion tonnes a year and unlike cereal most of that steel remains with us for a long time. It is

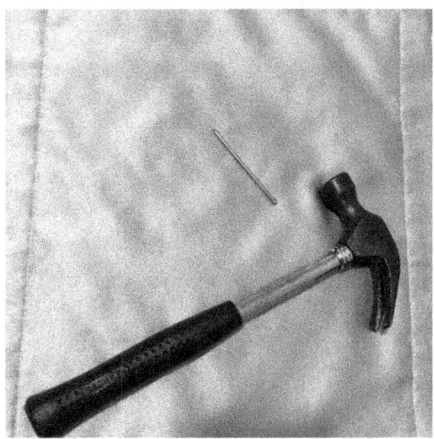

estimated that each of us use [per capita] 230 kg (*500 lb*) of steel per year. The main types of Iron are:

- Pig iron - crude iron formed in the first stage of iron production with 3-4% carbon
- Cast iron - Iron with high carbon content (2-4%) making it non-malleable and brittle. As the name implies it is cast using molds similar to plastics.
- Wrought iron - Nearly pure iron with almost no carbon content; It is malleable and can be shaped by hammering or rolling. It was historically important but became commercially insignificant after the availability of steel with the Bessemer process.
- Steel - Iron alloys with low (less than 1%) carbon and combinations of other metals such as chromium and nickel in stainless steel and chromium, manganese, molybdenum, nickel, tungsten and vanadium in tool steel.

Before the Bessemer process, steel was produced by *adding* carbon to wrought Iron in a very time consuming process lasting days. The Bessemer process produced steel by *removing* carbon from pig iron; The oxidation of impurities by air, a violent *exothermic* reaction requiring the special converter, sustained the high heat required for removing the carbon; The process was so fast that the pig iron was converted to steel in 20 minutes. It is only natural that we use it heavily as Iron is the most common element on earth with more than 32% of earth made up of iron. One important form of iron is stainless steel which contains around 11% of chromium along with iron and carbon. Iron smelting took time to develop as it requires very high temperatures. An article on The University of Pennsylvania Museum of Archaeology and Anthropology (*Penn Museum*) states:

"So what is so difficult about smelting iron? The ideal temperature for smelting iron is between 1100°C and 1400°C—much higher than required to smelt any other metal known in antiquity (e.g. copper can be smelted at temperatures below its melting point of 1083°C). While these temperatures will not melt iron—which melts at 1540°C—they will melt the non-metallic minerals (especially silica) found in iron ore, allowing them to drip away in the form of slag from the solid iron in a furnace. ... In short, compared to copper smelting, iron production requires richer ores, larger fuel supplies, closer controls over furnace air supplies and composition, and a great deal more post-smelting work before anything that even looks like a metal is produced."

Iron and steel is everywhere. The *Golden Gate bridge* contains 88,000 tons of steel; The *Empire state building* contains 60,000 tons of steel; Romania's *Palace of parliament* building used a whopping 700,000 tons of steel. Railway lines contain 50kg or more of steel per meter of track. India had 95,981 km of railway track in 2019. This translates to 4.8 Million tonnes of steel in Indian railway tracks alone. China, USA and Russia have even longer tracks. 1.5 to 2 million tons of steel is shipped to the appliance, utensils and cutlery market in the US every year according to the *American Iron and Steel institute*. It also says 75% of the weight of household appliances come from steel. China leads the world in steel production with almost a billion tonnes, half of the world production, in 2018 followed by India with around 100 million tonnes. Australia produces around a billion tonnes of iron ore every year followed by Brazil with around half a billion tonnes [in 2020]. A tonne of steel costs around $700 or only $0.70 per kg as compared to gold with a price of around $55,000 per kg or aluminium with a price of around $2 per kg making it one of the most affordable materials.

Iron metallurgy is one of our foundational inventions. Without knowing how to work iron, most of the inventions in the list such as steam engines, skyscrapers and rockets won't exist. What the printing press did to ideas, iron did to physical items. It has been with us for around 3000 years, much longer than the printing press. However ideas are more powerful than physical things; For instance, radio replaced wired telegraphy but the idea remained the same, encoding the *information* in a medium that travels fast. As very powerful and transformative ideas, timekeeping, calendar and money also rank higher than iron. Ships, made of wood, have been with us for longer than iron and steel. We are still living in the Iron age which will extend all the way to our species' existence, *appropriate for a species living on a planet with ⅓ of it Iron*. Our mastery of Iron and Nitrogen, the most common elements on earth and air respectively, are two of the turning points of human history leading to the ability to support large human populations (see the chapters on *skyscraper* and *fertilizer*) by increasing the productivity of the limited piece of land.

Knife

Earliest knives were made out of flint stones and predates history. Both the *Science Museum* in London and the *Metropolitan Museum of Art* in New York contain flint knives that predate history such as the *"Knife Blade, Dark Grey Flint, Mousterian, from Le Moustier, Peyzae, France, 10000-5001 BC."* and *"Bifacial Knife with Handle Ca. 2949 B.C. from Egypt"* that are 5,000-12,000 years old. The knife from Egypt looks distinctly modern including the handle except it is made of flint. Bronze knives started to appear around 3,500-5,000 years ago such as the beautiful *"Knife with Ram's Head"* from China during the Shang dynasty (ca. 1600–1046 BCE) or a bronze knife from the middle kingdom of Egypt (2055-1650 BCE) dated to around 3800 years ago. In fact the oldest bronze age object is a knife of *Majiayao Culture* from Gansu province in China near Chengdu dated to 5,000 years ago. *Mycenaean* daggers and swords from Greek islands such as *Karpathos*, *Rhodes* (Ialysos) and *Cyprus* (Enkomi) belong to the bronze age as well, dated to around 3,000-3,500 years ago. Italian student archaeologist Vittoria Dall'Armellina recently discovered the oldest known sword, a 5,000 year old bronze sword from *Kavak*, a settlement near the ancient Greek colony of *Trebizond* in what's now eastern Turkey, from the monastery on the Venetian island of *San Lazzaro degli Armeni* in Italy. In 2019,

19-year-old intern Nico Calman unearthed a 2,000-year-old Roman silver dagger with an iron blade at the archaeological site of *Haltern am See* north of Dusseldorf in Germany.

Knives grew in importance after the iron age starting 3200 years ago. The *British Museum* has a Late Cypriot Iron Knife dated to 3200 years ago (1200-1050 BCE) from Enkomi; It also has Iron swords from Iran and Netherhampton, England, both dated to 3000 years ago. One of the oldest iron daggers is found in the famous *Tutankhaman tomb*, produced with material from heaven literally, made of *meteorite iron* as humanity didn't learn about iron smelting till much later (see the chapter on *Iron metallurgy*); Tutankhaman's dagger at the *Egyptian Museum* in Cairo features a decorated gold handle and a gold sheath with a floral lily motif on one side and a feather pattern on the other; It is dated to 3,300 years ago, 100-300 years before other iron age knives. *Omura Shrine* in Hidaka near Sapporo, a city of 2 Million and the capital of the northern Japanese island of Hokkaido, in Japan has as the *shintai* ("object of worship") of the shrine a sword, the *National Treasure* gilt (gold plated) bronze tachi dated to 1,500 years ago. The legendary *Honjō Masamune* made by Japan's greatest swordsmith Masamune 700 years ago is considered the finest of Samurai swords (*katana*); It disappeared at the end of World War II when swords were surrendered to *Foreign Liquidations Commission* of the US; It hasn't been found so far as of this writing (June 2021).

Japanese, German and Swiss companies have dominated knife making in the world for centuries. Based in Nara near Osaka in Japan and founded by Samurai sword maker Shiro Kanenaga, *Kikuichi* has been producing handmade Japanese blades for over 750 years. They started out producing samurai swords and many of them are now considered *national treasures* in Japan; In 1868, Kikuichi switched over to making knives using similar techniques. *Zwilling J. A. Henckels* was founded by Peter Henckels in 1731 in Solingen between Dusseldorf and Cologne, Germany. Solingen is called the *"City of Blades"* as it has been renowned for the manufacturing of swords, knives and scissors for centuries. Another German great, *Wüsthof,* was founded by Johann Abraham Wüsthof in Solingen in 1814. Böker was founded by Heinrich

Böker in 1869 in Solingen, although they were making sabers from as early as 1829; In 1769, Johann Lauterjung founded *Puma*, another knife maker (no relation to the shoe maker) in Solingen.

Karl Elsener founded *Victorinox*, in Ibach-Schwyz near Lucerne in Switzerland In 1884. In 1891 he supplied the soldier's knife to the Swiss Army for the first time; He went on to develop the *Swiss Officer's and Sports Knife*, now the iconic Swiss Army Knife (*"Schweizer Offiziersmesser"*) in 1897. In 1893, *Paul Boéchat & Cie* (later *Wenger*, maker of *Genuine Swiss Army Knife*) was founded by Paul Boéchat at Courtetelle near Bern in Switzerland; It also received a contract from the Swiss military to supply knives. The two companies split the contract for army knives until Victorinox acquired Wenger in 2005. In England, Sheffield, midway between Edinburgh and London, was well known for knife making from the 14th century. Geoffrey Chaucer refers to it in *The Canterbury Tales* (*thwitel* is knife in the English of his time):

> Ay by his belt he baar a long panade,
> And of a swerd ful trenchant was the blade.
> A joly poppere baar he in his pouche;
> Ther was no man, for peril, dorste hym touche.
> A **Sheffeld thwitel** baar he in his hose.
> Round was his face, and camus was his nose.

Knives, and other blades such as swords, dagger and machete, rank high as a foundational human invention. Its importance should be compared to our innovations such as wheels, fire and agriculture. It helped our ancestors hunt big animals more efficiently, cut and cook them. It is so important in cooking our food, a basic necessity, that we hardly notice it at all. It is used in medicine and surgery as *scalpels*. It played an important role as a war weapon before firearms were invented. It has been with us from prehistory ranking it high.

Pot

"In the Kitchen" are found:

	£.	s.
36 Pewter dishes	12	5
1 copper pott	1	4
Brass kettles of several sorts	6	18
1 large iron kettle and 2 small ditto	1	5
3 iron potts	..	6
1 pr. large andirons	1	..
1 iron back	1	..
3 trammells	1	..
2 frying pans	..	15
3 spirits	..	18
2 gridirons	..	3
1 iron driping pan	..	12

... the list of livestock is thus resumed:

| 1 negro man | 30 | .. |

1 negro woman	25 ..
1 pr. of brass scales	.. 6

The *New York Times* of July 17, 1882, in the article with the subtitle, "INVENTORY OF THE POSSESSION OF A BOSTON MERCHANT IN 1712" lists the kitchen pots and pans that was in use in the late 16th to early 17th century America and likely in England. Pottery was a hunter-gatherer innovation that first emerged in East Asia between 20,000 and 12,000 years ago, towards the end of the Late Pleistocene epoch. Earliest known pottery is from the *Xianrendong Cave* near Taiwan in south-eastern China dated to around 20,000 years ago, predating agriculture by 10,000 years. Japanese *Jōmon period* ceramic pottery used for cooking seafood from around 15,000 years ago was discovered at the northern end of Japanese mainland near Tokyo. Pottery was widespread in East Asia (Russian far east, China and Japan) well before agriculture as containers for cooking food.

Grazia Deruta at Deruta near Rome in Italy is one of the oldest pottery makers in the world, producing *majolica* (tin-glazed italian pottery) from 1500. Deruta is well known for majolica, just as Solingen in Germany is for knives, with the tradition dating back to 1200. *Zwilling Henckels*, the well known knife-maker from Solingen founded in 1731, itself is one of the oldest cookware makers in the world. Luxury home appliances maker *Gaggenau Hausgeräte*, founded in 1683 by Louis William in Black forest region of Germany near Zurich, is one of the oldest companies making cookware. *Royal Tichelaar Makkum*, a Dutch pottery company in Makkum at the North sea near Amsterdam in the Netherlands, was founded in 1572. *Mauviel* in France, founded in 1830 at the Bay of Mount St Michael in English Channel near Paris, was well known for its tin-lined copperware.

Lodge, founded in 1896 by Joseph Lodge, is the oldest extant American manufacturer of cast iron cookware based in South Pittsburg near Atlanta in the US. *The Selden-Griswold Manufacturing Company* was founded in Erie, Pennsylvania in the US in 1865 by Matthew Griswold and his cousins, the brothers J.C. and Samuel Selden; It was one of the oldest and well known cast iron cookware makers in the US; "Griswold" is still one of the most valuable antique cast iron cookwares although the company itself went out of business in 1957. *The Wagner Manufacturing Company*, founded in 1891 by the brothers Milton and Bernard Wagner in Sidney near Chicago in the US, was another well known cast iron cookware maker in the US and was the first to produce Aluminium cookware in 1894. *Le Creuset*, well-known for its colorful, enameled cast iron dutch ovens, was founded in Fresnoy-le-Grand near Paris in 1925 by Belgians Armand Desaegher and Octave Aubecq.

In 1915, *Pyrex containers*, made of heat resistant glass, were introduced by *Corning glass*, founded in 1851 by Amory Houghton at Somerville near Boston in the US. *Non-stick* is often used to refer to surfaces coated with *polytetrafluoroethylene* (PTFE), a polymer of simple molecule CF_2-CF_2 and a well-known brand of which is *Teflon*. Teflon was discovered by Roy Plunkett at *DuPont* in Deepwater near Philadelphia in the US in 1938. PTFE is stable, non reactive and non toxic at temperatures below 400°C (752°F).

Aluminium and copper should not be used with acidic foods as acids can leach the metals. Recycled aluminium cookware in many developing countries contain harmful levels of lead in addition to aluminium itself. Older copper cookware may have tin or nickel coatings and should not be used for cooking. A popular way to combine the excellent conductivity of copper with the safety of steel is *copper base* or *tri-ply* cookware in which a base plate made from a layer of copper and aluminium is fused to a stainless steel inner layer.

Pots and pans rank as one of the most important inventions for one of our basic necessities, food. It has been with us for around 20,000 years, predating even agriculture. It is also one of the inventions we use directly or indirectly every day multiple times. Knife is more useful and ranks higher than pot as food can be cooked directly over fire as in *Tandoori* while knife has no good alternative. As a foundational invention predating history it ranks very high, in the top 20, in our list of inventions.

Furniture

In the *British Museum*, there is a sculpture of a man, with a shoemaker's *last* in hand sitting on a beautiful chair called *Klismos*. Uri Friedman writes in *Atlantic* on the *Klismos*:

In the fifth century B.C., the Greeks invented the klismos, which featured curved legs and a curved backrest, and which Rybczynski described to me as "one of the most beautiful chairs made by anybody." Ever. In his book, he argues that chairs "of equal elegance" to the klismos didn't emerge for more than 2,000 years, until the "golden age" of chairs in the 18th century, when a flurry of creative craftsmanship and global trade produced ornate items like the French Louis XV armchair and Chinese/English cabriole-legged furniture.

In ancient Greek art, "virtually everybody [is] sitting in a klismos chair. We have women, men, gods, and clearly important people, musicians, workers," Rybczynski told me. It was a comfortable, "democratic chair," not a

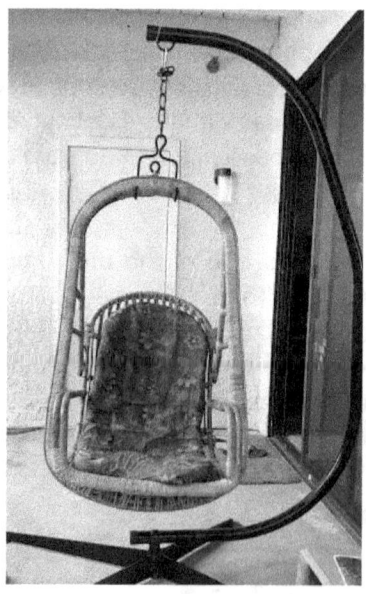

throne. The klismos is also mysterious: It appeared out of nowhere, with a design that was original rather than a variation on a past style, and then disappeared for millennia, only to reemerge as part of the Greek Revival movement in the late 18th and early 19th centuries.

The sculpture from Athens, Greece, *"Marble gravestone of Xanthippos, the shoemaker"* is dated to 2400 years ago. Wooden furniture found in the tomb of King Midas and his father Gordios, known for *Gordian knot* and king of Phrygia in Gordion, near capital Ankara in Turkey midway between *Black sea* and *Mediterranian sea* coasts, are dated to 2700 years ago. The *Metropolitan Museum of Art* in New York City has Egyptian chairs dated to around 3500 years ago such as the *Chair of Reniseneb* or the *Chair of Hatnefer*. The well known *Tutankhamun* treasures including his furniture such as the famous golden thrones and chairs are from this period as well. *Queen Hetepheres' Throne* from around 4300 years ago, a replica of which is at the *Harvard Museum of the Ancient Near East,* may be one of the oldest known pieces of furniture in the world. The oldest mattress, at 77,000-year-old, was discovered in the *Sibudu rock shelter* near Durban in South Africa at the Indian Ocean and was made of compacted layers of grasses and sedges. The *throne of King Dagobert*, at the *National Library of France* in Paris, is a folding stool of bronze, estimated to be from the early 7th century but with some 12th-century additions. Archbishop's chair in the *Canterbury cathedral* near London in England, a large stone seat, known as the *cathedra* or *Saint Augustine's chair*, is estimated to be from the early 13th century.

Monobloc chair may be one of the most useful and affordable inventions in the world. Manufactured in one piece from 2.5 kg (5.5 lb) of *polypropylene*, they're cheap, light, portable, waterproof, stackable and easy to clean and takes a few minutes to manufacture. In 1983, the first mass-produced monoblocs were brought to market by *Grosfillex*, founded in 1927 in Arbent, France near Geneva by Auguste, François and Jean Grosfillex. The *New York Times* traces its origin in *"Celebrating the Everychair of chairs, in cheap plastic":*

"The plastic part of that quest began in the 1950s, when furniture designers, like Charles and Ray Eames and Robin Day, made chair seats from newly invented plastics. The first all-plastic chair, the Universale, was designed by Joe Colombo in 1965. Three years later Verner Panton unveiled the Panton, the first chair to be made from a single piece of plastic. Sleek and sexy, it looked like the perfect industrial object, but had to be made by hand....

These developments culminated in the monobloc's launch in the 1980s ... while the humble monobloc played its trump cards of cheapness and usefulness to stealthily become the best-selling chair in history. The first monoblocs cost $50 each, but as more and more were manufactured, the price tumbled to less than $5 today."

IKEA, a manufacturer and supplier of ready-to-assemble furniture and homeware, is one of the largest furniture brands in the world with a revenue of 47.5 Billion dollars in 2020. IKEA was founded in 1943 by 17-year-old Ingvar Kamprad with its first store in Älmhult, Småland in Sweden near Copenhagen, Denmark. The acronym IKEA is made up of the initials of Ingvar's name (IK) plus those of *Elmtaryd*, the family farm where he was born, and the nearby village *Agunnaryd* where he was raised. IKEA is headquartered in Delft, Netherlands, and operates in 52 countries now.

Furniture ranks high as a foundational invention for humanity. It has been with us for as long as the wheel and very likely longer than writing. Now the worldwide furniture market is valued at more than 500 Billion dollars. There are more than a billion monoblocs alone in the world. Both knife and pot predate furniture, play a more important role in human history and rank higher. Given its antiquity, with it the time for high impact in human history, ubiquity and usefulness, furniture ranks as one of the highest ranked foundational inventions.

Toilet

In 2000, the United Nations came up with eight *Millennium development goals* with a target year of 2015. It included lofty goals such as "Eradicate extreme poverty and hunger" and "Reduce child mortality"; It also had targets for each goals, such as "Halve, between 1990 and 2015, the proportion of people whose income is less than $1 a day" and "Reduce by two thirds, between 1990 and 2015, the under-five mortality rate". How did the world do in these goals and targets, given the condition of the world today?

Contrary to the expectations of naysayers and cynics among us, the world did pretty well on these goals and targets, in some cases reaching the targets a few years earlier than 2015. For the "$1 a day" target, "More than 1 billion people have been lifted out of extreme poverty since 1990" and the world reached the goal by 2011 and extreme poverty fell by 68%, full 18% more than the target. For the "under-five mortality" target, "The global under-five mortality rate has declined by more than half,

dropping from 90 to 43 deaths per 1,000 live births between 1990 and 2015". After the success of the millenium goals, the UN came up with 17 *Sustainable Development Goals* to be achieved by 2030. One of the targets in "clean water and sanitation for all" goal is "By 2030, achieve access to adequate and equitable sanitation and hygiene for all and end open defecation, paying special attention to the needs of women and girls and those in vulnerable situations". *Open defecation* refers to "the defecation in the open, such as in fields, forest, bushes, open bodies of water, on beaches, in other open spaces or disposed of with solid waste". Open defecation spreads infectious diseases, diarrhoea (especially in children), adverse health outcomes in pregnancy, malnutrition, as well as increased vulnerability to violence for women and girls. In 2017, 673 million people (8.7% of humanity) still practised open defecation, down from 946 million in 2015; As late as 2000, In India 66% of the people practised open defecation that was down to 40% in 2015. Some countries such as Niger, Chad, South Sudan and Eritrea still have a prevalence between 60 and 80%. The target is to bring it down to zero around the world by 2030. Modern cities and skyscrapers couldn't evolve without the invention of toilets. Think about someone living on the 50th floor of the building in Manhattan or Singapore having to find a field nearby in a hurry and he would settle for living in the Amazon rainforest as a recluse.

Open defecation became a problem once people started living in dense settlements after the invention of agriculture, particularly when it could contaminate the drinking water supply. The earliest flushing toilets, water removing the waste, are found in *Minoan civilization* on the Mediterranean island of Crete in Greece around 5,000 years ago and the Indus Valley Civilization of present-day Pakistan and India, 4,600 years ago. The Romans treated going to the toilet as a social event with shared public toilets (no privacy barriers). In medieval Europe, the affluent used a *garderobe*, a protruding room with an opening for waste, suspended over a moat or river directly. The modern *flush toilet* was invented by John Harrington, godson of Elizabeth I, who invented a water closet with a raised cistern and a small downpipe through which water ran to flush the waste in 1592. In 1775 Alexander Cummings, a watchmaker in London, developed the S-shaped pipe under the toilet basin that stops the waste water from flowing back and keeps out the foul odours. John Snow, "Father of epidemiology", in London, England in 1854 famously traced the source of Cholera to a water pump supplying water from the Thames river contaminated with sewage. Pit toilets and outhouses were common across the world till the twentieth century. Growth of modern cities with skyscrapers in the twentieth century made toilets with proper sanitation systems very important. The toilet technology is also complicated by cultural norms. Squat toilets where people simply squat on the toilet is common in Asia, particularly India, Japan, China and Turkey in Europe; Western countries use the "western style" toilets with seating. (Japan is converting its public toilets to western-style for the 2021 Olympics visitors).

Founded in 1817 by Georg Horn, *Duravit* at Hornberg, Germany near Zurich is one of the oldest sanitaryware companies. *Kohler Company* was founded in 1873 by Austrian immigrant John Kohler at Sheboygan near Chicago in the US. Now it is one of the biggest sanitaryware companies in the world with a revenue of 7 Billion dollars (2019). *American Standard* traces its origin to *Standard Sanitary manufacturing company* in 1875 at Pittsburgh in the US. *Hansgrohe*, sanitaryware supplier to *Burj Khalifa*, the tallest skyscraper in the world, was founded by Hans Grohe in 1901 in Schiltach, Germany near its competitor Duravit. *Roca*, another multinational sanitaryware giant with a revenue of 2.3 Billion dollars (2019) was founded at Gavà, near Barcelona in Spain in 1917.

Toilet ranks high as one of the important inventions for improving human health and increasing life expectancy. Cholera pandemics were very common in the 19th century killing hundreds of thousands of people mainly due to fecal contaminated river water that was used as the drinking water source. However, till dense cities evolved open defecation was an option making toilet rank lower than knife or furniture; Modern toilets are only a few centuries old. Toilets are a prerequisite for skyscrapers and modern cities and serve an important biological function and human health, ranking it high in the list of inventions.

Religion

Intentional burial particularly with grave goods is considered one of the earliest signs of religion as it implies a belief in the *supernatural* [that is a belief without any physical evidence], a soul with people living after death. The earliest human burial is claimed to be in *Qafzeh cave*, at Nazareth in Israel midway between Beirut and Jerusalem, around 100,000 years ago. This would suggest a very early date of religion. However we don't see any evidence of intentional burials in Africa around this time or anywhere else which makes this claim very weak. As modern humans (*homo sapiens*) left our motherland, Africa, only around 80,000-60,000 years ago it is not our ancestors in any case and a different species of *homo* altogether. *Göbekli Tepe* ("belly hill") in Turkey near Aleppo in Syria being the first *temple,* religious holy site, in the world is also not fully supported by evidence. Many items in this site were dated to around 11,000 years ago. Well known pyramids of Egypt which were built in the Early Dynastic period of Egypt around 5,000 years ago are religiously important burial sites and make a much stronger claim to being evidence for the invention of religion. This gives the invention of religion a date around 10,000-5,000 years ago, around

the time of the invention of agriculture and a few thousand years before the invention of writing and wheels.

In any case, all the major extant religions were invented over the last 4,000 years which make them younger than the age of writing. Judaism's oldest manuscript, *dead sea scrolls*, are dated to around 2400 years ago. It's other *codices*, books made up of sheets of *vellum* or *papyrus* instead of paper as these books predate the widespread adoption of paper (see the chapter on *paper*), like *Leningrad Codex* and *Aleppo Codex* are much more recent, from only 1000-1100 years ago. Christianity is dated to around 2,000 years ago and Islam was founded around 1,400 years ago and *Qura'n* was written down shortly there after with one of the oldest codices, *Sanaa* palimpsest dated to around 578-669 CE. A *palimpsest codex* is a codex that contains a lower text, the original text that was erased to reuse the expensive parchments, below the visible upper text of the parchments in the codex; Lower text is reconstructed using x-ray imaging and computer analysis. Hinduism is probably 3,000-3,500 years old based on *Rigveda*'s age; However its age is harder to ascertain as the oldest manuscripts are only available in birch bark. (It decomposes easily unlike *vellum*). Buddhism and Jainism are around 2,500 years old.

Catholic church was a very powerful force in Europe during the middle ages that brought Europe into her *dark ages* by discouraging science and secular knowledge. Religion played a secondary role in European exploration of the new worlds, for spreading christianity to natives. Later many of the American colonies were established by religious minorities such as *Mayflower pilgrims* or *Pennsylvania Quakers* persecuted in European mainland. Islam played a motivating role in Arabs building an empire spanning Spain, North Africa all the way to India, whose conflicts are still simmering in places like the Indian subcontinent. As recently as the 20th century Hitler's hatred of Jews played a strong role in the second world war. Religious ideas were used to oppress half of humanity, women throughout history and in many Arab countries like Saudi Arabia even now. Hinduism was used to justify the caste system in India, affecting the lives of the majority, so called lower castes. Japanese religion with its emperor as God played a strong role in Japan's second world war fight including its *Kamikaze* fighters. In recent times terrorist groups such as Al Qaeda and ISIS have wreaked havoc and played a major part in history. History of America in the 21st century would have been very different without the *9/11 attacks* and definitely the unfortunate histories of Iraq and Syria. If religion weren't invented human history would have turned out very differently including possibly industrial revolution and enlightenment and other inventions happening much earlier in human history but we would also be missing our great pyramids, cathedrals, temples and mosques and religiously inspired literature, poetry, music and festivals.

Religion played a powerful positive historical role in enriching human culture; Architecture, literature, music, poetry and other cultural areas such as religious festivals. Many of the beautiful architectures are that of cathedrals, temples or mosques. Tanakh, Bible, Qur'an, Vedas and other religious texts encouraged early human reading and writing at least among a tiny elite. Many of their stories such as *Adam and Eve*, *Noah's Ark* or *War of Mahabharatha* are woven into the culture and common part of normal education. Often the most beautiful pieces of music in all cultures come from their religions.

Religion ranks very high in the list of inventions for its impact on human history and lives of all the people who ever lived. However its impact is waning in recent times. It could also partially be considered an anti-invention for its negative impacts such as crusades, established churches restricting science and education, oppression of women in countries such as Saudi Arabia to name a few. Modern conflicts in India-Pakistan, Ireland-Britain and the Middle East can be traced to religion as well. It also has historically been a powerful force responsible for our great cultural heritage, architecture such as the *Gothic architecture* of *Notre Dame cathedral* (it was severely damaged in a fire in April 2019), literature as in Bible, Quran or Vedas and music and poetry. Religion motivated many of the historical wars and military conquests such as the spread of Islamic empire from India in the east to North Africa and Spain in the west and crusades in the middle ages. However we all can agree to the powerful influence of this uniquely human invention and it gets its very high rank, in the top 20, because of its long history and ongoing influence.

Pen

It seems like the ballpoint pen has been with us forever; We take it for granted that we throw away 1.6 Billion of them every year in the US alone. In reality, it is as new an invention as the computers. The first ballpoint pen outside of Argentina was sold on Oct 29, 1945 in New York City. It was made by *Reynolds Pen Company* and cost $12.50, more than $180 in 2020 dollars. Since 1950, *Bic*, one of the first to manufacture ballpoint pens, has sold more than 100 billion ballpoint pens globally. It was first manufactured commercially by Hungarian-Argentinian inventor László Bíró, who fled Nazi Hungary for Argentina. The first *"birome"*, as it became known in Argentina, was released in 1943. László's brother, György came up with the non-liquid viscous ink which spread easily but dried quickly; This was as important an invention as Gutenberg's oil based varnish-ink for his printing press. Ballpoint pen itself (without the all important ink paste) was invented by John Loud of Weymouth in the US. He explains the ballpoint pen in his 1888 patent "PEN" [emphasis added]:

"My invention consists of an improved reservoir or fountain pen, especially useful, among other purposes, for marking on rough surfaces such as wood, coarse. wrapping-paper, and other articles-where an ordinary pen could not be used.

...When the ball L is pressed against a surface, the spring S yields, **allowing the ink to flow out of the tube around the ball on all sides to**

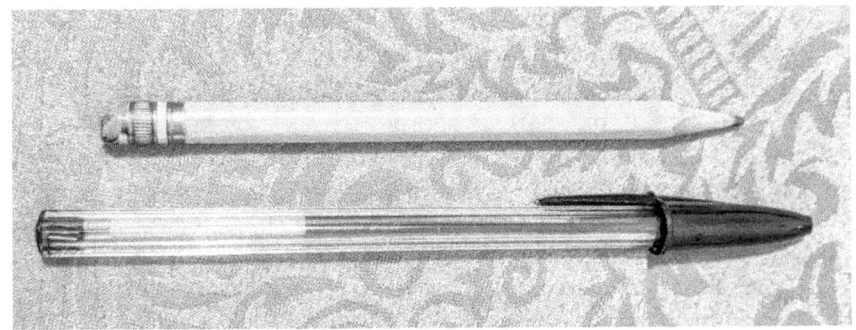

the point in contact with the surface to be marked, the amount of the flow and the width of the line being determined by the amount of play of the ball L inside the contracted mouth, which is in turn regulated by the distance between the opening b and the lower end of the screw C."

The new pen had a dramatic effect on the act of writing itself. BBC quotes Author David Sax:

"The ballpoint pen was the equivalent of today's smartphone. Before then, writing was a stationary act that had to be done in a certain environment, on a certain kind of desk, with all these other things to hand that allowed you to write. What the ballpoint pen did was to make writing something that could happen anywhere."

Before ballpoint pens were invented fountain pens were in common use. According to a *New York Times* article on 6 Feb 1927 [emphasis added]:

"An early attempt to devise a fountain pen has been disclosed by excavations in an Egyptian tomb dating back more than 4,000 years. The primitive instrument consisted of a section of reed the diameter of a lead pencil, about three inches in length and mounted on a long piece of copper. The nib of the pen is cut away to a fine point like an ordinary quill pen. **The narrow tube of the reed served to hold in reserve a small quantity of the writing fluid.**"

This shows primitive fountain pens may have been common in antiquity including a report about a *Fatimid Caliph* in Egypt in 10th century CE. Before fountain pens were invented, if people wanted to write with ink they had to use a dip pen, a sharpened stick or feather and later man-made metal nibs. The pen is dipped directly into ink and only holds enough ink to write a few words. The *Parker Pen Company* was founded in 1888 by George Safford Parker in Janesville near Chicago in the United States. The *Waterman Pen Company* was established in 1884 in New York City by Lewis Waterman. *Faber-Castell*, the great german pencil company founded at Stein near Nuremberg in 1761 by Kasper Faber started making fountain pens in the early 20th century.

Staedtler, another German great, founded in 1835 at Nuremberg, started making pencils in the 17th century. Nuremberg's strict guild rules that disallowed owning end-to-end pencil making by a single owner meant that the Staedtler company was not set up until 1835. Lead pencils have no lead. It is a mixture of graphite, known also as *black lead* or *plumbago*, and clay. **HB** stands for Hard Black, with hardness referring to clay and black the graphite content. A **2B** pencil will make a darker mark than B and **2H** is harder than H and will make a lighter mark. A *Washington Post* article explains the origin of *"lead pencils"*:

"In the early 1500s, a vast deposit of graphite was discovered in Cumbria, England. ... Because chemistry was a young science at the time, people thought graphite was a form of lead; hence the name given to pencils."

India ink, one of the oldest to be used for writing, was composed of carbon black, lamp black, and charred bones (called bone black), combined together with animal glue to create a block, which would then be re-liquefied by the addition of water. It was used in China from the 4th century BCE. Gutenberg created a new oil based ink that was more suitable for adhering to metal for his printing press (see the chapter on *book*). Modern inks are petroleum or soy based with synthetic dyes. Blue color is obtained with *triphenylmethane* compound dyes. Black color is achieved with *titanium oxide* (white ink) and carbon black.

Writing instruments, pens, pencils and ink, rank very high as one of the most important inventions affecting day to day life. They are one of the most common objects in the world, hundreds of billions of them, used by all the people including school children. Ink is critical for books and newspapers, ranked high due to their impact on human history; It is not ranked as high as paper and book as this invention's more popular forms, ballpoint pen, fountain pen and pencil are in common use only for a few centuries and Gutenberg's 'ink' is more of a varnish than writing ink (see the chapter on *book*). Even in the era of computers, they rank high as accessory inventions for books and a heavily used invention on their own.

Letter

"Unto my right well-beloved Valentine John Paston, squire, be this bill delivered.

Right reverent and worshipful and my right well-beloved valentine, I recommend me unto you full heartedly, desiring to hear of your welfare, which I beseech Almighty God long for to preserve unto his pleasure and your hearts desire. And if it pleases you to hear of my welfare, I am not in good health of body nor of heart, nor shall I be till I hear from you. For there knows no creature what pain that I endure, And even on the pain of death I would reveal no more. And my lady my mother hath laboured the matter to my father full diligently, but she can no more get than you already know of, for which God knoweth I am full sorry. But if you love me, as I trust verily that you do, you will not leave me therefore. For even if you had not half the livelihood that you have, for to do the greatest labour that any woman alive might, I would not forsake you. And if you command me to keep me true wherever I go, indeed I will do all my might you to love and never anyone else. And if my friends say

that I do amiss, they shall not stop me from doing so. My heart me bids evermore to love you truly over all earthly things. And if they be never so angry, I trust it shall be better in time coming. No more to you at this time, but the Holy Trinity have you in keeping. And I beseech you that this bill be not seen by any non earthly creature save only yourself. And this letter was written at Topcroft with full heavy heart.

Be your own Margery Brews."

This is the oldest known love letter in the English language written in February 1477 from Margery Brews to her fiance John Paston, now at British Library in London, England. *CURSUS PUBLICUS* ("the public way"), the Roman empire's postal service is the oldest known postal system in the world. It was created by Augustus Caesar around 2,000 years ago in the first century BCE. It was also the longest known postal system surviving all the way to the sixth century CE. It also spread to Eastern Roman Empire and in one form or another the Cursus Publicus subsisted in Eastern Rome until the fall of Constantinople (1453). According to F.K. Devos, writing in Public Affairs, "The most common cargo on the different vehicles was comprised of urgently needed war materials, parchment and papyrus, state revenue, express parcels of all kinds and officials and their families".

Interestingly, the oldest known letter in the world is a customer complaint; It is from a man named Nanni to another named Ea-nasir, written on a clay tablet 3750 years ago. We would do well to remember this, etching a letter in clay tablet, when we complain about navigating robotic dialogue for an hour and another hour waiting on hold for the "next available" human being with our phone company. The clay tablet is now in the museum with the description, "letter from Nanni to Ea-nasir complaining that the wrong grade of copper ore has been delivered after a gulf voyage and about misdirection and delay of a further delivery".

The *Correio-Mor do Reino* ("Royal Mail") of Portugal, founded in 1520 and available to the public from the very beginning, is the oldest extant public postal service in the world. *Royal mail* in England started for public service in 1635, although it started as a mail service for the crown in 1516. *Sanquhar Post office* (postcode:**DG4**), opened in 1712, near Edinburgh in the UK is the world's oldest (extant) post office in the world. Earlier, Franz von Taxis, a German noble in Italy, considered the founder of the modern postal system, set up a postal network in 1490 for the Holy Roman Empire that would eventually grow to cover all of Western Europe by the mid-16th century. The first postal service in the US started in February 1692; The USPS traces its roots to 1775 during the Second Continental Congress, when Benjamin Franklin was appointed the first postmaster general. *The Hinsdale, New Hampshire*, post office (zip code: **03451**), opened in 1816, is the oldest post office in the US. *India post* was started in 1774 by the *East India company* with the *Calcutta General Post Office* (GPO) opening on 31 March 1774; Madras GPO opened on 1 June, 1786. One of the highest post offices in the world is in Hikkim, Himachal Pradesh (PIN code:**172114**) operated by India Post at an altitude of 4,440 m (14,567 ft). India also has the largest postal network in the world with 156,721 post offices. Norway has the most expensive postage with 18 *Norwegian Kroner* ($2.18). India is one of the cheapest with it costing only 5 indian *rupees* ($0.07) with the cheapest option, that of sending a post card only costing 0.5 rupees ($0.007). The *Penny Black*, issued by the UK in 1840, was the world's first adhesive postage stamp used in a public postal system, a great invention in itself; It featured Queen Victoria and cost exactly a penny as the name implies and standardized the postage for the first time for letters. Its success led to the issuance of stamps in many countries in the 1840s and 50s.

Letter and the postal service, has a rich history and ranks as one of the greatest inventions of humanity. It still ranks as one of the most important services in the modern world. The US postal service delivers 429.9 million mail pieces every day. In 2018-19, India post delivered 5.76 Billion mail articles. It has also evolved into private delivery companies such as Fedex and UPS. Along with its modern forms such as email, Letter continues to exert influence as one of the longest and greatest inventions for human communications.

Shoe

"WASHINGTON, Sept. 4. - The discovery of fragments of bison-hide moccasins, bark sandals, throwing sticks and other articles in caves of Southern Oregon indicate that man may have inhabited the Pacific Northwest 10,000 years ago.

This was disclosed today by the Carnegie Institution in reporting on excavations made early this year in an expedition sponsored jointly by the institution and the University of Oregon and led by Dr. L. S. Cressman, Professor of Anthropology at the university.

Three caves were explored, one at Roaring Springs Ranch, about 100 miles south of Burns; one near Paisley, on the east side of Summer Lake, and the third in the vicinity of Fort Rock, about seventy miles south of Bend.

In the Roaring Springs cave, two throwing sticks, basketry and fragments of two bison-hide moccasins were found. At the Fort Rock cave more than seventy-five sandals made from shredded sagebrush bark were found under pumice near the mouth."

The *New York Times* of Sept 5, 1938 reports on the discovery of the oldest known footwear, sandals from *Fort Rock Cave* in the US, midway between San Francisco in the south and Seattle in the north, in the article, "Oregon Caves Yield Artifacts Of Habitation 10,000 Years

Ago; Particles of Bison Hide and Sagebrush Foot Coverings Are Dug Out of Pumice Attributed to Near-By Craters." Although 10,000 years looks relatively recent for shoes, it is still twice the age of wheels. In any case, the true origin of such a foundational invention for our species is likely to be closer to our home, *Africa*. First leather shoe, 5,500 years old, was found at the *Areni-1 cave* in the *Vayotz Dzor province* of Armenia, bordering Turkey and Iran. Roman soldiers wearing strap-style *Caliga sandals* are well known from the first century CE. Description on a Caliga in the *Museum of London* reads:

"One-piece, moccasin-style shoes and hobnailed boots were normal for outdoor and winter wear. Flip-flop style sandals were worn in better weather and around the home. Some boots were the strap-style caliga (military boot) or, as this example shows from AD 70 onwards, had intricate cut-out designs that would have shown off the wearer's coloured stockings."

In the *British Museum*, there is an exquisitely beautiful marble sculpture, "Marble gravestone of Xanthippos, the shoemaker", from Athens, Greece dated to 420 BCE; This *stele*, a grave stone slab, shows a bearded man sitting on a chair (*klismos*, see the chapter on *Furniture*) and holding up a shoemaker's *last,* a tool that looks like a foot blueprint used to make and repair shoes. This shows shoemaking was a common and sophisticated profession in Greece, 2400 years ago. Wooden footwear was used in places such as India (*paduka*) and Netherlands (*klompen*). The earliest known *klompen* in Amsterdam dates to around 1230.

Charles Goodyear invented *vulcanized rubber* in 1844 providing an important new raw material for footwear. Daniel Mills of New York working for Charles Goodyear Jr. (Son of Charles Goodyear) patented in 1869, "IMPROVEMENT IN SEWING-MACHINES FOR BOOTS AND SHOES", the shoe making process now called the *Goodyear welt*; It is still one of the best shoe construction methods after 150 years. The Goodyear welt process is a machine-based alternative to the traditional hand-welted methods. The shoes made with this method have prominent stitches, made by the *Goodyear welting machine*, circumscribing the shoe sole on the outside; This welt is so popular that some companies make fake "stitches" on rubber soles to mimic goodyear welts. (This is a good time to check your shoes). This is an expensive method but makes the shoes more durable and shoes can be resoled. There are alternative welts such as *blake welts* with stitches inside the shoe or simply cementing or glueing the sole to the upper part of the shoe as in sports shoes. According to the *Italian Shoe Factory* website, "Those who favour the Blake will tell you their shoes are less rigid and more comfortable to wear. The Goodyear, however, is hardier and better in wet weather". In 1975, Jerry Turner working at *Brooks*, founded in 1914 at Philadelphia in the US, introduced the first shoe to use the rubber substitute EVA (*ethylene vinyl acetate*), an air-infused foam that is now common in running shoes and flip-flops.

Eduard Meier, founded in Munich, Germany in 1596, is the oldest extant shoe company in the world. With a revenue of 37.4 Billion dollars, *Nike* of the US, founded in 1964 near Portland, Oregon, is the biggest shoe company in the world. Adidas, founded in 1924 near Nuremberg in Germany, with a revenue of 23.9 Billion dollars is the second biggest and the biggest footwear maker by volume, 448 Million pairs in 2019. *Bata*, based in Lausanne, Switzerland, is the second leading shoemaker by volume at around 200 Million pairs, selling footwear at affordable prices in countries such as India (also making their revenue look lower in nominal, non-PPP, terms). It was founded in 1894 at Zlín, Czechia near the Poland border, by Tomáš Baťa and his siblings. China is the world's largest footwear producer with 14.2 Billion pairs annually followed by India with 2.2 Billion. The global footwear market is estimated to be worth 365.5 billion U.S. dollars (2020). The common materials for shoes are leathers (natural and PVC/Polyurethane), rubber (natural and EVA), textiles and plastics.

Footwear ranks high as a foundational invention, comparable to clothing in its importance; It is not ranked as high, as the invention of clothes for the first time defined our species; Shoes are a natural corollary, an *incremental* invention. It has been with us longer than writing or wheels; However the latter were more *original,* not a derivative of a higher ranking invention, and more productive in generating descendant inventions. Every year, around 24 Billion pairs of footwear are produced in the world or 3 pairs of footwear per person, making it one of the most ubiquitous, useful and highly ranked inventions.

Stove

People have been cooking with open fires even before the invention of agriculture. Some three billion people still cook with rudimentary stoves that burn wood, dried dung or coal, together called *solid fuels*. Despite its romantic notion and bucolic charm, it is not a very healthy lifestyle. "Having an open fire in your kitchen is like burning 400 cigarettes an hour in your kitchen," says Kirk Smith, a professor of global environmental health at the University of California at Berkeley. Exposure to smoke from traditional cookstoves and open fires causes more than 4 million premature deaths, including more than 1.2 million deaths in India, every year. An estimated 700 million Africans (82%) cook primarily with solid fuels, 7% with kerosene, 5% with *Liquified Petroleum Gas* (LPG), and 6% with electricity. F.W. Lindqvist of Sweden invented and patented the first *kerosene stove*, with a distinctive burner vaporizing kerosene that I remember from my childhood, that doesn't leave black soot in 1892. It was called the *Primus stove* and was popular with explorers such as Roald Amundsen who took a Primus stove on his expedition to the south pole in 1911. It is still one of the most important

stoves in the developing world such as India (although starting to lose out to LPG recently, a good development); Kerosene stove is a much better improvement over open hearth with its incomplete combustion, particularly indoors. Mark Aldrich says about this intermediate step in stove evolution in his paper, *"The Rise and Decline of the Kerosene Kitchen: A Neglected Energy Transition in Rural America, 1870-1950"*:

"...focusing on the transition from wood and coal to electricity and gas in home heating and cooking misses a step. ... Especially for women in rural and farm households, kerosene provided an important bridge fuel to the newer age of gas and electricity. To ignore it is to ignore what was for many an important introduction to modern times."

First gas stove was invented by James Sharp at Northampton, north of London, England in 1826. It became popular in the late 19th to early 20th century with gas pipeline infrastructure available in Europe and the US. Gas pipelines deliver natural gas that is more than 95% *Methane* (CH_4). There is a report in The *New York Times* of Apr 24, 1887 with an interesting reference to gas stoves [emphasis added], titled "ECONOMICAL COOKING":

"About 300 ladies braved yesterday's rainstorm to attend the last of Miss Juliet Corson's lectures on **gas-stove cookery** at the Metropolitan Opera House assembly room. An elaborate domestic menu was served, and the ladies were not slow to express their surprise at the end of the seance that less than 10 cents' worth of gas had done all the boiling, baking, broiling, frying, and roasting with a minimum of trouble and no handsoiling whatever."

This shows that gas stoves were somewhat of a novelty even in big cities of the US towards the end of the 19th century. As late as 1919, stoves supported multiple fuels with gas, coal and wood. In the US 63% of the households used electric stoves and 33% used gas stoves in 2015, with 74.9 million households using electricity in a stove, cooktop, or oven. An important variation in gas stoves is the *LPG stove*. LPG is mostly made up of 3-carbon *propane* (C_3H_8) and 4-carbon *butane* (C_4H_{10}) as they are much easier to liquefy than natural gas and more energy dense for the same volume at 93.2MJ/m³ vs natural gas at

38.7MJ/m^3. Natural gas emits slightly less CO_2 than LPG, making natural gas the cleanest of the fossil fuels to burn. However methane itself is a potent greenhouse gas, 84 times more potent than carbon dioxide (GWP_{20}, *20 year warming*) and more toxic, making it very harmful if leaked. This makes LPG the overall better choice for the environment. LPG is very common in countries such as India without a gas pipeline infrastructure common in countries that require heavy winter heating. In Dec 1955, *Burmah Shell oil company* introduced the first LPG service in India as *Burshane* at Mumbai. 80% of Indian households have access to LPG now, though used in combination with other fuels including wood due to limited supply. LPG was isolated by Walter Snelling at the *US Bureau of Mines* in the US in 1910 and he also was the first to commercialize it. Gas stoves are the most common stoves in the world now with solid fuel (wood) still dominating Africa and rural areas of the developing world.

First electric stove was invented by David Curle Smith, of *Kalgoorlie* near Perth in the west, Australia in 1905. It became popular in the mid twentieth century with electricity widely available in the developed world. First Microwave oven was invented by Percy Spencer in 1945 at Raytheon in Newton near Boston in the US. Earlier in 1940, John Turton Randall and others at Birmingham University in England invented *cavity magnetron*, the critical component that produces microwaves in the oven. In 2015, 96% of the US households reported having a microwave oven. *Induction stoves*, although very efficient with fast heating, play a minor role likely due to the requirement for specialized magnetic pans.

Stoves rank very high as one of the foundational inventions as it addresses a basic necessity, food. Stoves rank relatively lower given their purpose and long history as it is an incremental invention of fire and modern stoves are only a few centuries old. Better stoves likely contributed to increased life expectancy in the 20th century, given how harmful open fire cooking indoors is. With an average household size of 4.9, there are around 1.5 Billion stoves in the world, making it one of our most ubiquitous and life giving inventions.

Road

"To make it easier to transport the heavy stones from one of these quarries, the Egyptians laid what may have been the world's first paved road. Research geologists mapping the ancient Egyptian stone quarries have identified **a seven-and-half-mile stretch of road covered with slabs of sandstone and limestone and even some logs of petrified wood.** ...

They said that pottery fragments at a quarry and a camp for the ancient stone workers, both discovered near the road, helped **date the site to the period of the Old Kingdom, about 2600 to 2200 B.C.**, when major technological advances were being made, but before Egypt's political zenith. The oldest previously known paved road, made of flagstone and dated no earlier than 2000 B.C., was in Crete. The Egyptian paved road, with an average

width of six and a half feet, ran across desert terrain 43 miles southwest of modern Cairo. "

The *New York Times* of May 8, 1994 describes the discovery of the world's first known road, 4500 years old, in Egypt. Given that the first wheel is dated to around 5500 years ago in Poland (see the chapter on *Wheel*) this is a reasonable time frame for the first paved roads. The article also mentions the second oldest, that of Crete, a greek island smack in the middle of the mediterranian sea not far from Cairo itself. Romans built a vast network of roads including the famous *Via Appia,* "Queen of the roads". At the empire's peak about 85,000 km (53,000 miles) of road connected the capital Rome with its far-away frontiers. The saying *"All roads lead to Rome"* was literally true; Twentynine major public roads radiated from Rome, including the Via Appia, the most illustrious example of Roman civil engineering skills. The *New York Times* describes it in an article:

"The Via Appia, which began here at the foot of the Palatine, was started in 312 B.C. by the Republican magistrate Appius Claudius and eventually covered the nearly 600 kilometers (370 miles) to Brindisi, making it the gateway to Greece and the Empire in the East. The Queen of Roads, as it was dubbed, became much more than just a road, the land on either side of it near Rome in particular, being lined with patrician villas and cemeteries, and in due course the site of the underground complexes of major Christian catacombs."

Edward de Smedt of *Columbia University* at New York, a Belgian-American is credited with inventing the modern man-made road *asphalt* in 1870. First modern asphalt road was laid at Newark near New York. In the US, bicycles provided the impetus for good roads with the *Good roads movement* started by bicycle enthusiasts. Earlier in 1816, John McAdam of the UK developed one of the modern methods of road construction called *McAdam* which simplified road construction. When motor vehicles became prevalent this method created a dust cloud and was fixed by spraying tar on the surface of McAdam road creating tar McAdam road or *tarmac*. According to *Asphalt Magazine*:

"The first record of an asphaltic road being constructed in the 1800s was from Paris to Perpignan, France, in 1852, using modern macadam construction with Val de Travers rock asphalt."

The US has 6.59 Million kms of road followed by China and India at 4.96 and 4.70 Million respectively; The US roads can go around the world (at equator) 165 times. World's highest road is through *Umling La Top* in Ladakh at a height of over 5.88 km, located 230km from *Leh* in India near the China border. The *Pan American Highway* (excluding *Darien Gap* of around 100km between Panama and Colombia) is the longest road in the world with a total length 30,000 km (19000 miles); It goes all the way from *Prudhoe Bay*, Alaska to the tip of Argentina in Ushuaia in *Tierra del Fuego* crossing both North America and South America in full. *Highway 1* in Australia is the world's second longest road and the longest national highway at 14,500km. Moscow is the city with the worst traffic in the world followed by Mumbai and Bogota. Among major countries, the Netherlands has the highest road density followed by France and Germany. Roads are expensive; To construct a 6-lane Interstate highway in the US it costs about $7 million per mile in rural areas, $11 million in urban areas.

Wheel is of no use unless roads, paved ones are available. It is reflected in the fact that natural selection never automatically evolved wheels on any animals unlike legs. Given the importance of wheels in human civilization, roads as the enabling invention for such an important invention ranks high on its own. It is ranked relatively lower for a foundational invention as it is a derived invention of wheels, a more complex, original and productive invention. Almost all of the 7.8 Billion people in the world use roads making it one of the inventions that heavily impact day to day life. For instance, in Europe, the average one-way commute time was 25 minutes in 2019. This means people spend around 5% of their waking lives on the road in Europe; Indians spend 7% of their lives getting to the office. Given the time we spend on roads it ranks high in the list of inventions.

PART 2

Thermometer

It may look surprising at first, that thermometer, a 'simple' invention is ranked so high, above all inventions of the industrial revolution such as steam engines. In fact, the quantity thermometer measures is very deep and complex with myriads of practical applications. *Temperature* is a macro measure of the *velocity distribution of molecules* particularly of gases or liquid. Higher the temperature of the fluid, higher the *average velocity*. Daniel Fahrenheit of Polish, German and Dutch descent invented the first modern thermometer in 1714 at Amsterdam, Netherlands, as an improvisation on his earlier alcohol thermometer of 1709. In higher temperatures mercury atoms have faster average velocity; This means the same quantity of mercury spans a bigger volume. This expansion of mercury is converted into a linear scale using a narrow tube in the thermometer. Galileo and a few others had created rudimentary thermometers earlier. Fahrenheit also invented a scale to go with his thermometer, *Fahreinheit scale*. In 1742, Anders Celsius at *Uppsala university* near Stockholm in Sweden developed the more intuitive *celsius* (centigrade) scale with 0 as the freezing point of water

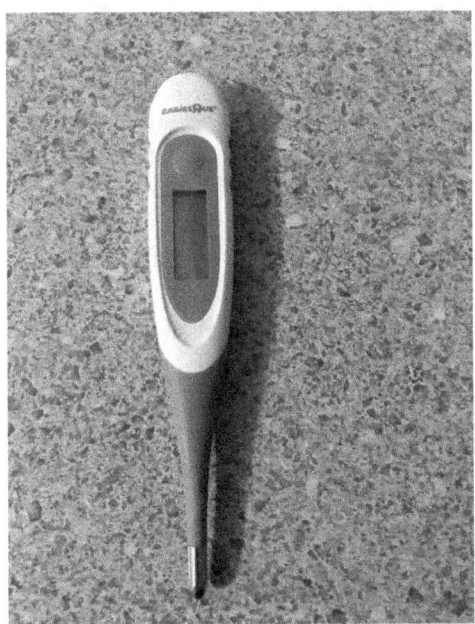

and 100 as the boiling point of water. William Thomson at the University of Glasgow in the UK came up with a theoretical temperature scale (*kelvin*) and explained the notion of "absolute zero", state of 'no motion' for the molecules (other than quantum mechanical vibrations due to *uncertainty principle*). In 2019, the temperature scale of the SI system was redefined using a fundamental constant of nature, *Boltzmann constant* using the two relations for energy, the two sides of $hv = kT$. It is now defined as:

"one kelvin is equal to the change of thermodynamic temperature that results in a change of thermal energy k T by $1.380\ 649 \times 10{-23}$ J."

That is, *kelvin* is a derived unit of - *kilogram, metre, second* - defining energy (*Joule*) and the Boltzmann constant. Main difficulty with the invention of the thermometer was understanding the concept of temperature itself. Once it was understood that the increase in the temperature increases the volume of liquids such as mercury, inventing the thermometer was a short step. However the full understanding of the temperature and heat had to wait for the development of thermodynamics in the 19th century. Interestingly, the study of temperature and heat led to quantum mechanics via Boltzmann's equation for *entropy*; Entropy is a measure of *disorder* ("messiness" or "randomness") in a system and is very important in modern physics and computer science; A glass broken into shards has more entropy than an intact glass; A random number, 2371, has more entropy than a number with a regular pattern, 1111. Boltzmann's equation expresses entropy as the number of possible different arrangements of a system [$S = k\ logW$, W being the number of different arrangements. This is where Boltzmann's constant for the current definition of temperature originates from]. One of the most profound laws of physics, the *second law of thermodynamics* flows from entropy. It says in any [isolated] system entropy always increases; That is it goes from less probable state to more probable state and never the other way around. This explains why scrambling an egg is easy, unscrambling one is so hard as to be impossible for practical purposes. Sir Arthur Eddington, well-known for confirming Einstein's General Relativity in 1919, dramatizes its importance [emphasis added]:

"The law that entropy always increases—the Second Law of Thermodynamics—holds, I think, the supreme position among the laws of Nature. If someone points out to you that your pet theory of the universe is in disagreement with Maxwell's equations—then so much the worse for Maxwell's equations. If it is found to be contradicted by observation—well these experimentalists do bungle things sometimes. But **if your theory is found to be against the second law of thermodynamics I can give you no hope; there is nothing for it but to collapse in deepest humiliation.**"

Humble thermometer and its theoretical offshoots have so much importance that even explains the specialness of the *origin of the Universe* itself and the *arrow of time*; *Big bang* is one of extreme low entropy for the obvious reason that it can only increase with time (obvious once we have the second law); In fact it is the only way past and future can be differentiated as all other physical laws are symmetrical with respect to time.

Invention of the thermometer and the study of temperature had an enormous impact on humanity including the development of steam engines, internal combustion engines and refrigeration. Its theoretical side ultimately led to the development of quantum mechanics. Quantum mechanics apart from being a supreme [and frustrating] human achievement in our understanding of the universe has also led to innumerable practical inventions, electron microscopes, solar cells, lasers and transistors to name a few; And via transistors to modern computers and mobile phones. As mentioned above, it also explains many aspects of the early universe and cosmology. In this sense, Thermometer, the concept and the measurement of temperature and heat, has been one of the most productive inventions of all time and deserves to rank very high. On a more practical side, It is useful in medicine and one of the most common devices used in medicine, as anyone who had to take care of a baby would know. Daily weather is reported on most of the radio and tv stations. It has been with us for more than 300 years, more than most other inventions of the industrial revolution including steam engines, plenty of time to make a strong impact on human civilization.

Steam Engine

When we think of the industrial revolution, one of the first inventions that comes to our mind is a steam engine. Steam engine was invented by James Watt at Glasgow in the UK in the 1760s with a crucial breakthrough on Thomas Newcomen's earlier design of 1712; Watt patented it in 1769 as "NEW INVENTED METHOD OF LESSENING THE CONSUMPTION OF STEAM AND FUEL IN FIRE ENGINES". Newcomen's *Atmospheric Engine* ("Fire Engine" in the patent) had to condense the steam in every cycle using a splay of cold water, the resulting vacuum pulled the piston down and did useful work; The cylinder [containing the piston] had to be reheated all over again resulting in lost efficiency, fuel and time required for the conversion of water into steam in every cycle. James Watt removed the need for this by removing the used steam to a separate *condenser* to cool the steam which kept the cylinder at the same operating temperature. It would be the difference between having to cool the pot every second and reheat it while cooking and keeping it at constant temperature all the time. James Watt's invention of a separate condenser was an improvement in efficiency, a beautifully ingenious one at that, instead of a brand new invention as the title of the patent itself points out. But it mattered, *a lot*.

It brought in the revolution in power and energy by making steam power much more efficient and commercially viable. Earlier Thomas Savery of England developed a steam pump in 1698 for pumping water from coal mines. This could be considered an early precursor for Newcomen's engine although it is not an engine in the sense it didn't have a working piston or cylinder. It created a vacuum by condensing steam in a container which pulled the water from the mine into the container; Then the new steam was used to push the water from the container up and out of the mine. Newcomen's engine cylinder shared its similarity with Savery's container, both doing their main work when steam is suddenly condensed in the working chamber creating the vacuum for atmospheric pressure to do its work, hence the name *atmospheric engine* for Newcomen's invention. Steam engines are external combustion engines, as in heated from a fire outside similar to how we cook, as opposed to internal combustion engines in cars.

Steam engines played a critical role in railways, ships and factories of the industrial revolution. It helped provide the coal required for the industrial revolution by replacing human muscle power with steam power to remove water from coal mines. Steam engines powered behemoths like the *Titanic* [SS in SS *Titanic* standing for steam ship] and *Lusitania*, actually steam turbines in the case of the latter. *British gas* website on Titanic's steam engines:

"Two reciprocating steam engines – with a combined output of 30,000 horsepower and each weighing 720 tonnes – and one low-pressure turbine powered the Titanic. They needed the steam produced by 29 boilers, each capable of holding more than 48 tonnes of water. To keep this system going, more than 600 tonnes of coal per day had to be shovelled by hand into the ship's 159 furnaces, which involved the 24-hour attention of over 175 labourers... Titanic's daily coal requirements would have supplied over 250 average UK homes for one year today."

Steamboats were important in the economic development of countries with long rivers like *Mississippi* in America. Modern electric plants are powered by steam turbines. Steam turbines have a boiler in which a fuel is burned to produce hot water and steam in a heat exchanger, and the steam powers a turbine that drives a generator. From the modern perspective, coal burning for steam turbines in electricity plants has been one of the biggest contributors of carbon dioxide in the atmosphere. IPCC special report 2018 says:

"Human-induced warming reached approximately 1°C (*likely* between 0.8°C and 1.2°C) above pre-industrial levels in 2017, increasing at 0.2°C (*likely* between 0.1°C and 0.3°C) per decade (*high confidence*)."

Steam engines rank very high for their role in the invention of railways and their critical role in the industrial revolution. Railways revolutionized the history of the countries, particularly big ones such as the US, Russia and India by providing connectivity to far flung areas of the country (see the chapter on *Railways*). Steam engines also for the first time provided power which is independent of human or animal muscle power [horse power as in *horsepower*]. It also provided the inspiration for internal combustion engines which power all automobiles. Steam turbines are used to generate the majority of the world's electricity and they accounted for about 48% of U.S. electricity generation in 2019. Steam engines also powered ships in the 19th and early 20th century making them much faster and not at the mercy of the wind and sail. For some time in the early 20th century steam engines were in the running for automobiles as well but they lost out to internal combustion engines. In an ironic twist, internal combustion engines are starting to lose ground to electric cars which are powered by electricity [stored in their batteries] generated using a steam turbine, an external combustion engine. It was also instrumental in the second phase of the industrial revolution powered by steam engines in textiles and mines instead of just man power, horse power or flowing water in rivers. Interestingly, modern nuclear power plants used to generate electricity or in aircraft carriers use a steam engine as well; Nuclear power is used to heat the water to produce the steam [that is for external combustion] which in turn powers the stream turbines. That is instead of using coal we use nuclear power for external combustion, everything else being a classic steam engine.

Railways

Modern railways commenced with the development of steam locomotives in England. First railway line opened in 1825 and connected the northern England towns of *Shildon* and *Darlington*; *Stockton and Darlington railway* was the first railway to use steam locomotives and carried passengers and coal from coal mines to ships. George Stephenson, *"father of railways"* invented the steam locomotive [building on some early models] used in the first train called *Locomotion*. This was followed in 1830 by an inter-city train, *Manchester and Liverpool railway* built by Stephenson which had a fixed time table for the first time and used steam power exclusively. First railway in British India started operations in 1853 near Bombay by the *Great Indian Peninsular Railway* formed by an act of British parliament. China's first train started in 1876 near Shanghai and was shut down in 2 years. Most of the rail network there was built after 1949 in communist china. An experimental train started in Russia in 1837 and it started in earnest in 1851 with the construction of the Moscow-St.Petersburg railway line. The *Baltimore and Ohio railroad* in the United States started operating in 1830. By 1869 US had an inter continental railways network connecting its East and West Coast greatly expanding the western settlement. In 1876, the *transcontinental express* from New York City reached San Francisco in 83 hours and 39 minutes, a journey which would have taken weeks or months earlier. *Trans-siberian railway* of 1916 in Russia from Moscow to Vladivostok with a length of more than 9,000 kilometers fulfilled a similar role in Russia. In fact, it connected Europe all the way to the far eastern end of Asia with connecting trains now going to China and North Korea

and spanning *eight time zones*. It was one of the amazing technological accomplishments of 19th century humanity. Now the US, China, Russia and India have the largest four train networks in the world; Railway tracks in each of these countries can go around the earth [at equator] a few times.

As an invention, Railways can be called the *"ships of the land"*. Railways played the role in human history of the last few centuries that ships played in earlier periods. Just like ships, railways were important politically, economically and militarily. It brought people together and exponentially increased the trade by connecting producers from far flung places to consumers all over the country. British could rule a vast country like India only using the railways. Railways played a very prominent role in the development of American west as portrayed in the *Westerns*. Railways unified the countries by replacing local time differing from city to city with standardized time with fixed time zones as railways required fixed time tables. Railways with rolling iron wheels on rails have much less friction than automobiles on roads, making them more fuel efficient. It makes it much cheaper and efficient to transport large numbers of passengers and huge amounts of freight.

One liter of fuel can move a tonne of freight 200 km (125 miles), as compared to a typical car fuel economy of 11 km per liter (25 miles per gallon) in 2019, which makes individual private car fuel economy around twenty times worse. Societies such as America which are heavily reliant on cars and trucks [including electric cars, a questionable alternative to climate change] are very inefficient and they are better served by moving to more efficient public transport systems and becoming less dependent on cars. Making matters worse, according to the *Texas Transport Institute*, highway congestion costs Americans $166 billion in wasted time (8.8 billion hours or a million *people-years*) and wasted fuel (3.3 billion gallons, enough to power 7 million cars for a year at the US average of 474 gallons per car) in 2017 which makes America's under investment in public transport including rail transit and dependence on cars a self-defeating choice in a society known for idiosyncratic choices for a developed country - no universal health care, employers being responsible for healthcare with the double whammy of a huge competitive disadvantage for employers while simultaneously making the people out of workforce due to an illness uninsured, subpar public school education system even in richest places such as the *Silicon Valley* and adhering to the imperial system of miles and gallons even after billion dollar space disasters (see the chapter on *Ruler*); This is after not including anachronisms like the candidate with 3 million more votes 'losing' the presidential election in a democracy, leading to the disastrous mismanagement of coronavirus and the death of half a million in 2020 under an inexperienced president. (Closer to home, *California high speed rail* connecting San Francisco and Los Angeles, a mere 600 km (380 miles), has been the butt of jokes as the rail that will be "ten years away and always will be").

Railways were particularly important in connecting large countries such as India and the United States in the 19th century, Russia in the early 20th century and China later in the 20th century. It increased the mobility of people along with innovations and ideas. It only takes eight days to complete the *Moscow-Vladivostok trans-siberian railways*, whereas before railways it would have taken weeks or months. Although airplanes nowadays can travel even faster, for the same reason as ships, trains can carry far more goods and people at a much cheaper cost. Railways helped governments reach far flung areas of the country faster and govern the country better. It also increased trade by making the transport easier and cheaper and reduced the price of commodities. As an example, Indian railways carry over 23 million passengers every day (and 3 million tonnes of freight); Indian railways carry more people in a single day than the population of all but the top 50 countries in the world. It serves as the vital vascular system of India in a country with a per capita GDP of 7,000 dollars in 2019 (OECD average is 46,000 dollars, 6.5 times higher). There were 144 Million airline passengers in India in all of 2019, less than what the railways carried in just a week.

Railways ranks very high due to its impact similar to ships. Given the time in history railways ranks lower than ships. Railways ranks higher than automobiles as it has two centuries of impact as compared to automobiles with only a century of impact. Although electricity is far more useful in day to day lives and led to multiple additional inventions in the 20th century, it has been widely available for less than a century even in developed countries (see the chapter on *Electricity*). For its pivotal role particularly in making far flung parts of huge countries accessible, Railways gets a very high rank as the *"ship of the land"*.

Internal combustion engine

In internal combustion engines, sudden expansion of high temperature and high pressure *gases produced by combustion inside a closed chamber* applies a direct force to a component of the engine directly, this component being a

- piston connected to wheels in an automobile
- turbine blade in a gas turbine powering a ship
- or interestingly, being forced through the nozzle of rocket engine at high velocity moving the rocket forward
- or even more interestingly, on a bullet in a rifle or on a cannon ball in a cannon forcing out the bullet/cannon ball as a projectile at high velocity

It contrasts with an *external combustion engine* like a steam engine in which the combustion occurs outside literally by burning a fire like in cooking. This combustion heats up the water in the boiler converting it to steam; Steam, being a gas, works similar to the gases in an internal combustion engine. This steam gets converted into water in a condenser for it to be used again in the work cycle. Internal combustion engines are typically much smaller and more efficient as they don't require this huge

boiler and condenser. However they need to use much more energy dense fossil fuels as the fuel. Internal combustion engines also can start to operate much faster as they don't have to boil the water to produce the working fluid, steam as they operate on air and gases produced from the burning of the fuel itself. They simply use air and gases produced during the combustion as the working fluid, the molecules of which push the piston or turbine doing useful work. Internal combustion engines in automobiles operate in discrete steps called strokes as in *two-stroke* and *four-stroke* engines unlike external combustion engines which operate continuously.

Belgian-French inventor Étienne Lenoir at Paris invented the first commercially successful internal combustion engine and patented it as, "Un moteur à air dilaté par la combustion des gaz" (literally "An air engine dilated by the combustion of gases" - patented in England as "IMPROVEMENTS IN OBTAINING MOTIVE POWER AND IN THE MACHINERY OR APPARATUS EMPLOYED THEREIN"), in 1860. This "two-stroke", uncompressed engine used *coke oven gas* as the fuel. The patent reads:

"This invention consists in the application and use of an inflammable gas mixed with a proper proportion of atmospheric air and lighted inside a cylinder by the aid of electricity, the expansion thereby produced acting upon the piston and imparting motion thereto which motion may be transmitted in any convenient and well known manner to a driving shaft."

Four stroke engines with the *Otto cycle* - intake, compression, expansion and exhaust - were built by Nikolaus Otto and Eugen Langen at Cologne in Germany with their 1876 engine marking the beginning of modern engines. Four stroke engines burning a liquid fuel, *gasoline*, were built first by Gottlieb Daimler and Carl Benz in Germany in 1885. These engines form the basis for current automobile engines with more than a billion cars in the world now. *Diesel engines* are an important variant of internal combustion engines, invented by Rudolf Diesel at Berlin in Germany in 1892. In diesel engines, hot, compressed air ignites the fuel directly without the need for a spark plug that ignites the fuel in gasoline engines. In short, diesel fuel explodes itself under heavy pressure in each cycle, requiring much sturdier engines to withstand the pressure. Diesel engines are more fuel-efficient, 15% more energy than petrol for the same volume, and require a cheaper, less refined fuel, diesel. (Diesel is costlier in the US for economic reasons like higher tax on diesel unlike most other countries).

Internal combustion engines, and steam engines that burn coal for that matter, have had an interesting impact on the planet as they burn fossil fuels with a lot of carbon from old organic materials. Average global temperature on earth has risen by 1°C since 1880. In case this feels trivial, *Nasa earth observatory* explains what happens when it goes the other way by just a few degrees [emphasis added]:

"A one-degree *global* change is significant because it takes a vast amount of heat to warm all the oceans, atmosphere, and land by that much. In the past, a one- to two-degree drop was all it took to plunge the Earth into the Little Ice Age. **A five-degree drop was enough to bury a large part of North America under a towering mass of ice 20,000 years ago.**"

As a crucial invention powering automobiles and airplanes it deserves to be ranked above automobiles and airplanes. It greatly reduced the size of an engine making it possible to actually build a heavier than air airplane for the first time due to its efficiency derived from its use of energy dense fuels. Both automobiles and airplanes appeared in human history shortly after the invention of efficient internal combustion engines making them look almost inevitable. As we have seen time and time again, the main difference between external and internal combustion engines is one of efficiency, as they operate on the same principle, that of producing hot, expanding gases which do useful work, but it mattered, a *lot*. If internal combustion engines were never invented we would only have had comparatively heavier modes of transport which can comfortably accommodate the huge boiler of a steam engine such as ships and trains. Although the automobile industry is moving to electric cars with batteries in a hurry as explained in the chapter on batteries, It will take decades before airplanes or rockets run on batteries or solar power and internal combustion engines will keep their high rank in the list of supreme human inventions.

Automobile

Towards the end of the 19th century there was a huge environmental crisis brewing in big cities around the world. Great cities such as London and New York were literally drowning in horse manure from hundreds of thousands of horses used to carry people around in horse drawn carriages and buses. Henry Ford's *Model T* solved this problem called, *"The Great Horse Manure Crisis of 1894"*. Although it sounds incredulous, when automobiles were introduced it was a big improvement not only in human convenience but for urban pollution as well. In 1903, Henry Ford founded *Ford Motor Company* at Detroit in the US and introduced the Model T in 1908. The Model T, at just $850 ($24,000 in 2020), revolutionized the world and made cars affordable for normal people for the first time. It hit a million vehicles by 1914 (and 10 million by 1924), consigning horse carriages and "great manure crisis" to the dustbin of history. It all started in 1886, when German inventor Carl Benz at Mannheim near Stuttgart in Germany, got the first patent for a motor car using an internal combustion engine; His patent, "FAHRZEUG MIT GASMOTORENBETRIEB" ("Vehicle with gas engine operation") - machine-translated from original German - reads:

"The present construction is primarily intended for the operation of light wagons and small boats, such as those used to transport up to 4 people.

The accompanying drawing shows a small tricycles-style car, built for 2 people. A small gas engine, regardless of the system, serves as the driving force. The same gets its gas from an apparatus to be carried, in which gas is produced from ligroin or other gasifying substances. The cylinder of the engine is kept at the same temperature by evaporating water...."

Carl Benz produced the first commercially available automobile in 1888. His wife Bertha Benz's iconic drive the same year, now celebrated as the *Bertha Benz memorial route,* is the start of mainstream automobiles. *Mercedes-Benz* and all cars originated from his patent, Bertha Benz's dowry and her iconic drive. A BBC article says:

"So much of Bertha's 194km round-trip drive between the Benz family home in the city of Mannheim and her mother's place in Pforzheim, typifies the fearlessness of automotive pioneers. ... In an era before road maps and GPS, Bertha had only rivers and railroad tracks to guide her to her mother's home. Imagining her jostling over blackened cobblestones in a buggy with wooden wheels and a 2-hp, four-stroke engine, I realised how brave she was. *Maybe a little crazy, too*. And maybe that's the reason her plan succeeded."

It is the invention with one of the most day to day impacts as private cars in developed countries and as public buses in developing countries. There are now more than a billion automobiles in the world making it one of the ubiquitous and widely useful inventions in our day to day lives. Automobiles changed country landscapes with interstate highways, *autobahns* and a maze of overpasses which characterize modern industrialized countries. It has also changed the landscape and life of cities in developing countries such as India with congested roads with massive traffic and air pollution. Long waits in congested freeways are common in big cities in developed countries with no proper public transit (*San Francisco Bay Area* in the US for instance). However, contrary to people's intuition that automobiles are dangerous due to fatal accidents, it may have historically saved people's lives. A scholar comparing horse carriage accidents with modern automobiles says [emphasis added]:

"...we obtain the average fatality rate of **18.8 per 100 million miles of horse travel**. In contrast, the rate for the motor vehicle years between 1925 and 2000 is **8.1 deaths per 100 million miles**. Were we to assume that Americans never adapted to automobile use but nonetheless traveled modern distances (about 2.75 trillion miles annually (U.S. Bureau of Transportation Statistics 2001: Table 1-29)) by horse-drawn vehicles, we would expect 517,000 deaths annually. Even if modern medical treatment could save the lives of half of these casualties, horse-related injuries would cost at least 200,000 lives more than the current average annual fatalities from motor vehicle travel. According to such conjecture, **the automobile may be responsible for saving the lives of at least 10 million Americans since 1950.**"

Automobile ranks as one of the high ranking inventions although it has been with humanity only for slightly more than a century. If we consider clothing, fire as foundational inventions occupying the top layer of human inventions and printing press, map and ship as *classic foundations* occupying the second layer, the next layer just below these two layers is occupied by the inventions of the industrial revolution: steam engine, railways, electricity and automobiles, the foundational inventions of a technologically advanced society. Automobiles are ranked below internal combustion engines as the latter is a prerequisite for automobiles. Railways have been with us for almost double the time as automobiles and have played a bigger role in our history and deserve a higher rank. Automobiles are ranked higher than Electricity as it has a longer history, with electricity being widely available in the US only after 1935 and in India only at the end of the 20th century (see the chapter on *Electricity*).

Electricity

Electricity drives the modern economy and daily life. Nowadays we can't imagine life without electricity. Let's do the impossible and imagine a world without electric energy for a moment to understand how much impact it has in our lives.

- For light we will have to use some form of chemical energy such as the *kerosene lamp*, which is inefficient and creates severe indoor pollution.
- For appliances with a motor such as the washing machine, we need to use an internal combustion engine as we do for automobiles using natural gas or gasoline.
- Our TV and computers can't use transistors as they depend on electricity flow in a single direction nor even vacuum tubes for the same reason. Our 1s and 0s will be represented by some mechanical contraption like a *semaphore* in railways going up and down. Given the size of these 'bits' these devices will be much bigger than a modern computer and way bigger than even ENIAC which used vacuum tubes. This means we couldn't have afforded a computer for the simple reason we don't have big enough houses. This would also be driven by mechanical wheels and gears powered by a transmission from an internal combustion engine powered by chemical energy from fossil fuels. Even if we

had a big enough house, we wouldn't be able to afford the gas bill from these behemoths.
- In communications, we would neither have telegraph nor telephone as they depend on electric signals travelling close to the speed of light. We will have to use light energy in some form of optical fibres which directly send optical pulses of 0s and 1s somehow converted from our dots and dashes of Morse code at the transmitter.
- A telephone would be even more complex in this light based system. We have to convert voice into a light signal of varying intensity using some mirror contraption or as digital encoding of the analog voice signal.
- Electromagnetic [Radio] waves won't exist as they are produced by accelerated electrons flowing through a conductor [that is electricity but with a much higher frequency than a standard AC current of 50/60 Hz]. Strictly speaking light won't exist either as it is an electromagnetic wave produced the same way but we will ignore it in this thought experiment.
- We won't have mobile phones as we don't have the radio waves and our mechanical transistors are too big to make a mobile phone circuit.

This thought experiment tells us that without electricity we will have a very low quality of life just like our ancestors of the 19th century and earlier. We will be using a dim kerosene lamp with indoor pollution and risk of fire, a mechanical computer and TV [with an optical broadcast] the size of a huge building with a big gas bill, internal combustion engine motors in our kitchen and no radio or mobile phone. Electricity is the foundation for almost all modern inventions such as light, household appliances, Radio, TV, movies, computers, internet and so on. Electricity is critical for large scale industrial production; Without electricity, industrial production will have to switch to internal combustion engines or steam engines requiring vast amounts of fossil fuel and associated pollution as during the industrial revolution of the 18th and 19th centuries.

The work of two Italians, Luigi Galvani at *University of Bologna* near Venice in 1791 and Alessandro Volta at *University of Pavia* near Milan in 1800 made electricity more than just a curiosity for the first time.

Father of modern electricity is Michael Faraday of London in England with his invention in 1831 of *electromagnetic induction*, the principle behind electricity generators and transformers, which converts the motion of an electrical conductor such as metal coils in the magnetic field into electricity. Nowadays, huge coils inside equally big magnets are moved using some form of mechanical energy such as water from a dam or falls or steam from a coal powered plant. In 1882, Thomas Edison at Menlo Park near New York City used his pearl street station generator to illuminate *Manhattan* (main island of New York City) and the rest is history. In 1888, Nikola Tesla in New York City patented, "DYNAMO-ELECTRIC MACHINE", the alternating current (AC) used in mainstream electricity now instead of Edison's direct current (DC). (However, very long distance power lines such as the 1360 km (846 miles) power line from *Celilo station* near Seattle to Los Angeles use DC). In the 1880s and 1890s public lighting started appearing throughout the world in a hurry and electric trams started to replace horse-drawn carriages. In Germany the first electricity company, *Allgemeine Elektricitäts-Gesellschaft,* was created in Berlin in 1883. Far in the east, the first electric trams of India appeared in Chennai [Madras at that time] in 1895. Household electricity started appearing in the 1920s throughout the world. In Europe, England set up the national grid in 1926 and it linked the nation by 1935.

Electricity is one invention we will immediately feel if it were to go missing from our lives. Without electricity there are no lights, computers, phones, internet or any of our household appliances, at least in the forms we use them now. This means it is one of the inventions with a strong multiplier effect. Another important effect of electricity has been to make our energy non polluting, resulting in much better air quality and increased life expectancy. However electricity (for household use) has been with us for only a century as compared to many modern inventions. In 1990, 57% of the population didn't have access to electricity in India. Even in the US, 90 percent of the rural homes didn't have electricity in 1935. In fact, rural electrification is one of the less talked about but more long living achievements of FDR. In India 240 Million people were still without access to electricity as of 2015. However its importance should increase in the 21st century with developing countries moving towards full access to electricity and electricity deserves its very high ranking.

Light bulb

"The object of this invention is to produce electric lamps giving light by incandescence, which lamps shall have high resistance, so as to allow of the practical subdivision of the electric light.

The invention consists in a light-giving body of carbon wire or sheets coiled or arranged in such a manner as to offer great resistance to the passage of the electric current, and at the same time present but a slight surface from which radiation can take place.

The invention further consists in placing such burner of great resistance in a nearly perfect vacuum, to prevent oxidation and injury to the conductor by the atmosphere. The current is conducted into the vacuum-bulb through platina wires sealed into the glass."

Thomas Alva Edison, the greatest inventor of all time as can be seen from his frequent appearances in this list and 1,093 patents to his name, at Menlo park near New York City in the US invented the light bulb in 1879. When we see the satellite image of earth at night at *Our World in Data* (a beautiful site from the *University of Oxford* for people

who love to think in data and facts), we are struck by the places which have advanced human societies and the places which are still living a few centuries behind others. We can clearly see Europe, North American East and West Coast, India, Eastern China, Japan, Coasts of South America and Australia burning bright. We can also see almost all of Africa, the interior of Australia, interior of South America and Mongolia and Central Asia showing almost no light. The difference between North and South Korea is stunning. (If a computer is nearby it is worth taking a look and coming back to read the rest of the chapter). It is interesting that artificial light is a proxy and the foundation for our civilization and clearly one of the most important inventions of humanity. Although there were other sources like kerosene lamps or whale oil before electric light, they were very expensive, particularly relative to the average wages at the time and very *inefficient*, not very bright in terms of lumens. Electric light brought us modern nightlife and extended our waking lives, making humanity more productive. For one, I am sitting at home writing this at 9 PM at night with the aid of an artificial light and my daughter is doing her homework with the help of the same light. Two centuries back we simply couldn't have afforded the light even if it provided enough lumens. Steven Pinker says in his excellent book *Enlightenment Now* based on research from economist William Nordhaus [emphasis added]*:*

"A Babylonian in 1750 BCE would have had to labor fifty hours to spend one hour reading his cuneiform tablets by a sesame-oil lamp. In 1800, and Englishman had to toil for six hours to burn a tallow candle for an hour... In 1880, you'd need to work fifteen minutes to burn a kerosene lamp for an hour; **in 1950, eight seconds for the same hour from an incandescent bulb**; and in 1994, a half-second for the same hour from a compact fluorescent bulb - a 43,000-fold leap in affordability in two centuries."

It is not just about the quality (*lumens*) or affordability (*amount of labour required*) of the light. Electric light is much safer than open flame systems with serious fire hazards such as gas or kerosene lamps and has saved many lives over the century. There was a fire in the *Vienna Ring theatre* in 1881 that killed 794 people after a gas lamp caught fire.

The *New York Times* reports on the next day Dec 9, 1881 [emphasis added]:

> "The ring theatre, formerly the Comic opera-house, where Sarah Bernhardt recently performed, took fire at 7 o'clock to-night, just before the beginning of the opera ' Les Contes d'Hoffman.' **The fire was caused by the fall of a lamp on the stage**. The house was tolerably full, and the loss of life is very great. ... The scene was terrible, the flames shooting up through the roof and eventually gutting the entire building."

Another serious fire in Chicago at *Iroquois theatre* from an electric arc lamp killed more than 600 people in 1903. Earlier *Brooklyn theatre fire* in 1876 killed around 300 people after a gas border lamp ignited the stage scenery. These incidents serve as the reminder that electric light has been a life saver by enhancing fire safety and reducing indoor pollution. It also would have been very dangerous to venture outside at night without street lights or in dim gas lights (One of my distant relatives was bitten by a snake when he went outside at night and died as late as in the 1980s India). This shows why rural electrification has been a humanitarian, life saving mission in developing regions such as India or Africa or indeed in the US in the time of President Franklin Roosevelt (FDR) in the 1930s and 40s.

Electricity ranks very high in this list because of electric lights. If people are given only one choice from electric amenities they choose electric lights over everything else. I should know, as I grew up in a house where the only electric amenity was an electric incandescent lamp [and later a fluorescent tube] with power shared from one of our neighbours. It should be considered one of the most important inventions which brought us to the modern times with its impact on quality of life, safety and economy. Light bulb gave us an extra 12 hours which increased human productive lives vastly [reading books, school children able to do more schoolwork, shops open late at night, factories can run in multiple shifts to produce a critical vaccine]; Part of this enhanced productivity goes into new inventions, making electric light a foundational invention powering other inventions.

Fertilizer

"The year 2013 marks the 70th anniversary of the **Bengal Famine which resulted in the death of an estimated 1.5 to 3 million children, women and men** during 1942-43. ... Famines were frequent in colonial India and some estimates indicate that **30 to 40 million died out of starvation** in Tamil Nadu, Bihar and Bengal during the later half of the 19th century."

M. S. Swaminathan, architect of India's *agricultural revolution* wrote this in *The Hindu,* an indian newspaper, about India's food situation a century ago. It doesn't happen in modern India or almost anywhere else in the world now. Of course, quite a bit of blame for the famines should go to british colonial policies treating Indians as second class citizens, in particular British Prime Minister Winston Churchill's callous attitude

during the *Bengal Famine*; He famously asked how, if the shortages were so bad, Mahatma Gandhi was still alive. Modern India is a democracy, in fact the biggest democracy in the world which definitely matters. There is another equally important reason for this modern miracle.

One of the greatest inventions in the world most people are not aware of happened in one of our basic necessities, food. This invention is responsible for the lives of almost half the people living on our planet at this moment, 3-3.5 Billion people. *Haber-Bosch process* for converting inert nitrogen (N_2) in the air into *Ammonia* (NH_3) which has Nitrogen in a form consumable by plants and animals. The main reason this is required is because two nitrogen atoms like each other so much that the N_2 molecule has one of the strongest chemical bonds known. Nitrogen atoms prefer this arrangement so much that most of the explosives contain nitrogen atoms in some other form which on the first chance they get, try to get back to their preferred inert N_2 form, violently releasing a lot of energy in the process. This also means nitrogen molecules in air don't participate in any chemical reactions making them useless in biology. For nitrogen to be useful it needs to be converted to an active form like Ammonia whose nitrogen atoms can be used to build organic molecules including proteins, DNA and enzymes, very important for all life on earth. Fritz Haber at University of Karlsruhe near Stuttgart and Carl Bosch at *Badische Anilin und Sodafabrik* (BASF) in *Ludwigshafen on the Rhine* near Mannheim in Germany are credited with discovering this process which is responsible for modern fertilizers in 1909. This reaction is so important that nearly 50% of the nitrogen found in our body originated from the Haber–Bosch process running in an industrial plant. Before this process was invented we had to rely on nitrogen fixing bacteria in the family *Azotobacter*, living in the roots of *legumes* (pea family) which simply couldn't support a population of 8 Billion people. This process is credited with eliminating famines in the modern world and supporting such a large population in much smaller farmland by increasing land productivity.

The Haber-Bosch process uses high pressure of 200 times the atmospheric pressure and high temperature of more than 400°C and iron as the catalyst to combine nitrogen and hydrogen [from a source like Methane] to make Ammonia. Haber invented and proved this at laboratory scale with *Osmium* or *Uranium* as the catalyst. Bosch at BASF scaled up this process to work at industrial scale. BASF remains the largest chemical producer in the world with 72 Billion dollars in revenue (2020). The book *Full Planet, Empty Plates: The New Geopolitics of Food Scarcity* says [emphasis added]:

"The transition was dramatic. Between 1950 and 1973 the world's farmers doubled the grain harvest, nearly all of it from raising yields. Stated otherwise, **expansion during these 23 years equaled the growth in output from the beginning of agriculture until 1950.** The keys to this phenomenal expansion were fertilization, irrigation, and higher-yielding varieties, coupled with strong economic incentives for production."

As the invention responsible for giving life literally to billions of people and eliminating famine it deserves to be ranked very high. If ever there was a magic potion for eliminating famine and hunger this is it. It can be called the *"printing press of agriculture"* in the sense it completely transformed modern agriculture by increasing land productivity exponentially. 157 million tons of Ammonia was made in 2010 using this process (For comparison, the world produced 170 million tons of sugar in 2019 and 380 million tons of plastics). As an invention affecting the first of the three basic necessities it ranks very high. It also gets a minor boost from its role in explosives. As fertilizer accelerated food production and sewing machines accelerated cloth production both rank very high in the list of inventions. Fertilizer ranks just above the sewing machine as food is a more important necessity than cloth for survival and clothes are reusable, so we need less of them. Electricity and electric light ranks higher than both of these inventions for reasons explained in the sewing machines chapter.

Sewing machine

"Day after day, the grim drudgery of digging for bodies had progressed at Rana Plaza. Talk of rescuing survivors had faded. This was the recovery phase, and what was being recovered were corpses, the numbers spinning remorselessly forward: 700 dead became 800, then 900, with no end in sight. By Friday morning, the number had pushed above 1,000."

The *New York Times* reported on the *Rana Plaza disaster* this way, 17 days after the tragedy. In 2013, Rana plaza, an eight story building in Bangladesh capital Dhaka collapsed, killing 1134 textile workers making clothes for major world brands like Walmart, Gucci and Versace. Food, shelter and cloth are our three basic necessities. Nowadays we take our clothes for granted as it is very easy for us to buy clothes with a click of a button or pick it from a supermarket from hundreds or thousands of choices for as little as the price of a single meal, 10-20 dollars in the US. However, unlike fruits or grains, clothes don't grow on trees [or plants]. Each one needs to be made by a person somewhere in the world; We only come to know about them when a tragedy like that of Rana plaza

occurs. Behind every piece of clothing we wear there was a person stitching that cloth and there was a sewing machine, one of the greatest inventions of all time. French Tailor Barthelemy Thimonnier invented the first working Sewing machine in 1830. The story has it that he was almost killed by other tailors fearing unemployment, *Luddites* of England style. Elias Howe in the US invented the first practical sewing machine with the crucial *lockstitch*; Lockstitch is the mechanism to lock an upper thread using a threaded needle with a lower thread, *bobbin thread*, in the bobbin making the familiar machine stitch pattern. This is accomplished by a rotating hook in the bobbin catching the upper thread and looping it around the lower thread. (If all this sounds confusing, it is time to look at one of the many sewing videos on the *internet* and come back; One video is in the *bibliography*). This is the heart of the invention in the sewing machine. For this reason Elias Howe is the main inventor of the sewing machine. The *Science Museum* in London has one of the original Elias Howe's sewing machines from 1846 with the all important lockstitch on display; This was brought to England from America by his brother Amasa Howe when they tried to sell their machines in England. Isaac Singer at New York mass produced the first commercially successful sewing machine in America, the very famous *Singer machine,* pooling inventions from multiple people including Elias Howe in 1856. In a few years, this brought a sewing machine to almost every household which needed one as Singer's company introduced the now well known financial innovation to go along with it, monthly installments (**EMI**) instead of paying the full cost upfront for the first time. (Elias Howe later won a lawsuit and royalty from Singer company).

Sewing machine, along with globalization, made the clothes so cheap and affordable that *Guardian* says [emphasis added]:

"For the consumer, of course, this has all meant that while prices of everything else except communications have risen, clothes cost less. **In 1900, 15% of a US household's income was spent on clothing**. In 1950, it was still 12%. Even as late as the early 1990s, major purchases of clothing – a suit, a dress, a coat – marked a special occasion or a rite of passage. But by 2004, the total amount spent by households on clothes had dropped to just 4%. By 2010, according to the US Bureau of Labour Statistics, **clothing cost the average American family only $1,700 (£1,017), 2.8% of their income**. And for that money the consumer gets much more. Cheap no longer means nasty; it just means affordable. In 1997, the average woman in the UK bought 19 items of clothing a year; in 2007, she bought 34."

By bringing down the price of clothing, one of the basic necessities of human life, sewing machines played a major role in improving the quality of life. It increased productivity of cloth making on the same scale as the printing press increased book printing over human scribes. In the 1840s before this invention a shirt and pants cost close to 500 dollars in today's prices. Our forefathers were able to only afford new clothes once a year or so. Its importance is on the same scale as electricity and it has been with us for longer than electricity, by almost a century. However, electricity was a platform invention which spawned multiple inventions in turn. Hence electricity ranks higher than sewing machines despite being a relative luxury as compared to a sewing machine. For instance, in many Indian and the US villages sewing machines reached first, not electricity (see the chapter on *Electricity*) which reduces the human impact of electricity compared to sewing machines. Even with that electricity ranks higher as the meta-invention which in turn created so many inventions. Major textile inventions from the industrial revolution such as the Spinning Jenny can also be compared with the sewing machine. However no single invention in textiles is as original, played as big a role as the sewing machine, or continues to be as relevant. For instance, Spinning Jenny is an incremental invention originated from the millenia old *spinning wheel* and was replaced by spinning technologies such as spinning mules and modern open-end spinning (see the chapter on *Spinning Jenny*). The humble sewing machines and the people using them to make our clothes across the world are something we should be thankful for everyday when we wear our clothes.

Vaccine

The Coronavirus epidemic which started in Dec 2019 has killed 2.4 Million in a little more than a year [*Feb 2021*]. At this rate it would have killed 240 Million in the 21st century. Fortunately for us we now have vaccines for this dangerous epidemic. There was a disease which actually killed 125% of this hypothetical number, 300 Million people, around 90% of the US population now, in the 20th century making it a much more tragic incident for humanity over the long term. *Our world in data* has a harrowing graph where it shows that as high as 18% of the 18th century Londoners, almost one in five, dying of this disease. Steven pinker emphasizes this in his excellent book *Enlightenment Now*:

> "scientific knowledge eradicated smallpox, a painful and disfiguring disease which killed 300 million people in the 20th century alone. In case anyone has skimmed over this feat of moral greatness, let me say it again: scientific knowledge eradicated smallpox, a painful and disfiguring disease which killed 300 million people in the 20th century alone."

Edware Jenner of England invented the first vaccine, an inoculation with the related but weaker cowpox virus, for *smallpox* in 1798. He tested his vaccine by inoculating eight-year-old James Phipps

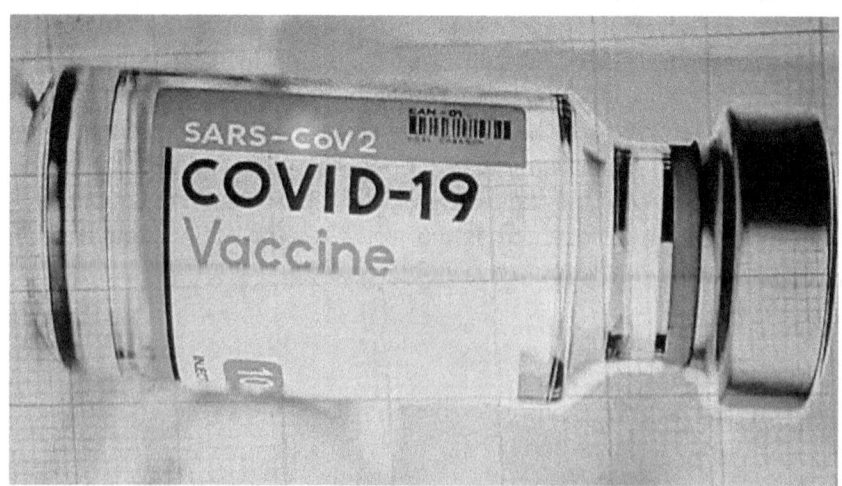

with cowpox pustule liquid recovered from the hand of a milkmaid, Sarah Nelmes, who was infected with cowpox. This is functionally equivalent to using the live attenuated virus causing the smallpox disease as the vaccine. This weaker form of virus or bacteria causes our immune system to learn about the disease without making us very sick with the severe disease. It builds antibodies that can fight the normal, stronger form of the microbe. Interestingly, Jenner's vaccine was invented before the *germ theory of diseases* was known and he didn't know about viruses at all; It would have been a mystery to Jenner as to why it worked. Louis Pasteur of France put the vaccines on solid foundation by developing the germ theory of diseases in 1877. He proceeded to develop many vaccines of his own; Louis Pasteur developed the first live attenuated bacterial vaccine for *chicken cholera*; He developed another vaccine for a dangerous virus disease that spreads to people from dogs via dog bites, *rabies* in 1884. He also founded the *Pasteur institute* in Paris in 1887 which has been a prominent institution in the fight against microbes for over a century; It was the first to isolate the HIV virus in 1983; In 1921, the *Bacille Calmette-Guerin* (BCG) vaccine for *tuberculosis* (TB) was developed at the Pasteur institute by Albert Calmette and Camille Guéérin. It is administered to babies in high risk countries such as India but is not part of the standard vaccines in developed countries like the US. Emil von Behring at *Halle on the Saale* near Dresden developed the vaccine for *diphtheria*, a major killer of children at that time, in the 1890s. Leila Denmark in Atlanta in the US (among others), a pediatrician who lived to 114 years and 60 days and practised till 103, developed the *pertussis* (whooping cough) vaccine. Jonas Salk at the *University of Pittsburgh* near Philadelphia in the US invented the first vaccine for *polio*, a debilitating disease that affects the brain particularly of children, in 1952. Albert Sabin, a polish-american at *Cincinnati's Children's Hospital* near Chicago in the US developed the oral polio virus in 1961 which played the key role in eradicating polio all over the world. In 1966 WHO started the global smallpox eradication campaign and achieved its goal on May 8, 1980. Maurice Hilleman working at *Merck* in West Point near Philadelphia in the US developed the *MMR vaccine* in 1971 from earlier vaccines, some of them he himself developed. Maurice Hilleman was responsible for developing more than 40 vaccines, including for measles, mumps, hepatitis A, hepatitis B, meningitis, pneumonia, Haemophilus influenzae bacteria, and rubella. WHO lists vaccines for 23 common diseases [excluding smallpox] and CDC lists 14 diseases in *"Diseases You Almost Forgot About (Thanks to Vaccines)"*. Many coronavirus vaccines of 2020 such as from Pfizer or Moderna use a novel technology based on *mRNA* which directly encodes viral spike protein instead of using a live attenuated virus or viral proteins.

Vaccines have been nothing short of a miracle invention eliminating deadly diseases such as Smallpox, Polio, Measles and Whooping cough. Steven Pinker says again in *Enlightenment Now* on the discovery of the Polio vaccine:

"On April 12, 1955, a team of scientists announced that Jonas Salk's vaccine against polio—the disease that had killed thousands a year, paralyzed Franklin Roosevelt, and sent many children into iron lungs—was proven safe. According to Richard Carter's history of the discovery, on that day "people observed moments of silence, rang bells, honked horns, blew factory whistles, fired salutes, . . . took the rest of the day off, closed their schools or convoked fervid assemblies therein, drank toasts, hugged children, attended church, smiled at strangers, and forgave enemies."

It is hard to overestimate the importance of vaccines. They are second only to fertilizers in life-giving ability and transforming human life for the better. (If you have a computer nearby, take a look at the pictures of children in *iron lungs*, leg braces or smallpox scars; Last one is not for the faint of heart). The single biggest achievement of humanity in the 20th century is arguably the elimination of smallpox virus completely from the face of the earth. Remember we are claiming this in an extraordinary century for humanity when we discovered *relativity* and *quantum mechanics*; We connected almost all of humanity to the electricity grid; We invented radio, tv, computers, internet and mobile phones; We sent people to the moon. Vaccines vastly increased the number of children living to see their first and fifth birthdays. Given its life-giving importance it ranks high among modern inventions.

(Image credit: NBC News)

Antibiotic

Leprosy (*Hansen's disease*, in honor of Armauer Hansen who discovered the *bacillus* that causes the disease in 1873, in Bergen, Norway) is a horrible disease. It is caused by a bacterium called *Mycobacterium Leprae*; This bacteria attacks the victim's nervous system, in particular the nerve endings; People affected by it lose fingers and toes from injuries and infections of unnoticed wounds as the nerve endings connecting them are affected by the bacteria and don't transmit signals to the brain; It historically had huge stigma and people with the disease were isolated into *leper colonies* in many countries in the world. Now there is a magical medicine intervention which can cure this disease completely, *antibiotics*. WHO provides these antibiotics [combination of *dapsone*, *clofazimine* and *rifampicin*] for free as well. For someone affected by this disease being born 100 years back or now can make all the difference. (It is worth looking at the images of people with deformed extremities from Leprosy to really appreciate the importance and impact of the antibiotics). Antibiotics did the same many times over for many bacterial infections. Normal surgeries and childbirth were much more dangerous before the age of antibiotics due to the risk of fatal infections. *Washington post* quotes Author Eric Lax:

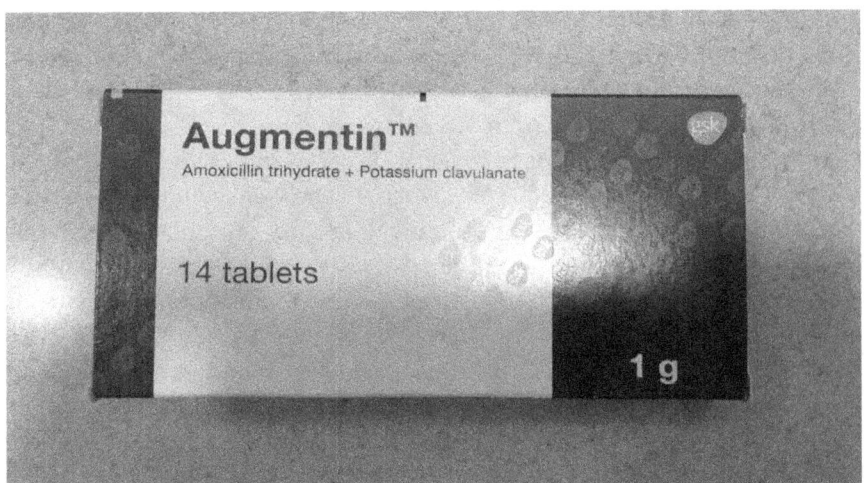

"Of the 10 million soldiers killed in World War I, about half died not from bombs or shrapnel or bullets or gas but rather from untreatable infections from often relatively minor wounds and injuries."

In 1928, Alexander Fleming at St. Mary's hospital in London invented *Penicillin*, the first antibiotic from the fungus *Penicillium rubens*. Howard Florey of Australia and Ernst Borin, a German born British (with eerie similarity to Young Einstein) at *University of Oxford* isolated the active ingredient, *2-Pentenyl Penicillin* (Penicillin-F), in 1940 that made penicillin mass producible. More powerful and commonly used penicillin, *Penicillin-G* contains *Benzylpenicillin*. Penicillin (Penicillin-G) was widely used in World War II by allied forces with the US producing more than *650 billion units* by 1945. Penicillin prevents the bacteria from synthesizing *peptidoglycan*, a molecule in the cell wall, providing its strength. In 1945, British doctor Robert Cochrane, who worked in *Tamilnadu*, a state in south India near Sri lanka, at the time first used *dapsone* for leprosy, laying the foundation for the modern treatment.

Most commonly prescribed antibiotic *Amoxicillin* was developed by *Beecham* (now *GlaxoSmithKline*) in the 1960s. Amoxicillin is a derivative of penicillin; This in turn was derived from *Ampicillin*. Ampicillin only differs from Penicillin-G by an extra *Amino* [NH_2] group. Amoxicillin itself differs from Ampicillin by an addition of the *hydroxyl* [OH] group in the functional benzene ring of Ampicillin [and Penicillin-G]. These additional groups make them more lipid soluble and able to pass through the cell walls of *gram negative bacteria*, destroying their cell walls. It addresses the main limitation of original penicillin that was only effective against *gram positive bacteria*. (Gram negative bacteria have an outer lipid membrane protecting the peptidoglycan layer of the cell wall, making it harder for penicillin to reach the layer). These semisynthetic penicillin derivatives work against a broader range of bacteria. These breakthrough antibiotics started in 1958, when Beecham scientists invented a process for the Isolation of *6-APA* [6 aminopenicillanic acid] from penicillin; Semi-synthetic penicillins such as Amoxicillin could then be synthesized from 6-APA. The major commercial source of 6-APA is still natural Penicillin-G. Penicillin and its derivatives, together called *ß-lactam antibiotics* for their main functional group of four atoms [three carbon and one nitrogen] ring in the middle, are a major group of antibiotics in use now. Interestingly, bacterial resistance works by bacterial cells producing an enzyme *ß-lactamase* which breaks down this ring in the antibiotic [arms race between us and the bacteria]. Macrolides [*azithromycin*] and Fluoroquinolones [*ciprofloxacin*] are the two other major groups of antibiotics that work quite differently from penicillins and *cephalosporins* (ß-lactam antibiotics). Fluoroquinolones are controversial as they have serious side effects including nerve damage. Latest WHO list of critically important antibiotics lists around 100 antibiotics.

It is estimated that Penicillin alone saved on the order of a few hundred Million lives. It played a very important role in reducing childhood mortality. Before antibiotics, 90% of children with *bacterial meningitis* died, for instance. Among those children who lived, most had severe and lasting disabilities, from deafness to mental retardation. More than 270 Million antibiotic prescriptions are written in the US alone for a population of 330 Million or 836 per 1000 people (2016). Antibiotic resistance is an important issue which makes some of the antibiotics, particularly ß-lactam antibiotics, ineffective. (This is the reason it is important to finish the antibiotics course even after the symptoms have gone away).

Antibiotics easily rank as one of the most important inventions for their life saving impact [compare world war I deaths from infections to world war II with Penicillin]. Along with vaccines they have produced a medical miracle in the 20th century and one of the main reasons for increase in life expectancy and reduction in child mortality. They are the reason we don't see leper colonies all over the world now. Vaccines rank higher than antibiotics as they work on viruses as well and typically provide long term immunity. Viruses are as big a threat to humanity as bacteria as we know from the ongoing coronavirus epidemic (March 2021). In general, prevention is better than cure and vaccines provide the prevention and antibiotics provide the cure. Vaccines also erased many diseases from the face of the earth such as smallpox and polio [with only very few cases in remote areas]. That said antibiotics come second only to vaccines in their life giving importance and rank well above other modern inventions of humanity such as radio, television or computers.

Bleach

"ST. PETERSBURG, Sept, 16. - The cholera epidemic which, originating in Southern Russia, has claimed already upward of 100,000 victims, is stretching its way across Asiatic Russia, and to-day was officially declared to threaten the Province of Amur, in Southeast Siberia, and separated by the Amur River from Manchuria.

The reports now in possession of the Sanitary Bureau show a **total for the season of 182,327 cases, with 83,613 deaths.** These include the early returns for the week ended Sept. 10 and the revised figures for the preceding week. Complete reports for the week of Sept. 4 to Sept. 10, inclusive, are lacking, but the totals for the seven days at hand are 7,559 cases, with 3,557 deaths. The totals for the preceding week are 18,330 cases, with 6,187 deaths."

In an article, "CHOLERA DEATH TOLL IN RUSSIA 100,000", The *New York Times* of Sept, 17, 1910 provides the grim statistics of a cholera epidemic [emphasis added], one of the many around the world

till the early 20th century. *Vibrio Cholerae*, a comma shaped bacterium, was a serial, mass murderer in the world in the 19th and early 20th century; It spread through contaminated drinking water and devastated the world through a series of pandemics starting with the first one in 1817 that started in *Ganges river delta* near Calcutta in India. (Technically there is still an ongoing Cholera pandemic, seventh overall, that started in 1961 according to WHO). It is an acute diarrhoeal disease that can kill within hours if left untreated due to the rapid loss of large amounts of fluids and electrolytes. In those times, life giving water also used to take away hundreds of thousands of lives. During the first epidemic (1817-1835) in Paris, the *mortality percentage* was 31-34%, that is it killed one out of every three who contracted the disease; It killed 5.6% of people in the age group 20-39, healthiest adults in the population. As recently as 2010 in Haiti, only 1,000km from the US, a Cholera epidemic that originated from the UN peacekeepers spread to hundreds of thousands and killed more than 10,000 people. A simple invention, adding *chlorine* to the water using bleach, became the magic bullet that saved millions of lives over the last century putting this invention on a par with vaccines and antibiotics.

In 1854 during a cholera epidemic, John Snow, "father of epidemiology", in England proved with experimental data that cholera spread from contaminated water. One of the first known (one-off) uses of chlorination was in 1897, when bleach solution was used to disinfect a water main in Maidstone near London in the UK, following an outbreak of *typhoid*. The first continuous chlorination of water was in 1902 at *Middelkerke* near Brussels in Belgium. In the UK, the first use of bleach was in 1905 in Lincoln near Manchester after a typhoid epidemic. In 1908 in Chicago in the US, George Johnson instituted chlorination by adding *chloride of lime* (crystalline mixture of *calcium hypochlorite* with alkaline *calcium chloride*) to contaminated river water. Chlorination of a river water supply to Jersey City near New York in the US in 1908, led to a litigation that publicized the chlorination effort and led to the spreading of the process across the country. It spread in the US and Europe in the early 20th century and spread to countries in Asia such as Japan in the second half of the century.

In 1785 French scientist Claude Berthollet in Paris discovered *sodium hypochlorite*, the first commercial bleach, named Eau de Javel ("Javel water"). Charles Tennant and **Charles Macintosh** in Renfrewshire near Glasgow in the UK in 1798 discovered *calcium hypochlorite* ("bleaching powder") as an alternative for Javel water. In 1820, French chemist Antoine Labarraque in Paris discovered the disinfecting power of hypochlorites, crucial for the chlorination of water. Chlorination controls typhoid, cholera, amœbic dysentery, bacterial gastroenteritis, shigellosis, salmonellosis, Campylobacter enteritis, Yersinia enteritis, Pseudomonas infections, schistosomiasis, giardiasis and viral diseases such as hepatitis A. Although chlorine is by far the most commonly used disinfectant, other chemicals such as *ozone* and *chlorine dioxide* have also been used for many years. According to the CDC:

"...The occurrence of diseases such as cholera and typhoid dropped dramatically. In 1900, the occurrence of typhoid fever in the United States was approximately 100 cases per 100,000 people. By 1920, it had decreased to 33.8 cases per 100,000 people. In 2006, it had decreased to 0.1 cases per 100,000 people (only 353 cases) with approximately 75% occurring among international travelers. Typhoid fever decreased rapidly in cities from Baltimore to Chicago as water disinfection and treatment was instituted. This decrease in illness is credited to the implementation of drinking water disinfection and treatment, improving the quality of source water, and improvements in sanitation and hygiene."

Life Magazine, in 1997, called the purification of drinking water "probably the most significant public health advance of the millennium." Bleach, more broadly chlorination of water, saves millions of lives every year; It ranks on par with vaccines and antibiotics with much broader impact over multiple water-borne microbes. According to WHO, even now Cholera sickens approximately 2.9 million people every year, and kills 95,000 mainly in areas where water is still untreated showing the danger of water borne microbes and what it could have done to us without this life saving invention in water supplies around the world.

Radio

"Those who have been saved, have been saved through one man, Mr Marconi... and his marvellous invention."

Herbert Samuel, then Postmaster General of England, remarked in April 1912. On 15 April 1912, on her maiden voyage from England, *RMS Titanic*, a huge 200 Million dollars (in 2021 dollars) ship hit an iceberg and sank into the north Atlantic ocean more than 1,000 km (600 miles) from her destination, New York City literally in the middle of nowhere. More than 1,500 people lost their lives in one of the worst disasters in maritime history. However there is a silver lining in this tragic story. Had this disaster happened 15 years earlier in 1897 all on board the Titanic, 2,224 people, would have died. The *Chronometer* had been invented centuries earlier which meant Titanic knew her longitude [and latitude which is much easier to calculate with a *sextant* as it doesn't require the time of the day]. She needed a way to communicate that information to other ships nearby. There was just such a new technology onboard the

Titanic called *wireless telegraphy*; This enabled the Titanic to inform *RMS Carpathia*, a nearby ship, about the tragedy. Carpathia came within two hours of the sinking and saved 706 people's lives thanks to this new technology, radio.

Radio is a product of *Maxwell equations* in 1865 by James Maxwell at *University of Cambridge* and he predicted *electromagnetic waves* based on these equations. Heinrich Hertz at *University of Karlsruhe* near Stuttgart in Germany detected these waves experimentally in 1888. In 1896, Italian Guglielmo Marconi at London in England, patented the first commercially successful wireless telegraphy system based on radio waves. Then the radio technology started to be used for another spectacularly successful but completely different use, *public radio broadcast*. First radio station in the world was KDKA, (a *call-sign* not an acronym), in Pittsburgh in the US started on Nov 2, 1920 with the broadcast of the 1920 *Harding-Cox presidential election results*. New York City radio station WEAF broadcast the first paid radio commercial in 1922. Now there are at least 44,000 radio stations in the world covering more than 70% of the world population.

Radio waves are generated when electrons in a conductor are accelerated back and forth with the frequency of the wave corresponding to the frequency of the back and forth motion of the current. In a typical AC current, the direction of the current changes 50/60 times a second, that is it has a frequency of 50/60 Hz. If the frequency of the oscillation becomes high enough, above 20 Khz it produces *radio waves*. This is the base carrier wave and information is encoded in the wave by modulating some aspect of this wave. In the *Amplitude Modulation (AM) radio* it is modulated by changing the *amplitude* [hence power, power goes as the square of the amplitude] of the wave as a function of the information. In *FM radio* it is modulated by a [a relatively small] change in the carrier wave frequency itself. FM radio stations are allowed to deviate around 75 Khz (0.1%) around their base frequency which is between 88-108 Mhz. FM radio works better for this reason as the change in the power of the radio wave from degradation due to distance, reflection and so on will change the AM radio's useful signal as it affects the amplitude but not the frequency.

Radio waves rank high as one of the greatest inventions of the 20th century playing a pivotal role in human communications. Telegraph for the first time provided humanity with the ability to send information much faster than a physical carrier can travel, be it a person on a horse, *pony express* style, or a train. It used electrical signals close to the speed of light [around 90%] to send information but with physical wires. Radio removed the need for these physical wires and communicated information literally at the speed of light, the maximum possible speed in the universe due to *special relativity* of Albert Einstein. As illustrated by the Titanic story above, this freedom from physical wires proved very transformative and saved many lives. It is vital for ships and airplanes particularly in the form of air traffic control communications with the aircrafts. It played a pioneering role in its offshoots TV, space missions, satellite communications and mobile phones. In this sense it has been the *"electricity of communication"* laying the foundation for a flurry of great inventions in communications. It brought in the era of realtime news instead of waiting for a day to see the news in the next morning's newspapers. It is hard to pick between telegraph and radio as both were pioneering inventions producing massive changes in human lives and history. Going from 0 [for practical purposes our normal speeds] to 90% is a bigger change than 90 to 100% in our communications speed. However the requirement of a physical wire is a very serious limitation particularly if you have to communicate from a ship or from a spaceship going for moon landing; A satellite can't communicate with earth using a telegraph unless we use 36,000 km wires to keep it in geosynchronous orbit. Telegraph is also more expensive due to the wires and energy inefficiency due to the resistance in metal conductors. Radio also started the era of *public media broadcast* which telegraph didn't. Radio is still a very relevant invention but telegraph and its offshoots such as landline telephone are fast becoming history. (This is not to deny their historical importance or the transformative role it played in human history; Telegraph wires were humanity's *"umbilical cords of fast communication"*). Radio beats telegraph by a whisker even though telegraph had a 50 year lead time and played a critical historical role by introducing a bigger "jump" in our society.

Telegraph

Telegraph is one of the greatest inventions of humanity for a particular reason. Just like steam engines freed people from animal [including their own] muscle power vastly increasing the power at humanity's disposal, Telegraph freed the transmission of information from physical travel of animals or people vastly increasing the speed of information transmission. *Pony express* was the fastest form of communication in the 1860 United States between her east and west coasts; It took 10 days from New York to San Francisco and literally used Ponies [Horses]. Electricity travels at around 90% of the speed of light in electrical conductors like copper wires (*270,000 km/s*, seven times around the earth in a second). It would only take 20 milliseconds for electricity to cross the distance between New York and San Francisco (*4670 km/2900 miles*). Once electricity was discovered people immediately realized there was a quick way to pass information from one place to another very quickly if we could find a way to encode the information in the electric current and have a physical wire connecting the places.

The simplest way is a binary code as in on or off [0 or 1 in computers] which in the *Morse code* is encoded using — [dash]and ·[dot] as in the well known SOS [··· — — — ···]. Interestingly *Mayans* used — and · to encode all their numbers in a base 20 system (see the

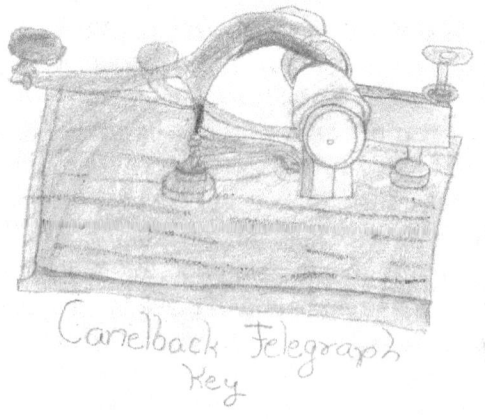

Camelback Telegraph Key

chapter on *Numbers*). In telegraphs, dash and dot were encoded by a long current and short current respectively. For comparison, modern computer memory simply stores a 0 and 1 using a capacitor in charged and discharged state. In some sense, old telegraph cables across the Atlantic are no different from the current optical cables carrying internet traffic across our oceans. They both carry information as binary code almost at the speed of light in vacuum. The transition from telegraph to telephone back to the Internet went from digital to analog and back to digital again. For the internet, computers interpret the binary signal and display it in human readable form as text, image, audio or video. For the telegraph an electromagnet with a metal armature converted the electrical signal into a human readable form as sound or as text. The metal armature, attracted by the magnet, hit the magnet and made the classic telegraph *clicks* and in some cases a pencil at the end of the arm made symbols corresponding to dashes and dots on a rolling paper tape.

Samuel Morse, of Morse code fame, in New York City in the US invented the telegraph in 1837 with help from many pioneers. In his patent of 1840, "IMPROVEMENT IN THE MODE OF COMMUNICATING INFORMATION BY SIGNALS BY THE APPLICATION OF ELECTRO-MAGNETISM", Morse also described the now well known Morse code ("Example 3" in the patent). It was a very efficient code in the sense it assigned fewer symbols to more common letters like E and more symbols for less common letters like Q. In England, William Cooke, an officer of Madras Army on leave at London and Charles Wheatstone at King's College, London invented the telegraph independently in 1837 but used a more complex code representing the letters of the alphabet directly. It has a modern analogy in computers in that some of the early digital computers like ENIAC used decimal code and ultimately lost ground to more modern binary codes.

It transformed all the information based industries particularly newspapers and financial services. Six New York City newspapers in 1848 founded the *Associated Press* to share the expense of gathering and transmitting news by telegraph. In 1867, the *New York Stock Exchange* introduced the *stock tickers*, telegraph devices that reported the sale of stocks. It was also a critical war technology in the 19th century and was used for improving the safety and efficiency of the railways. *American Civil War* was the first major war to benefit from the telegraph as a military communication technology as well as for war news reporting. Telegraph unified countries and continents; First transcontinental telegraph line was completed by *Western Union* in America in 1861. *Siemens AG*, founded in 1847, setup telegraph connectivity in Germany, Russia and other neighbouring nations. In British India, the first experimental telegraph service started in Calcutta in 1850. It connected all major cities such as Bombay and Madras by 1854. It also connected the continents for the first time permanently with the completion of the first *transatlantic cable* in 1866. With the emergence of widespread telephones, particularly mobile phones with internet in the developing world, telegraph services became obsolete. In 2013, India stopped its telegraph service bringing the telegraph era of the world to an end.

Telegraph is easily the *single biggest invention in the 19th century*. Telegraph's high ranking comes from the fact, that for the first time it freed the speed of information from the speed of physical travel and eliminated the need for physical travel just to convey information making the transfer of information both much faster and cheaper; It was an exponentially powerful breakthrough in human communications which led to all the modern communication technologies that purely send information such as telephone, radio, TV and email. It was at once a highly original and an "obvious" invention; Once electricity and magnetism were invented it was a short step to use it to communicate information to far away places. It was a very productive invention generating many important descendant inventions in this list including telephone and TV (John Baird broadcast his first TV through telegraph and telephone wires; see the chapter on *Television*). As it requires electricity to work it ranks below electricity. Its importance and historical impact is very similar to that of radio. Wireless telegraphy using radio eventually replaced electric telegraphs. (For reasons for radio's higher ranking over telegraph despite telegraph's longer history see the chapter on *Radio*).

Telephone

"The method of, and apparatus for, ***transmitting vocal or other sounds telegraphically***, as herein described, by causing electrical-undulations, similar in form to the vibrations of the air accompanying the said vocal or other sound, substantially as set forth."

Alexander Graham Bell of Salem near Boston in the US was granted patent no. 174,465, *IMPROVEMENT IN TELEGRAPHY* which is very clearly readable for a patent; The quote [with emphasis added] above from this patent succinctly describes what a telephone is; A clear offshoot of telegraphy that transmits sound signals using modulations of electric currents in the telegraphic wires. Telegraph was the first invention used to send information faster than physical carriers such as people, horses or trains. It meant, for the first time, information can travel in less than a second across continents instead of taking many days and months at the rate of 100 KM/hour (60 miles/hour) or less. Sending pure

information was also much cheaper than sending physical matter for long distances particularly if they were living matter, men or horses. The Internet can be thought of as a telegraph including binary code which can send arbitrary information such as images, audio and video. Theoretically we could have built the internet on top of the telegraph system as well, we just need a mechanism to encode [and interpret] arbitrary binary data in Morse code dots and dashes even as simple as literally interpreting dots and dashes as 0 and 1s. It is interesting to note here that John Baird used telegraph/telephone wires to transmit television in 1927.

However [classic] telegraph could only send textual information and not audio. Audio is particularly important as then we could talk to each other via a 'telegraph' system. Once a telegraph system was invented it was a relatively straightforward step and there was a strong incentive as well for the telephone. All it required was a way of encoding speech as an electrical signal. In fact the patent and the quote from it mentioned above for the first successful telephone clearly states "the method of, and apparatus for, transmitting **vocal or other sounds telegraphically**".

Alexander Graham Bell of England, a speech expert from a well known family of speech experts who came to the US [via a canadian detour] to train the teachers of *Boston School for Deaf Mutes* was a natural person to be the first to invent such a system for encoding speech. He invented the first telephone in 1876 by inventing a *transmitter* which used a magnet and a membrane with an iron piece as the armature [arm pulled by the magnet] and a *receiver* which used a magnet and an iron disc armature which recreated the sound. Instead of magnets, modern microphones in the transmitters use an electric capacitor with one of its two plates acting as a movable diaphragm; Voice causes the diaphragm to vibrate, varying the capacitance of the capacitor and the resulting current, encoding the voice signal in the current. Modern receivers still work the old way though; They use magnets to vibrate a movable diaphragm and reproduce the sound accurately. Elisha Gray of Chicago in the US almost simultaneously invented a telegraph, again pointing to the inevitability of voice in telegraphy. First commercial *telephone exchange* in 1878 was started by George Coy at Newhaven, midway between Boston and New York City, in the US. Early telephone exchanges were manually operated with the caller literally telling the operator the number to call and the operator completing the circuit from the caller to the receiver manually. First automated exchange was built by Almon Strowger at La Porte near Chicago in the US in 1892. In 1885, American Telegraph and Telephone company (AT&T) was formed and dominated the telephone industry in the US in the 20th century. By 1900, the US had more than half a million phones; By 1948, it increased to 30 Million phones. In 1960, England had more than 5 Million phones for 52 Million; Germany had more than 3 Million phones for around 55 Million. In contrast, India only had less than half a million telephones for a population of close to 500 Million in 1964; Telephones were very slow to penetrate in developing countries due to high infrastructure costs. This is the reason telegraphy was very popular in developing countries until very recently.

Telegraph is a more foundational invention with an exponential increase in human productivity due to lightning fast communications [literally, as lightning is electric current as well]; This brought human society's quality of life to a distinctly higher level than where it was before. Compared to telegraphs, the telephone was an incremental invention and natural evolution which would have come about sooner or later in the 19th century after the 1830s. This clearly ranks telegraphs much higher than telephone. Telephones also evolved into modern cell phones, which are now more numerous than 7.8 Billion people on earth and increased human productivity particularly in developing regions of the world such as Africa, India or Indonesia. Its importance can also be compared to newspapers. Newspapers have become an institution in itself, unlike telephones, as upholders of facts and democracy; Telephones play a much more personal role in society. However the majority of people in history, particularly in the developing world, were illiterates who could still use telephones ranking the telephones slightly higher. As compared to computers, telephones have been with us longer and play a much more critical role in emergencies like natural disasters and rank above computers.

Flying shuttle

Let's start with a quick introduction to weaving for us, modern city dwellers of the 21st century:

- Fiber - First cotton is picked and separated from its seed manually or using an instrument such as *Cotton gin* that has a chapter of its own in the list.
- Yarn - Cotton fibers are spun into threads using an instrument such as *Spinning Jenny* that has a chapter of its own as well.
- Fabric - Yarn needs to be woven into fabric using a loom. A piece of fabric consists of two sets of threads, one vertical and one horizontal, going across in a grid pattern. Current chapter is about this step, weaving.

Weaving in a loom itself is an interesting process:

- In weaving, longitudinal threads are called *Warps* ("turn or bend" as in warped spacetime) and cross threads are called *Wefts* ("weave").

- To make a fabric using hand you would take a weft thread and weave it up a warp thread, down through next and then up the next thread and so on till the end of the fabric, that is the end of warp threads. However it would be a painstakingly slow and time consuming process.

- In a loom it is done much more efficiently. Warp (longitudinal) threads are grouped into two alternating groups, with all odd

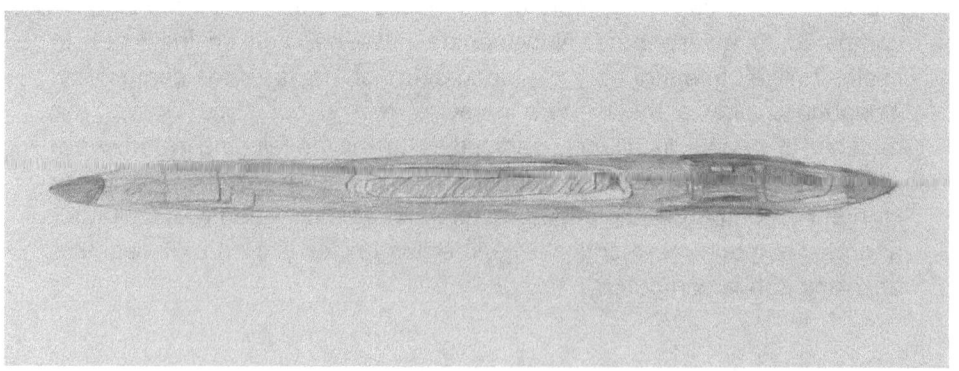

threads in the first group and and all even threads in the second group; Each warp thread is threaded through the eye of a *heddle* in its group, odd or even; Heddles are a series of vertical strings with eyes at the center for threads to pass through.

- In one weaving step, odd heddles are raised by a shaft mechanism making space, called *shed*, between the two groups of warp threads, as all odd threads are now above even threads. This is the most important step to grasp to understand weaving.

- A *shuttle*, a bobbin holder, containing the bobbin of weft (cross) thread is thrown across the *shed* all the way to the end of the fabric, completing a weaving step.

- The beauty of this process is, now the weft thread is below all the odd warp threads and above all the even warp threads creating the required up and down pattern of weft thread across all the warp threads, saving time in the weaving step by creating the up and down pattern of weft thread, one line of the cross-grid pattern, all at once.

Readers paying attention would notice that in the next step even heddles have to be raised, with odd and even warp groups alternately being raised in consecutive weaving steps.

This gets us all the way down to the *shuttle*, the bobbin of weft thread thrown across the shed in one weaving step. Although more efficient than a process without the loom, it still has some limitations. If one weaver has to be able to throw the shuttle back and forth across all the warp threads, and hence the width of the fabric as obviously the fabric starts at the first warp thread and ends at the last warp thread, it can't be longer than the width of the weaver's two hands. Otherwise he will have to get up from the loom, go and collect the shuttle some distance away and come back, slowing down the weaving and tiring the weaver out. Another option is two have two assistants at each end throwing the shuttle across to each other through the *shed* like two kids playing a game of shuttle throw and catch. However these assistants would need to be paid increasing the cost of the fabric.

Flying shuttle, also called *wheeled shuttle* (a more appropriate name), invented by John Kay at Bury near Manchester in England in

1733 automated this shuttle playing game by making the shuttle fly across the shed all by itself, hence the name. It is accomplished by having a track across the shed and shooting out the shuttle from either end using a mechanism that's controlled using the weaver's hand, and connected to the ends by a string from the control; If you have seen weaving videos (and if you haven't, it may be a good time particularly if you have lost track of the explanation by now), it is the thing the weaver pulls to either side with his hand in weaving steps making the shuttle fly to the side in the direction of pull. The modified shuttle has wheels for it to "fly" on the track quickly.

Edmund Cartwright at Lincoln near Manchester in England invented the *power loom*, the mechanized weaving machine, in 1784 although it took decades to be useful given the complexity of weaving. Modern advanced power looms are shuttle less. *Rapier* ("pointed sword") *loom* works similar to sewing machine needles with the rapier, a long needle-like stick, taking the weft thread across all warp threads in each weaving cycle. It can be thought of as a shuttle being carried across the warp threads by a stick, a *stick shuttle* instead of a *flying shuttle*. The *Jacquard loom*, invented by Joseph Marie Jacquard at Lyon in France near Geneva in 1804, is an interesting combination of textiles and programming. It is not a loom at all but a loom "accessory" for automating the weaving of complex patterns by customizing the warp threads to be lifted in a step instead of all odd threads or all even threads. This *beautiful invention* foreshadowed many concepts in computing including stored programs for automation and binary numbers. It used punched cards, later literally borrowed for computer programming in the 20th century, to *program* a pattern and reproduce it on weaving fabric automatically in a loom. Modern air jet looms carry the weft thread across using a pressurized air jet (*"flying weft"* as it were) instead of a shuttle or a rapier.

Flying shuttle automated the throwing of the shuttle across the shed back and forth and increased the productivity of weavers at least two fold by removing the need for the assistants while being able to weave much wider fabrics. This was one of the founding steps of the industrial revolution in textiles, impacting human history for around 300 years. For its founding influence, it ranks with other foundational inventions of the industrial revolution.

Spinning Jenny

"The advent of the mechanical cotton picker has aroused the ancient debate, are machines the friend or foe of mankind? Ancient kings destroyed the spinning wheels so that human labor would not be displaced; the nineteenth century witnessed riots against cartwright's spinning jenny by outraged laborers who thought it an abomination of the devil and even today men are unable to conceive an adjustment to cotton pickers and other devices which ostensibly displace man in industry.

...The attack on the spinning jenny was indeed a mistake. Men who used slow hand labor were unable to purchase expensive shirts and clothing but with the coming of this great invention men and women were able to live more comfortable, healthful lives. It has always been so; machines have produced untold benefits to mankind.

It is our knowledge of social adjustments that is rusty and decrepit at times. Hear the voice of such static thought from the lips of a prominent Southern politician who proposes to meet the change by legislating this machine out of existence. Natural science surges forward to new glories; machines do not displace labor, they produce new labor and bring leisure and comfort to mankind. Would that our knowledge of social science and government were as progressive!"

One Francis Joseph Bassett of Ridgefield, Connecticut wrote to The New York Times of Sept. 21, 1936 in a letter titled, "MACHINES vs. MEN", about the mechanical cotton picker soon rendering human pickers obsolete and unemployed. This debate should be familiar to us in this age of computerized automation and algorithms (along with tall claims of "artificial intelligence" that doesn't even have the common sense of a three year old yet) displacing workers and the need for universal basic income for all the would be idle citizens with no work to do. If history and spinning jenny is any guide, machines, our inventions, have proved beneficial to humanity and caused no long run unemployment as people, ever inventive, have turned to more productive work one rung higher; If you can't be the cotton picker, you can be the (mechanical) cotton picker operator.

Unlike other inventions in the list, many of the modern readers know Spinning Jenny was very important but don't quite understand what exactly it does. It improved the efficiency of an invention which existed for thousands of years called *Spinning Wheel*, charka, popularized by Mahatma Gandhi. Spinning wheel is [still] used to make yarns from natural fibres like cotton which is one of the first steps in the textile manufacturing process. Spinning wheel simply twists the yarn [to make it into a thread] when the spinner turns the wheel. Turning a wheel is a much easier operation than twisting the yarn which is far more painstaking [as the human hand can't rotate the full 360 degrees unlike the wheel, we have to keep stopping midway and "unrotate" the hand]. Spinning wheel also can be automated, for instance by using the foot (*treadle*) or water power as in the *water frame*. Spinning Jenny simply increased the productivity of that wheel by making it spin the yarn on more than one spindle, anywhere from 8 to 100 or more. This meant one could spin hundred times more yarn in the same time using the same power. James Hargreaves at Oswaldtwistle near Manchester in England invented the Spinning Jenny in 1764, making the transition to industrial scale yarn spinning. Richard Arkwright at Preston near Manchester in England patented the *Spinning frame* in 1769; Spinning frame can be considered as a spinning wheel powered by horses or falling or flowing water; It is called the *water frame* when powered by the latter. Samuel Crompton at Bolton near Manchester in England in 1779 invented the *Spinning mule*, the next logical step that combined the spinning jenny and water frame, hence the name *mule*; It automated the wheel rotation in Spinning Jenny using water power. This for the first time freed it completely from human muscle power and enabled one operator to spin more than 1000 spindles at the same time, bringing in true industrial scale textile manufacturing.

We have seen that incremental, efficiency inventions matter, a lot. One previous example is the printing press of Gutenberg which made the copying of the manuscript more efficient, fast and cheap by making it independent of human labour. Another is James Watt's steam engine which "simply" introduced a separate condenser to the existing *Newcomen engine*, the efficiency jump making all the difference in its widespread adoption. Spinning Jenny is another incremental, efficiency invention, many times efficient over the spinning wheel, which changed spinning from a cottage industry to an industrial factory operation.

Spinning Jenny, and its successors such as Spinning mule, rank high as an important step in ushering in the industrial revolution of England. Flying shuttles and sewing machines wouldn't have been of much use if yarns for them couldn't be produced at scale. Looking at one level below, cotton gin won't be required if all that clean cotton can't be turned into yarn fast enough. As one of the three foundational inventions in the industrial revolution of textiles it ranks high in our list of inventions.

Cotton gin

If there were a Moore's law during the late 18th to early 19th century American south, it would be for cotton production instead of transistor density and the size of transistors. After the invention of the Cotton gin, the yield of raw cotton doubled each decade after 1800. In fact a hallmark of a good invention is it improves human productivity, not how complex it is. A good invention enables people to "do more with less", in less time, less resources or with less workers. It is true of *Flying shuttle* that enabled a weaver to weave wider fabric without assistants or *Jacquard loom* that enabled complex weaving patterns human assistants couldn't even do with consistency and speed; It is true of skyscrapers that made limited land in places such as Manhattan and Singapore more productive, that is support more people; It is true of fertilizers that made the same agricultural land yield more food. Before cotton gin, the average cotton picker could remove the seeds from only about one pound of short-staple, harder to remove seed but more available variety, cotton per day. After the invention in 1793 by Eli Whitney at Savannah near Atlanta in the US, hand-cranked gin could remove the seeds from 50 pounds of cotton in a single day, a fifty fold improvement in

productivity. Modern ginning machines operating on electricity can produce 400 pounds in an hour, or 10,000 pounds a day.

Cotton gin is also the story of why strong intellectual property protection matters for inventions to flourish. Eli Whitney, arguably as inventive as Thomas Edison, spent so much time trying to protect his invention from people's piracy and state governments actively subverting the patent, he hardly had any time left for further inventions during that time. Edison, living in a time of stronger IP rights, could keep churning out invention after invention - light bulb, movies, phonograph, battery and on and on - without worrying about other people and governments stealing his hard work. In the 21st century, most of the inventions happen in the US and Germany and not in India or Egypt for the same reason. An inventor in India would be in a situation similar to Eli Whitney with everyone including governments stealing his invention and he will be left struggling to make ends meet. Further corroborating this theory, Whitney did have time for inventions after he gave up protecting his patent, inventing standardized, interchangeable parts for muskets, the precursor to the modern assembly line developed by Henry Ford for building automobiles in the early 20th century.

Cotton gin was an ingenious but simple device. It had a rotating drum with hooks *pulling on cotton lint, separating it from seeds in the process* instead of separating seeds from lint as we (or 18th century slaves in the American south) would do manually with our hands, a *crucial insight* it turns out. Stephen Yafa explains the Cotton gin in his book "Cotton: The biography of a revolutionary fiber" [emphasis added]:

"Ten days later Whitney came back with a working model of his cotton gin—crude, but essentially governed by mechanical design principles that are still in operation today. **No engineers since have reinvented a better gin; they've simply built better versions of Whitney's original.** Legend has it that Whitney's inspiration came as he was roaming the plantation grounds pondering how to solve the problem and paused to watch a cat hunt down a chicken. At the last moment the chicken fled and the cat's lunging paw came away with only a few feathers. **Why then try to separate the seeds of**

upland cotton from its fibers? Why not instead build a device to separate the fibers from the seeds? Small difference, huge implications. If Whitney's machine could allow the fibers to be pulled away while creating a barrier that held back the seeds, the claws could exert enough force to yank the fibers free.

That elegantly uncomplicated premise led Whitney to build an apparatus that duplicated the motions of the slaves who cleaned cotton manually. He fashioned a mesh sieve or hopper with narrow slits in it running lengthwise to do the work of the hand holding the seed. On the surface of a drum rotating around the hopper he duplicated fingers pulling off the lint by attaching wire claws that protruded through the slits; these hooks grabbed the lint and wrenched it away from the seeds, which were held in check by the tight mesh. A cylindrical brush swept off the freed lint. This hand-cranked contraption, the cotton gin, was, in Whitney's words, "an absurdly simple contrivance"—and unfortunately for Whitney, he was right: it was all too easy to copy."

Cotton gin again proves the recurring theme in the book - that simple, incremental inventions, matter, *a lot.* Cotton gin ranks high as one of the three foundational inventions - Flying shuttle, Spinning jenny being the other two - of the industrial revolution in textiles. From less than 200,000 pounds in 1792, America's cotton production became 35,000,000 pounds in 1800, an astonishing 175-fold increase. In the long term, due to the productivity increase from the cotton gin, cotton was the leading American export from 1803, a decade after the invention, to 1937. It played an important role in human history; In England it provided the raw materials for the industrial revolution; In the US, it contributed to the economic growth of the south and on the darker side, expansion of slave labor for cotton farming and eventually to the US civil war. For its impact in human history and industrial revolution, it ranks very high in our list of inventions.

Newspaper

Relation aller Fürnemmen und gedenckwürdigen Historien ("Account of all distinguished and commemorable stories") is considered the first newspaper in the world; It was published by Johann Carolus of Germany in 1605 as a German weekly newspaper in Strasbourg, now in France just beyond German borders near Stuttgart. It was only appropriate that the country where the printing press was invented would start *'the press'*. *Courante uyt Italien, Duytslandt, &c.* ("Current Events from Italy, Germany, etc.") was the first newspaper to be published in a newspaper style format; It was a dutch newspaper published in Amsterdam, Netherlands in 1618. *Daily Courant* was the first daily newspaper in the world and was published in London, England; It was published by Edward and Elizabeth Mallet in 1702. Asia's first newspaper was *Hicky's Bengal Gazette or the original calcutta general advertiser* in 1780 by James Hicky of Ireland in Calcutta, India. *Publick Occurrences Both Forreign and Domestick* was the first newspaper in

the Americas; It was published in 1690 in Boston, US by Richard Pierce and Benjamin Harris.

Newspapers started along with the earliest democracies in the modern world and have become a pillar of these democracies. Newspapers improved the average intelligence [think graphs, statistics] and awareness of the general populace. In this age of social media and fake news one can readily understand the role mainstream newspapers such as The *New York Times*, *Washington Post* or The *Guardian* play in countering them and spreading factual information. Newspapers and literacy played a positive feedback loop with newspapers requiring literacy and in turn newspapers playing as the first useful application for the general platform of literacy. Newspapers enabled modern nation states by diffusing information from the government in the capital to all corners of the country. The invention of the telegraph transformed the newspaper industry by providing uptodate information from all over the country as the news could now travel close to the speed of light. *Associated Press* (AP) was formed in 1846 by New York City newspapers to share the cost of sharing news by telegraph. Paul Reuter founded *Reuters* in 1851 as a news wire agency in London. *Agence France-Presse* (AFP) was founded earlier in Paris, France in 1835 as *Havas* and is the world's oldest news agency; AFP predates telegraph and used trains and carrier pigeons for news collection and later switched to the telegraph. Now another invention, the internet is impacting newspapers' survival by impacting their advertising revenue as most of the [digital] advertising revenue is vacuumed up by the likes of Google and Facebook. It has also led to the decline of paper based news readership particularly in developed countries impacting paper subscription and advertising revenue.

There is a reason authoritarian regimes like China control the newspapers heavily. If conditions in regions like *Xinjiang* were to be reported free and fair by [chinese] newspapers both the international community and domestic population would rise up and create problems for the authoritarian communist party rule. We can contrast the situation in Xinjiang with the recent migration crisis of 2018 with children in cages in the US; In the US it led to widespread reporting by newspapers and the resulting outcry put an end to this abhorrent practice. The main difference is the freedom and independence of the newspapers. Another point illustrates this effect; India, a vibrant democracy has more than 118,000 newspapers while China, an authoritarian regime has around 2,000 newspapers. Antagonism between authoritarians and newspapers goes a long way. In 1722, James Franklin, publisher of one of the first american newspapers, New *England Courant* [and brother of Benjamin Franklin], was jailed by the colonial government for publishing controversial articles.

Newspapers rank very high in the list of modern inventions for creating modern democracies and as a powerful pillar of current democratic societies. Political freedom is one of the fundamental components of *quality of life*, given a choice one would choose to be born in India or Botswana over China or Saudi Arabia. In some cases, it is literally a matter of life and death, if you are a *Xinjiang* Muslim for instance. In fact, the *Washington Post*, the well respected newspaper in America recently adopted *"Democracy dies in darkness"* as their slogan. It is no coincidence that the very first modern democracies were formed around the time of early newspapers. They existed in a symbiotic relationship from the early days. First amendment to the *US constitution* was for the *freedom of the press*. Newspapers were the first modern public broadcast mechanism before radio, TV or the internet. With its rich 400 years heritage and ample time to impact human history, it has played a much bigger role than inventions such as radio in society. Newspapers enabled the modern economy by spreading information about prices of commodities and stock prices to people all over the country. In some sense they can be called *"ships of information"* [for the same reason railways can be called *"ships of the land"*] for their role in making the world a global village. As the first broadcast media, it was the pioneer for other broadcast media such as radio and TV. They have the power to shape the views of the people and bring in political revolution. The printing press and paper were prerequisite inventions for the newspaper and rank higher than newspapers. For their importance, newspapers rank just after the top inventions of the industrial revolution and above the inventions of the *computer revolution*.

Glass

There is a beautiful glass *chalice*, called *Lycurgus cup* from the 4th century CE in *British Museum*'s Roman collection; It bears a scene of *King Lycurgus* tied up in a tangle of grapevines and appears jade green when light is shone from the front and blood-red when it is lit from the back (you should see it for yourself to appreciate how amazing the effect is). Recent research suggests that Romans achieved this amazing feat by using fine silver *nanoparticles*. It is one of the greatest achievements in human art and engineering. Even without the nanoparticle wizardry, it is amazing to think that glass, this beautifully transparent material is made from sand found in the beaches and deserts, an opaque crystal with no interesting optical features. One of the most common glasses, making up 90% of all glasses in the world, *soda-lime glass* is made by melting together sand [*Silicon Dioxide* called Silica], soda ash [*Sodium Carbonate*] and lime [*Calcium Oxide*] roughly in the ratio of 75%:15%:10%. Soda is added to reduce the melting point of sand and lime is added to counteract the effect of soda that would make the resulting glass water soluble. The mixture needs 1425-1600° C [2600-2900° F] for it to melt and produce the glass.

It is claimed that glass making dates back to 4,000 years ago in Mesopotamia based on a sample of translucent blue glass from the *British Museum*; It is also claimed that many ancient civilizations such as India, Egypt and Persia had glass for more than 3,500 years based on some archeological artifacts. However it would have been very hard to produce the high temperature, more than 1400° C, required for glass making in ancient times; *Iron metallurgy* with its similar requirement for the high temperature only started around 3200 years ago (see the chapter on *Iron Metallurgy*). Based on this high temperature requirement, It is more likely that glass making started in the *Iron age* around 3,000 years ago. This confusing historical picture makes it hard to find the exact origin of glass. In any case, glass making was already widespread around 2,000 years ago in Roman times. In the middle ages, *Venice* in Northern Italy was a centre of glass making; Venetian glasses from the Adriatic Sea island of *Murano* around 1300 CE are well known, particularly *Cristallo glass* made with *Manganese Oxide*; It is extremely thin and colorless and one of the finest glasses ever made.

Glass is a so-called *amorphous solid*; This means it has short range order but doesn't have the long range solid crystal structure as in metals and salts. This contributes to it not absorbing energy in visible light and becoming transparent. In particular photons in the visible light spectrum don't have enough energy to cover the *band gap* of electron energy levels, making them pass through without being absorbed. This big band gap of glass also makes glass a very poor conductor of electricity and heat. It is very similar in this regard to *diamond* which is also transparent as it doesn't absorb visible light due to large energy gaps; Diamond also happens to be a poor conductor. Diamond can be contrasted with *graphite* that is also a crystalline form of carbon but with a movable planar sheet structure that makes it conductive and black. Diamond and glass are similar in one more important respect; Both are formed from materials which are opaque, coal and sand respectively, and are amazingly transparent.

Glass is one of the most common items around us after steel and plastic. It is used as glass containers for storing food and for drinking; Mirror is a common household item. Glass is used in eyeglasses, microscopes, telescopes and camera lenses. Homes and Office buildings have glass windows and doors; *Tempered glass* is used in automobile windows. Glass is also used in optical fibers that carry internet data across the continents. Glass is important in medicine for instance, for vaccine vials; Glass is also important in laboratories for test tubes, flasks and beakers. Glass is used in lighting appliances such as bulbs and fluorescent tubes; Glass is critical in many electronic appliances such as cell phones and TVs.

Glass has also been culturally important as stained glass windows in cathedrals, particularly in *Gothic architecture*. Cathedrals such as *Notre-Dame de Reims* in France [that unfortunately burned down in April 2019], *Westminster Abbey* and *Canterbury Cathedral* in England showcase this beautiful stained glass art form. *Metal oxides* provide the color in these glasses, with bright ruby red from *Copper Oxide*, sapphire blue from *Cobalt Oxide*, green from *Iron Oxide*, yellow from Sulphur and purple from *Manganese Oxide*.

Glass is an invention with one of the most widespread uses in our daily lives. It is used in kitchen utensils, windows and modern skyscrapers. Glass is a very *productive* invention, producing multiple high ranking descendant inventions in the list including the *microscope*, the *telescope*, *eyeglasses* and the *light bulb*. Microscopes have been critical in germ theory of diseases that saved billions by now. Telescopes have been fundamental in foundational questions such as earth's place in the solar system and the age of the universe itself [Hubble telescope with its reflecting mirrors]. It historically played a very important role in cathedrals and mosques that is an important part of human cultural heritage. It was the most commonly available container during the historical period. Its importance is comparable in some aspects such as household use to plastics; However glass has been with humanity for much longer time and influenced human civilization longer; Glass has also been more productive; This makes the glass rank much higher than plastics despite the latter's versatility including as textile fibers (see the chapter on *Plastic*).

Anesthetic

It is unthinkable to do a medical operation such as *open heart surgery* without an anesthetic. A dentist from Boston, US, William Morton was the first to demonstrate a general anesthetic in 1846. He demonstrated *diethyl ether* at *Massachusetts General Hospital* in Boston with help from surgeon John Warren. It has transformed horribly painful medical procedures into memory less, pain less procedures of modern times. In 1847, James Simpson of the UK, demonstrated *chloroform* as an anesthetic. On April 7, 1853, Queen Victoria of England gave birth to her eighth child with the successful administration of chloroform, a primitive form of modern "epidural" procedure, making chloroform a very popular anesthetic. In the early 20th century, Chloroform use was discontinued after it was associated with cardiac arrests (*ventricular fibrillation*, useless beating of the heart that doesn't pump the blood) and much better anesthetics were developed. As diethyl ether was inflammable, safer ether derivatives such as *Sevoflurane* have replaced its use now.

In 1973, John Glen of England developed *propofol* ("milk of amnesia") at *AstraZeneca*, a company that traces its origins to *Astra* founded in Sweden in 1913 and *Imperial chemical Industries* founded in

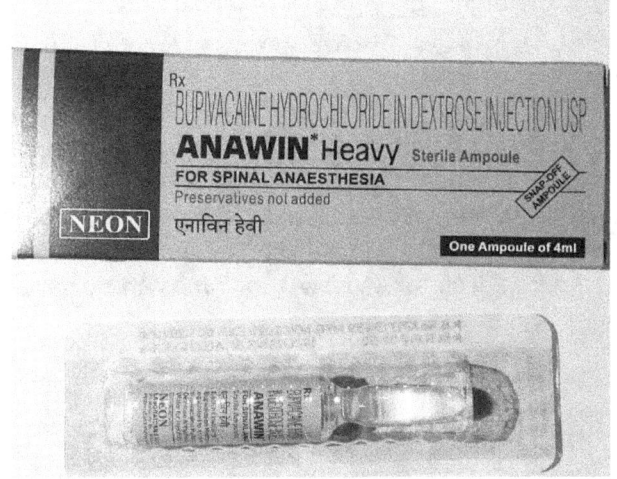

England in 1926; It was approved for public use in 1986 and has become the most commonly used modern anesthetic. Now, propofol is administered around 60 million times per year in the United States; Considering the US population of 330 million, that means one out of every 5.5 people a year goes under propofol every year. *Sodium thiopental* [so-called *truth serum*], the most common anesthetic before Propofol's invention, was developed by Ernest Volwiler and Donalee Tabern of the US in 1934 at *Abbott Laboratories*, founded at Chicago in 1888 by Wallace Abbott. *Sevoflurane*, a fluorinated *isopropyl ether*, was invented by Bernard Regan and others in 1975 and it was developed commercially by Japanese company *Maruishi* and Abbott in the US in the 1990s. *Procaine*, a local anesthetic, was first made in 1905 by Alfred Einhorn, a German chemist, who named the substance "*Novocain*"; This was used commonly in dental procedures. *Lidocaine* (cylocain or lignocaine) was invented in 1943 by Nils Löfgren at the University *of Stockholm* in Sweden; It is replacing Novocain as the hypoallergenic alternative in dentistry. On the 50th anniversary of the invention of Ether, Dr. David Cheever of Boston said [emphasis added]:

"What victim of surgery, who, under ether, sinks into a calm and dreamless sleep, during which his abdomen can be cut open, his bowels taken out, handled, and replaced, his nerves cut, his veins or arteries tied, and his skin sewed up, and who is made so completely oblivious as to ask on awakening, 'Are you not ready to begin?' but concedes with gratitude, on realizing the result, that **this is the greatest discovery ever made for the happiness of mankind?**"

Local anesthetics such as Both Procaine and Lidocaine work by disrupting sodium channels in local nerve cells disrupting the signalling to the brain from the area. General anesthetics such as propofol and the inhaled anaesthetic sevoflurane inhibit the transmission of feedback signals from the frontal cortex. A *New York Times* article says:

"More recent research on injectable anesthetic drugs like propofol suggests that they interfere selectively with certain neurotransmitters and with the interaction between the cerebral cortex (where thought and perception resides) and the deep part of the cerebral hemispheres known as the thalamus, which acts as some kind of gateway between the cerebral cortex and the rest of the brain. In other words, it is by no means certain that all of your brain is "asleep" when you are anesthetized."

In 2018, Propofol dominated the (global) market at 33.83% and sevoflurane at 21.17%. Despite the market dominance few companies are willing to make propofol because it is complicated to manufacture, leading to its shortage sometimes; It's an *emulsion*, a combination of two liquids that don't blend together chemically, making it harder to manufacture and bacteria can contaminate it more easily than other drugs. John Glen, inventor of Propofol, explains why emulsion is required:

"There's a Catch-22 when it comes to anesthetics. In order for a drug to get into the brain, it needs to be a fat-soluble compound. But if a compound is fat-soluble, then it becomes very difficult to get it into water-based solutions that can be injected into the veins. Fortunately, in the early 1970s, a technique was developed to combine these fat-soluble compounds with a chemical called a surfactant to get them into water. This allowed us to go back and look through all the promising compounds we'd previously passed over because they weren't water-soluble. One of those was propofol."

Anesthetics rank high for alleviating human suffering and saving millions of lives through critical surgeries. After basic necessities of life, food, clothing and shelter, eliminating *avoidable human suffering* ranks as one of the most important policy goals for our society and anesthetic has played an important role in this goal. It has been with us for 175 years, longer than railways, electricity and vaccines. However, unlike these inventions it is not used by everyone in day to day life; It is not as productive, in terms of descendant inventions; It hasn't played as big a role as vaccines in saving lives particularly in reducing infant mortality. For these reasons, it ranks high, in the top 50, and ranks below foundational inventions of the industrial revolution and life-giving inventions such as vaccines.

Computer

"PHILADELPHIA, Feb. 14--One of the war's top secrets, an amazing machine which applies electronic speeds for the first time to mathematical tasks hitherto too difficult and cumbersome for solution, was announced here tonight... This resolver of difficult problems is what computing experts call a 'digital' counter. **Basically it does nothing more than add, subtract, multiply and divide. It does this by generating very accurately timed electrical impulses at a speed of 100, 000 per second. and can do one operation every twentieth pulse, thereby adding, for instance, at the rate of 5,000 per second...** In a matter of seconds it does what trained computers hitherto have required weeks to perform. The instrument contains 18,000 vacuum tubes, occupies a room 30 by 60 feet and weighs thirty tons. It took thirty months to build, cost about $400,000 and required 200,000 man-hours of work."

The *New York Times* of Feb 14, 1946 announced ENIAC, the first electronic, programmable, general-purpose digital computer, to the world this way in the article, "Electronic Computer Flashes Answers, May Speed Engineering", that amazingly accurately describes a computer [emphasis added]. In every clock cycle, a modern computer can do an

operation from a fixed repertoire of arithmetic, logical and control flow operations. Its speed comes from how short this clock cycle typically is. The computer I am writing this book on [An old computer bought as refurbished for around 200 dollars including monitor and keyboard, so it is a conservative estimate of typical speed] has a processor with 3.1 Ghz clock cycles. That is the processor can execute in a second 3.1 billion operations from this fixed repertoire of around 1000 instructions. Arithmetic operations typically include addition, subtraction, multiplication and division; Logical operations include AND, OR and NOT; Control operations include jumping to a particular location in a sequence of instructions, called a *program*. We use the same alphabet to write any book; In the same way these basic instructions are used to implement any process which is *computable*. Amazing power of the computers come from:

1. This *idiot savant* executing these instructions extremely fast.
2. The different *programs* [sequence of instructions] that make it do a variety of things from displaying a spreadsheet, rendering a web page, predicting the weather to computing most relevant news items for *you* in a news feed such as Yahoo homepage, Google News or, on the flip side, spreading fake news on Facebook.

Apart from the speed due to fast transistors, a computer can be implemented completely in any medium which can support these fixed instructions, in particular mechanical devices. Such a computer was designed by Charles Babbage at the *University of Cambridge* in England in 1837, called an analytical engine although it wasn't built. *Jacquard loom* invented by Joseph Marie Jacquard at Lyon in France near Geneva in 1804 for weaving complex patterns was remarkably prescient with "programs" written in punch cards and binary numbers (1 - warp thread up and 0 - warp thread down, see the chapter on *flying shuttle*).

Alan Turing at the *University of Cambridge* in England, among others, provided the theoretical foundation of computers in 1937 with his paper, *"On Computable Numbers, with an Application to the Entscheidungsproblem"*. He defined a mathematical computer now called the Turing *machine (TM)*. A TM contains an input tape and an output tape and is in a certain internal state at any one time; It operates in a sequence of discrete steps; At every step, it reads symbols from the input tape and writes symbols on the output tape and changes its internal state. When the machine enters a special HALT state it finishes computation and the output of the computation is found in the output tape. A simple algorithm can be: read a number, say 2 from the input tape and output 4, *square of 2*, the square of the number in the input at halt; this TM computes the *square* function: f: x -> x².

Universal Turing Machine (UTM) is a particular TM which is more like a Turing Machine Simulator; It can read a TM description, say for computation of square and mimics the behavior of this machine in its operation and produces the square of the number in its output tape; If it is given a description of cube computation it mimics the behavior of the TM doing the cube computation. In general, given any computable function, UTM can mimic its behavior, that is, compute that function. In fact, the definition of a computable function is one that can be computed by a TM and hence by the UTM. A machine that can compute any function UTM can compute is called *Turing Complete*. In fact a *computer* is any machine that is equivalent in its power to UTM, that is *Turing Complete*; In particular, our digital computers are *Turing Complete*. [In a very accurate sense due to George Cantor and Turing we won't go into], Computable functions correspond to integers but all possible functions, including non-computable ones correspond to real numbers. As the integers are a negligible portion of the real numbers there are vastly more non-computable functions than computable ones.

ENIAC [Electronic Numerical Integrator And Computer] was built at the *University of Pennsylvania* by John Mauchly and Presper Eckert in 1945 and it was *Turing Complete*. John Neumann described the stored-program architecture [program description is stored as data in memory instead of physical rewiring] in 1945 EDVAC design report. First stored program computer Manchester Baby was built in 1948 at the *University of Manchester*.

Computer is easily the *single biggest invention in the 20th century given its effect on human productivity in a few short decades.* Modern society is operated by computers, from electrical grids to banking and air traffic control to subways to reservation systems. Billions of people have a computer in their pockets in the form of a smartphone, ranking it very high.

Transistor

Transistors can do two things very well:

- Switching, that is it can represent a 0 or 1 and very quickly switch between them. For instance, a RAM memory chip stores a binary bit [0 or 1] in a tiny capacitor's charged state with billions of capacitors in a very small chip.

- Second they can be combined together in ways to implement *logic gates* which operate very fast, a NOR gate for instance. A gate operates on two bits of input and produces one bit of output. NOR gate is a so-called *universal gate* which means we can implement any other logical operations such as AND using these gates.

Transistor functions exactly like a vacuum tube does, but only much faster, much smaller and way more power efficient. These devices send

a current in one direction readily and in the opposite direction not at all [hence the name *semiconductor*] and they can be controlled by a much smaller current flowing through their 'control'. This means they can be used to distinguish two states required for 0 and 1, one of current flow and one of no flow. This can be easily switched by changing a small control current. This is the principle behind amplification as well; In amplification the signal at the control [say a radio signal] is amplified by the main current in analog fashion by copying the signal continuously. In the digital world only two states need to be distinguished, current and no current which makes it in some sense easier and more precise. As an aside, this is the reason in theory quantum computers can work much faster [apart from *entanglement* and its exponential power due to *superposition*]. An electron spin state can be switched between ½ and -½ much much faster than a transistor current switching.

John Bardeen, Walter Brattain and William Shockley invented the transistor at Bell Labs in the US in 1947. Mohamed Atalla and Dawon Kahng also at Bell Labs invented the MOSFET transistor in 1959, a transistor used in almost all modern computer chips. Transistors underlie all micro chips including calculators and personal computers in particular CPU and computer memory. Modern chips have billions of transistors, with size as small as 7nm [15 times smaller than human hair width] in an area of around 100 square mm, around ¼ of the size of the postage stamp. A vacuum tube is typically a few hundred square mm in size; That is a transistor takes up the area of one-billionth of a vacuum tube, making devices made with transistors billion times smaller and consume a billion times less energy or better for the same functionality. It is also far more reliable as there is no heating coil in the transistor and can operate much faster, on the order of nanoseconds; That is it can switch from 0 to 1 or vice versa in a few nanoseconds.

Jack Kilby of *Texas instruments* and Robert Noyce of *Fairchild semiconductors* independently built pieces of the inventions that would form the integrated circuits (IC) in 1958 and 1959 respectively; Noyce's monolithic ICs with transistors and other electrical components on the same silicon wafer paved the way for mass production of chips by etching circuits directly. *Photolithography* is an important technique used to create these tiny billions of transistors in a chip by companies such as Intel or *Taiwan Semiconductor Manufacturing Company* (TSMC). It was invented in 1955 by Jules Andrus and Walter Bond at Bell labs. After these inventions and that of MOSFET that allows much higher transistor density, the number of transistors in chips started to increase exponentially. Intel founder Gordon Moore made the observation known as *Moore's law* that *the number of transistors in a chip doubles every 2 years*. Transistor size went from 10 micrometers to 7 nm (a factor of 1000) or smaller nowadays. In 1971, Intel *4004 processor* had 2,300 transistors in 12 square millimeter area, with a transistor 'size' of 10 micrometers; Now the new chips with 7nm process can have more than 100 Million transistors in a square millimeter or 1.2 Billion transistors in the same 12 sq mm, almost a million times (1000x1000) more transistors. This also means million times smaller devices with more efficient power consumption.

Transistors rank very high as the prerequisite for all modern electronic inventions such as superfast digital computers and mobile phones. Now almost all household appliances and machinery such as automobiles and elevators include an integrated circuit; Billions of transistors are sitting in the chips in all our appliances surrounding us, computer, mobile phone, TV, washing machine, oven, cars and toys. A computer made up of vacuum tubes, ENIAC weighed 30 tons and a clock speed of 100,000 cycles per second, an order of 10,000 times slower than modern computers. An invention like a mobile phone with its small size would have been unthinkable without transistors. As we have seen many times earlier, it is an incremental invention over the vacuum tube, doing the same thing - switching - but much much faster. A transistor can switch [go from 0 to 1 or 1 to 0] a billion times a second, that directly corresponds to the speed of the device such as a computer built out of these switching elements, making the modern digital miracles possible.

Internet

In 1968, Senator Edward Kennedy of Massachusetts, US sent a telegram congratulating Bolt, Beranek and Newman (BBN) company in his state on its contract to build the "Interfaith Message Processor". Interestingly enough, the story of the Internet starts from these Message Processors although it had nothing to do with religious reconciliation unless we include the websites on interfaith dialogue, among millions of other things, that would eventually be built on them after 30-40 years. Much earlier in 1961, Leonard Kleinrock at Massachusetts Institute of Technology, near Boston, US pioneered the *packet switching* technology underlying the internet that contrasts with *circuit switching* used in telephone networks. In packet switching, snippets called *packets* from different messages are interleaved and sent through the physical network and later reassembled by receiving computers and put in proper order into constituent messages. This is only possible with the development of computers that can do this computation for splitting and re-stitching at either end. Earlier telephone networks used circuit switching that used the physical network exclusively for one telephone call till that user was

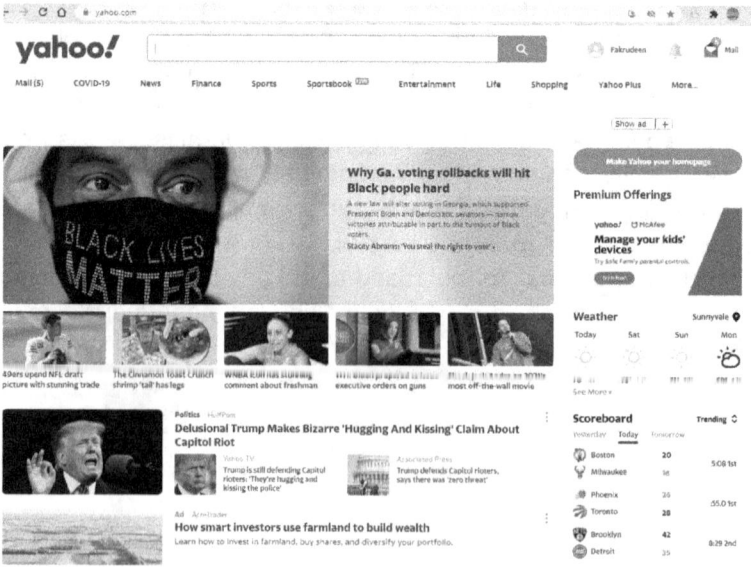

finished; Then the network, *circuit*, was switched over to the next person waiting to use the network. Circuit switching was simply the evolution of manual switching done by early telephone exchanges, automated using electro mechanical switches such as *Strowger switch*. This could work without much sophistication, in particular without computers, but had a serious limitation by current standards; Only one call could be active through one end to end physical wire connection (path) in the network. It is as if I am connected to *yahoo.com* no one else can connect to it till I am done with Yahoo; Even worse, if my path to Yahoo shares some physical connections with *google.com*, users can't connect to Google either. In contrast, packet switching is very efficient with multiple users sharing the physical paths in the network simultaneously. Once computers and packet switching were invented, the Internet was only a matter of time. The US Department of Defense's ARPA created the ARPANET (Advanced Research Projects Agency Network) in 1969. The first node of ARPANET, *Interface Message Processor* (IMP), was at the *University of California, Los Angeles* run by Kleinrock himself and the second node was at the *Stanford Research Institute*. Vinton Cerf at *Stanford University*, US and Robert Kahn at DARPA, Washington DC, US created the foundational protocol called TCP/IP in 1974. Paul Mockapetris at Los Angeles, US invented the *Domain Naming System* (DNS) in 1983. Tim Berners Lee of England working at CERN in Geneva, Switzerland invented, *world wide web* (www), three important pieces of technology that made the growth of the internet explosive bringing it to mainstream use in 1989:

1. HTML - Hypertext Markup Language, formatting language of the web, instructing computers how to display the web content of a web page to the users
2. URI - Uniform Resource Identifier, an 'address' used to uniquely name and identify a web page on the web
3. HTTP - Hypertext Transfer Protocol, technology used by computers to retrieve web pages from the web

Lee also set up the world's first website that is still running at: *http://info.cern.ch/hypertext/WWW/TheProject.html*. *Stanford Linear Accelerator Center* (SLAC) launched the second website and first in

North America in 1991. Yahoo, one of the first successful commercial web services providing directory listing, news and email launched in 1994 when only 3000 websites existed. Now there are more than 1.8 Billion websites in the world. Google was launched in 1998 and Wikipedia started in 2001. Email was invented by Raymond Tomlinson at BBN, a contractor for ARPANET, near Boston, US in 1971 including the famous @ symbol. This is the same BBN that got a contract from ARPANET in 1968 for building its network nodes, Interface Message Processors (IMP).

When we type a name such as *www.yahoo.com* in a web browser it is resolved into a low level Internet Protocol (IP) address such as *74.6.231.20* by the naming system called Domain Naming System (DNS). Then the web browser uses this address to connect to the machine at *74.6.231.20* using the protocol HTTP built on top of the foundational network protocol TCP/IP (roughly think of TCP/IP as the phone line and HTTP as two people talking to each other over the phone line). It retrieves the content written in the formatting language HTML and formats the content as per instructions in HTML and displays it to the user. Most important construct in HTML is the links which allow one to navigate to another webpage by clicking on the link, hence the name *hypertext*.

The Internet ranks very high as a powerful platform for both personal and broadcast communications. The Internet has completely transformed communication, commerce, entertainment, news and society itself in 30 years. The Internet has become very important for commerce and its importance will likely continue to increase. Online shopping accounts for 8.9% of all retail sales worth 744.1 billion dollars in the US. It accounts for 23.1% of all retail sales in China. Online banking is very prevalent with 34% using mobile banking that ultimately relies on the internet in the US. In India, services such as railway reservation are critically important with its reservation website booking 83% of all tickets booked. In 30 years, it has become a critical foundation of modern society along the lines of electricity, telecommunications or transportation and ranks very high despite its short history.

Compass

" On this day, at about nightfall the needle varied to the N. W., and on the following morning still more so."

Christopher Columbus' journal under the date of Tuesday, September 13, 1492 has this innocuous looking sentence that would create serious panic among his crew. On October 12, 1492, Columbus landed on the *new world*. One of the critical instruments that enabled his journey was the compass. Its importance in the journey is found in an interesting way; Although he didn't know it at the time, Columbus had independently discovered the phenomenon of *magnetic declination*; The magnetic north of the earth doesn't point to the north pole but is at an angle that is small in most places; It was not widely known to European explorers at that time. For Columbus' crew, the compass seemed to stop working and it created fear and panic. It was a matter of life and death as they were in the middle of nowhere in the Atlantic. In particular Columbus used a method of navigation called *dead reckoning*. In dead reckoning, the navigator finds his position by measuring the course and distance he has sailed from some known point. Starting from a known point, such as a port, the navigator measures out his course and distance from that point on a chart, pricking the chart with a pin to mark the new position. Each day's ending position would be the starting point for the next day's course-and-distance measurement. Course was measured by a compass; Distance was determined by a time and speed

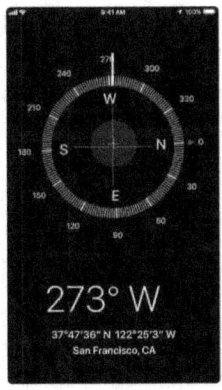

calculation: the navigator multiplied the speed of the vessel by the time traveled to get the distance. Without a working compass they simply couldn't navigate and reach land. The fears of Columbus' crew were not unfounded; In 1291, Vivaldi brothers from Genoa, Italy disappeared in the sea near Africa trying to find a route to India. For early explorers in the middle ages, a broken compass would be the equivalent of us getting lost in a spacecraft with a broken radio antenna.

Compass is possible because earth is 32% iron and itself is magnetic. However there is no particular reason earth's magnetic field should align with the axis of rotation, that is magnetic north doesn't necessarily point to the north pole; It is for the same reason that it would be a surprise if earth were to orbit in a perfect circle instead of an ellipse as circle is a configuration with much lower *entropy* (more organized, less disorder) among all ellipses (see the chapter on *Thermometer*); It would almost be a proof for a divine designer. Interestingly, earth is very close to a circle with an eccentricity of only *0.0167* (the *foci* of elliptical orbit is only 1.67% off from its center). The *National Oceanic and Atmospheric Administration* (NOAA) provides magnetic declination for any place on earth. For Sunnyvale, California, USA (37° 24' 01" N and 122° 1' 00" W) it is "$13° 8' E \pm 0° 21'$ changing by $0° 5' W$ per year"; **That is true north is almost 1/6 of the way east (90° E) of what the compass shows.**

First compasses were invented in China around 200 BCE and were used for only religious reasons, in fortune-telling by means of *geomancy*, with a soup spoon on a square base. It was first used for navigational purposes in the 11th Century in China and in Europe in the late 12th century. Alexander Neckam of England in his book *De naturis rerum* ("On the Natures of Things") written around 1190 makes some interesting claims including that "The loadstone, if placed on a woman's head in her sleep, made her confess her adulteries". More importantly he refers to the compass in Europe for the first time [emphasis added]:

"The sailors, moreover, as they sail over the sea, when in cloudy weather they can no longer profit by the light of the sun, or when the world is wrapped up in the darkness of the shades of night, and they are ignorant to what point of the compass their ship's course is directed, they touch the magnet with a needle, which (the needle) is whirled round in a circle until, when its motion ceases, **its point looks direct to the north.**"

This clearly shows that compass was common around the time of Neckam's writing. In 1269 Petrus Peregrinus de Maricourt of France, in his well-known *Epistola de magnete* ("Letter on the magnet"), described a floating compass for astronomical purposes as well as a dry compass for seafaring and included illustrations of the compass. Compass was known in the islamic world around this time; Al-Ashraf Umar ibn Yusuf, sultan of Yemen, wrote on the compass around 1293. One of the manuscripts of this treatise is preserved in the *Egyptian National Library* in Cairo. He describes its use as the *Qibla indicator* for Mecca during islamic prayer. Modern smartphones use a compass (*tri-axis magnetometer*, a micro electro mechanical sensor) that uses, instead of the magnetic needle, *Hall effect*, in which "electrons racing through a conductor in the presence of a magnetic field are forced to one edge of the conductor, creating a voltage difference whose orientation is perpendicular to the magnetic field".

Compass ranks very high for its role in the discovery of the Americas and discovery of the route to India. Both of these changed the world in profound ways including British colonialism and the founding of the United States. It was critical for the first circumnavigation of the earth by Magellan's voyage, the expedition that named the ocean between Asia and the Americas *Pacific*. Until GPS was invented recently it was very important for exploring even inland areas like forests ranking it very high in historical importance. Compass is ranked relatively lower than maps and ships for three reasons. The Compass is less than a millennia old unlike the latter with more than 2,000 year histories; It has also been replaced by GPS in modern navigation; Unlike telegraph, also an obsolete technology now, it wasn't very productive; Telegraph was directly responsible for an array of important inventions such as telephone and internet that replaced it; Compass wasn't responsible for GPS as they work very differently. This makes Compass rank high, in the top 50, and below both the inventions of the industrial revolution and the computer revolution.

PART 3

Firearm

"For 31 straight years, Louisiana has reported the nation's highest murder rate."

A *New York Times* story on murder rate in Louisiana in southern united states starts with this line. One of the reasons cited is firearms are the weapon used in 84% of murders (compared with the US national average of 74%, an astonishing number in itself, guns kill 3 out of 4 people murdered in the US). Guns are the most efficient killers of people. A *University of Pennsylvania* study shows that 33% of gunshot victims taken to hospitals die [For comparison, only 7.7% of stab wound patients die]. Another study from *international red cross* corroborates this finding by showing a similar ratio of wounded to that of killed in armed conflicts.The very efficiency of guns and its wide availability were what made two world wars a very bloody affair with the second world war alone killing 70 Million people worldwide [70M is almost 80% of the current German population or 21% of the current US population]. By comparison, nuclear bombs accounted for around 200,000 or 0.29% of the deaths. (To be fair, they were invented in 1945 towards the end of the world war; The story could have been different if they were invented earlier). For their efficiency in extinguishing lives, firearms rank very high as an [anti-] invention. First guns were probably cannons. Gunpowder itself originated from China where it was being used in fireworks. The oldest surviving firearm is the *Heilongjiang hand cannon* dated to 1288. Richard Gatling at Indianapolis near Chicago in the US in

1862 invented the first hand cranked machine gun that could fire up to 400 rounds a minute. Hiram Maxim at London invented the *Maxim Gun*, the first *automatic* machine gun operated with its own recoil, in 1884. *Sturmgewehr 44* (literally "assault rifle" 44) was the first assault rifle produced in 1943 by Germany for World War II. The AK-47, *Avtomat Kalashnikova*, was invented by Mikhail Kalashnikov in Russia in 1947 based on Stg 44; It is the most popular firearm in the world with around 100M of them.

If you are like me who finds all the gun terminology very confusing, here is a quick summary:

- In *automatic* (or *fully automatic*) firearms, the firearm continuously loads and fires as long as the trigger is pressed.
- In *semi-automatic,* the firearm continuously self-loads but the trigger needs to be released and pressed for it to fire every time.
- An *assault rifle* is *selective-fire*, that is can operate in either of the two modes above and uses *intermediate* cartridges, smaller than cartridges used in classic rifles of early 20th century, before the invention of assault rifles.
- A *pistol* is any handgun.
- A *revolver* is a pistol with a spinning cylindrical chamber with multiple bullets.

In 1836, Samuel Colt at Hartford, mid-way between Boston and New York, patented the revolver. *Colt Paterson* was the first revolver which started the handgun revolution. It is superseded by *semi-automatic* pistols that self-load the bullet from the magazine using recoil energy but the trigger still fires the bullet. The *Salvator-Dormus pistol* by Karl Savator of Austria in 1891 was the first semi-automatic pistol. In 2020, *Glock 19* ["9 mm"] was the best selling pistol in the world. According to *Glock*, Over 65% of federal, state and local agencies in the United States use Glock pistols. The *9×19 mm Parabellum* invented by German George Luger at Berlin in 1902 is the world's most widely used cartridge.

In some sense, the action of the firearm is the same as that of an *internal combustion engine*, particularly a diesel engine that requires no external spark (see the chapter on *Internal combustion engine*). Suddenly expanding gas from a chemical reaction [a controlled explosion] pushes a piston or a projectile at high velocity. However, in the case of the diesel engine, fuel at high pressure explodes by itself. In case of a firearm a sudden force explodes gun powder, typically an unstable nitrogen compound such as *saltpeter* (Potassium Nitrate) with carbon fuel (charcoal) and stabilizers with its nitrogen atoms eager to get back to their stable, inert nitrogen bonds in nitrogen molecules (N_2) in the air (see the chapter on *Fertilizer*). Gunpowder is 75% potassium nitrate by weight. In general, it is a good bet that if a [chemical] explosion happens it involves nitrogen compounds eager to form stable Nitrogen molecules and vice versa; if a chemical equation produces nitrogen molecules it likely involves a violent, exothermic explosion. As a non firearm, non dynamite example, airbags inflate in collisions when *Sodium Azide* explosively converts to nitrogen gas. A simplified equation for the burning of gunpowder is:

$$10KNO_3 + 8C + 3S \rightarrow 2K_2CO_3 + 3K_2SO_4 + 6CO_2 + 5N_2$$

Barrel length affects the exit velocity and the range of the bullet. Still a *9mm* [width of the barrel] pistol bullet, with a barrel length of 12 cm to impart its momentum, can cover 2 km at the rate of more than 1,300 kmph, faster than typical airplane speed; Its effective [kill] range is a more modest few hundred meters. A sniper rifle can kill at a distance of 3.5 km or more [real record by a canadian soldier in Iraq in 2017]; With a barrel length of 60 cm it has a muzzle velocity of 3,600 kmph [4 times typical jet speed], it can cover many miles easily.

Firearms rank very high as a life-taking invention for the same reason, fertilizer or vaccines rank very high. Guns killed people on an industrial scale in wars, at the rate of 5,00-1,000 rounds a minute for many assault weapons. This technology vastly accelerated the efficiency of the killing by freeing it from human muscle power similar to what steam engines or internal combustion engines did in a more positive context. In the US alone, there are almost *400 Million* guns, more than the US population of 330 Million; Americans make up 4.4 percent of the world but own 42 percent of the guns. 31% of mass shootings worldwide from 1966 to 2012 were committed by Americans. It is critically important for defense and law enforcement in a world proliferated by firearms. For its impact on human history as a potent anti-invention including world wars, colonialism and elimination of native populations in Americas and Australia by European settlers and transforming these continents it tops the second half of this book.

Nuclear Weapon

Nuclear weapons take its place in our list of inventions as the truly frightening one with the potential to end our species completely. It is another anti-invention included for its current and future impact on human society. First nuclear weapon was dropped on *Hiroshima*, Japan on Aug 6, 1945 by the United States under president Harry Truman with a yield of 15 kilotons of TNT (63 TJ) and killed at least 66,000 people; Second was dropped on *Nagasaki* on Aug 9, 1945 with a yield of 21 kilotons (88 TJ), killing at least 40,000 people; both are conservative (low) estimates. There are now 13,400 Nuclear weapons in the world that together can annihilate all of humanity. First weapon was developed under president Franklin Roosevelt in a special world war II project called the *Manhattan project*. When Robert Oppenheimer, who led the team that designed the bomb, first saw the successful test he was quoted as saying the verse from *Bhagavad Gita*, *"Now I am become Death, the destroyer of worlds."* According to the Bulletin of the Atomic Scientists:

"As of mid-2017, we estimate that there are nearly 15,000 nuclear weapons located at some 107 sites in 14 countries. Roughly, **9400 of these weapons are in military arsenals**; the remaining weapons are retired and awaiting dismantlement. Approximately 4150 are operationally available, and some 1800 are on high alert and ready for use on short notice."

The US and Russia own the bulk of these weapons with around 4,000 active warheads apiece. The US has around 3,570 strategic warheads (*"end of the world"*) and around 230 tactical (lower yield, short term battle use) warheads such as the *B61 gravity bomb*. Even current tactical warheads can be 25 times as powerful as the Hiroshima bomb that had only around 15 kT of TNT. China follows at a distance from these two at around 200-300 weapons.

In an atomic bomb a naturally radioactive element like *Uranium-235* is isolated from its predominant Uranium-238 isotope and kept in the weapon in such a way that during the explosion it would reach *supercriticality*, an amount which will sustain an exponential growth of nuclear chain reaction. Supercriticality threshold is around 52 Kg for Uranium-235 and 10 Kg for *Plutonium-239*, two most common nuclear weapon elements. It is very expensive to maintain these weapons; It costs 44.5 Billion dollars to maintain the US nuclear arsenal (2021) which is more than the full *state department* [US Foreign ministry] budget. From 1940 through 1996, the US spent nearly $9.33 trillion (in 2021 dollars) on nuclear weapons and weapons-related programs. This is more than what the biggest state in the US, California, spent on its public education with combined federal, state, and local funding for 94 years. According to the California Department of Education, "The total overall funding (federal, state, and local) for all K–12 education programs is $98.8 billion, with per-pupil spending of $16,881 in 2020–21". To add to that, education is an investment in the future of the country and nuclear weapon cost is simply the cost for running in place in an arms race with no benefits to society. The *International Campaign to Abolish Nuclear Weapons* (ICAN) that won 2017 Nobel peace prize says:

"In a city like Mumbai, India, with population densities in some areas of 100,000 people per square kilometer, a Hiroshima-sized bomb is estimated to cause up to 870,000 deaths in the first weeks. A 1-megaton bomb could promptly kill several million people, with the death toll rising over time. A 12.5-kiloton nuclear explosion in a New York shipping yard would produce casualties more than one order of magnitude greater than those inflicted in the September 11 terrorist attacks. Blast and thermal effects would kill 52,000 people immediately. Another 238,000 would be exposed to direct radiation from the blast. The fallout would expose a further million and a half people. In total, more than 200,000 would die."

It is truly an existential risk without being a hyperbole and ranks in importance close to humanity's foundational inventions mainly because it has the ability to send all of us to the stone age at the least and make us start over this book with a different set of inventions. That is assuming *homo sapiens* will survive a nuclear war; It is more likely that earth and the evolution itself has to start over with another species of *intelligent ape*. However it is hard to rank it as it hardly has any impact in day to day life and even in conventional wars. This ranks it below our foundational inventions like fire, modern foundations like printing press, inventions from industrial revolutions such as steam engines and electricity. The US attacking nuclear weaponless states such as Iraq and Libya with impunity has provided further incentives for countries such as North Korea and Iran to accelerate their development of nuclear weapons as an ultimate deterrent against the US and other invasions. However there are many encouraging signs; Total active warheads have come down from a high of 64,000 in 1986 to 9,220 in 2017; There is a nuclear arms reduction treaty between the US and Russia (*New START*) that limits the number of deployed strategic nuclear warheads to 1,550 per side. Abolishing all nuclear weapons could be one of the greatest gifts early 21st century humanity, our generation, can give to our children and grandchildren.

Eye glasses

"It is **not yet twenty years** since there was found the art of making eyeglasses, which make for good vision..."

This is the first known reference about eye glasses from an Italian priest in 1306 CE [emphasis added]. His reference dates the eye glasses to after 1286. It is not known when exactly the first eyeglasses originated or who made them. It must have been common in Italy around 1350 as the painter Tommaso da Modena depicted eyeglasses in his painting at a convent in *Treviso* near Venice in Italy. Interestingly it was an improvisation on the part of the artist as the person in the painting, a priest called *Cardinal Hugh of Provence*, had died even before the eyeglasses were invented! One of the first figures to be associated with the invention of spectacles was the thirteenth century English friar Roger Bacon, who was based in Paris and outlined the scientific principles behind the use of corrective lenses in his book *Opus Majus* (literally "Great work"). Prepared in 1267 at the request of Pope Clement IV, the book is a collection of ideas, an encyclopedia of knowledge embracing all science known at the time. According to the book *A history of Western Philosophy:*

"Part five of the book, which deals with optics, is thought to be the section that best illustrates Bacon's own work. He begins with the physiology of eyesight, the eye, and the brain, and goes on to discuss the conditions of

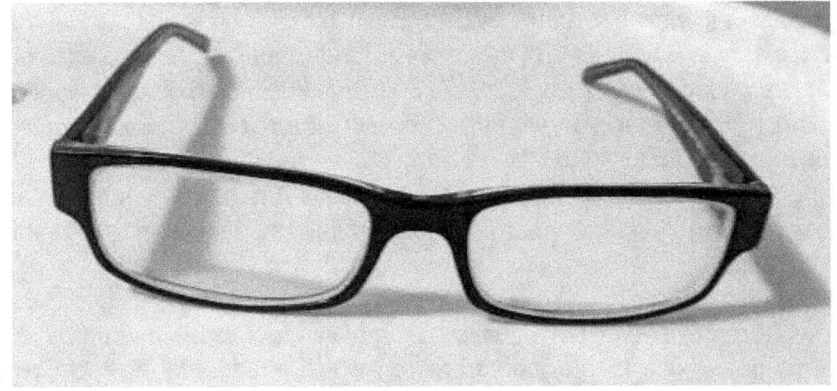

seeing: light, distance, position, size... He goes on to discuss direct vision, reflected vision, and refraction."

In part six, he discusses experimental science and the list of possible future inventions, eyeglasses among them, the first reference to eyeglasses. Much earlier, there is a greek story of Archimedes, antiquity's great scientist, using a concave mirror that focuses sun's rays to burn Roman ships to save his city *Syracuse*, at the southern tip of Italy in the mediterranian overlooking Africa around 212 BCE; At least one experiment in 1973, reported in The *New York Times*, could set a ship on fire using his method and came to the conclusion that this "burning glass" is plausible in theory.

Once books became widely available in the next century after Gutenberg's printing press, demand for spectacles increased exponentially. *Worshipful Company of Spectacle Makers*, a livery company (*trade guild* in modern parlance) was founded in 1629 in England which means it has spread from Italy to all of Europe and was a very common item in Europe in the early 17th century. **Established in 1777 by William Fraser,** *C.W. Dixey & Son* **of England, is the world's oldest independent eyewear company.** They claim "We have served seven successive Kings and Queens of England, the Royal Houses of nine nations, and the emperors of China, France and India."

According to the *Vision Council of America*, approximately 75% of adults use some sort of vision correction. About 64% of them wear eyeglasses, and about 11% wear contact lenses, either exclusively, or with glasses. Over half of all women and about 42% of men wear glasses. Adolf Gaston Eugen Fick, a German ophthalmologist, working in Zurich, Switzerland invented the *contact lens*; Fick's article on contact lenses was published in German in the journal *Archiv für Augenheilkundo* in March 1888. One out of six Americans wears contact lenses, translating into approximately 45 million people in the United States.

Corrective lens power is measured in *diopters*. It measures the optical power of a lens or curved mirror, which is equal to the reciprocal of the focal length (**1/f**, **f** being focal length) measured in *metres*. It was introduced by French Ophthalmologist Ferdinand Monoyer in 1872 at Lyon, France near Geneva. The *Snellen Chart* used for measuring vision acuity was invented by Herman Snellen, a professor of ophthalmology at the *University of Utrecht* near Amsterdam, in the Netherlands. If, from a distance of 20 feet, one can read a line that "normal" sight would enable a person to read at 20 feet, his sight is termed 20/20. In the Snellen Chart, big E at the top corresponds to an acuity of 20/200 and most charts go to 20/15 or 20/10. In terms of diopters, 20/400 corresponds to -4.0 diopters, 20/200 (line 1) corresponds to -2.5 and 20/25 (line 7) corresponds to -0.25 with 20/20 (line 8) corresponding to 0.0. *Hyperopia* (farsightedness) and *presbyopia* (inability to focus particularly in old age requiring reading glasses) require corrective lenses with positive diopters.

EssilorLuxottica, formed with the merger of Essilor of France and Luxottica of Italy in 2018 is the biggest prescription lenses and frames maker in the world with a revenue of 14.4 Billion in 2020 and sells close to a billion pairs of lenses and frames every year. It owns *LensCrafters*, the largest eye glasses company in the world. It also owns many popular brands such as *Ray-ban* and *Vogue*. Essilor, founded in 1849 as Société des Lunetiers ("association of spectacle-frame makers"), had 42% of the world's prescription lens market share followed by *Hoya vision* of Japan at 12% and *Carl Zeiss* of Germany at 8%. Hoya vision was founded in Hoya near Tokyo in Japan in 1941. History of Carl Zeiss, the great optical instruments company, is mentioned in the chapter on Microscopes. *Bausch and Lamb* (now *Bausch Health*) is one of the oldest eyeglasses makers in the US founded in Rochester, New York, in 1853 by German immigrants John Bausch and Henry Lomb.

Eye glasses, as the founding invention for *lenses*, ranks very high; Lenses, as a very productive invention, led to the invention of microscope, telescope, photography, movies and television. Eye glasses have been with us for more than 700 years which increases its impact and ranking. The Global prescription lens market is estimated to be $31.94 billion in 2019 and is expected to reach $52.07 billion by 2027. As an invention affecting half the world population, 4 Billion people, it also has a high impact in modern society ranking it high.

Microscope

Microscope was a critical invention in the science of *Biology*. Robert Hooke at the *Royal Society of London* in England, then newly formed in 1660, used the Microscope to come up with his *Cell theory* in 1667, the birth of modern biology. Later it was critical to the study of microorganisms such as bacteria and yeast and to the *germ theory of diseases*, the birth of modern medicine. The first *compound microscopes* (ones that used two lenses) were invented by father-son duo Hans and Zacharias Janssen in the Netherlands around 1590 but the evidence is not conclusive. The earliest microscopes could magnify an object up to 20 or 30 times its normal size. Cornelis Drebbel of the Netherlands is also credited with independently building a compound microscope around 1620. Antonie van Leeuwenhoek, "father of Microbiology", at Delft near the Hague in the Netherlands built his own powerful microscopes with only one lens around 1675 [based on his paper to the *Royal Society*] and studied the microorganisms for the first time. His

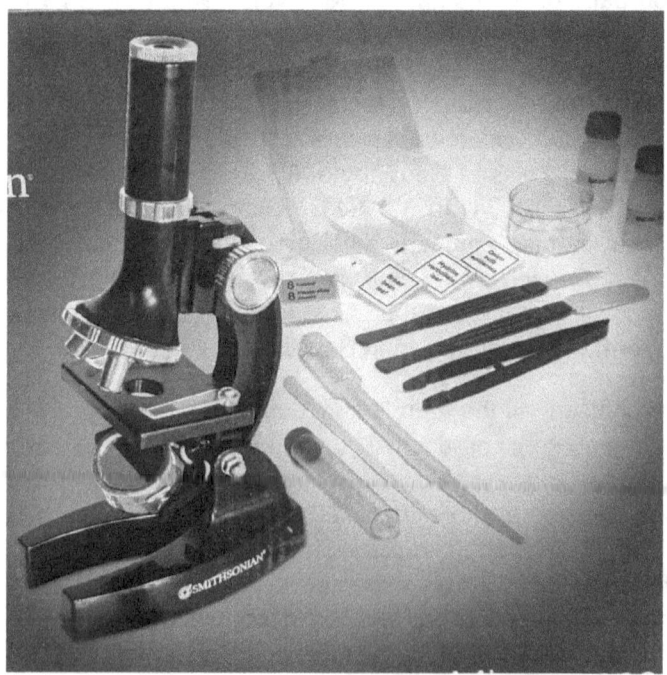

high-quality, hand-ground lenses could magnify an object by up to 200 times. Hooke published *Micrographia* in 1665, an astonishing collection of copper-plate illustrations of objects he had observed with his own compound microscope. We can almost say that it built modern biology and only Darwin's *natural selection* can claim a bigger role than a microscope. In 1846, Carl Zeiss founded *Carl Zeiss AG* at Jena near Dresden in East Germany, to mass-produce microscopes and other optical instruments; It is now an optical giant in the world with a revenue of 7.7 Billion dollars in 2019-20.

Ernst Ruska and Max Knoll at *Technical University of Berlin* in Germany invented the first *Electron Microscope* in 1931. The *resolution capacity*, ability to resolve the detail of the object being examined, of light microscopes is limited by the wavelength of light (*400-700 nm*). In quantum theory both particles and waves are one and the same; The role played by photons in optical telescopes is played by electrons in *Tunneling Electron Microscopes* (TEM). However, electron's wavelengths are much smaller, *0.01 nm*, than electromagnetic waves (*photon* waves) resulting in much better resolution for the magnified images. *Scanning Electron Microscope* (SEM) works similar to a tiny CRT tube by scanning in lines and analyzing the secondary electrons from the specimen itself. Electron Microscopes can magnify up to 50 million times; With a magnification of a million, a grain of salt will be as tall as the *Eiffel Tower* and with a magnification of 50M, it will be twice as tall as *Mount Everest*.

X-ray crystallography is another form of Microscope invented by Max von Laue at *Ludwig Maximilian University of Munich* in Germany and further developed by father-son duo William and Lawrence Bragg of England. It was used by Rosalind Franklin at *King's College London* in determining the double-helix structure of DNA in 1952. It is currently the most favoured technique for structure determination of proteins and biological macromolecules. A paper by Smyth and Martin explains:

"The aim of *x-ray* crystallography is to obtain a three dimensional molecular structure from a crystal. A purified sample at high concentration is crystallised and the crystals are exposed to an *x-ray* beam. The resulting diffraction patterns can then be processed, initially to yield information about the crystal packing symmetry and the size of the repeating unit that forms the crystal. This is obtained from the pattern of the diffraction spots. The intensities of the spots can be used to determine the "structure factors" from which a map of the electron density can be calculated. Various methods can be used to improve the quality of this map until it is of sufficient clarity to permit the building of the molecular structure using the protein sequence. "

There are now other forms of Microscopes such as *Scanning probe microscopes* and *Field Ion microscopes*. In Scanning probe microscopes, a (microscopic) probe literally probes the specimen (as a blindfolded person would do to an elephant by touch) and produces the surface pattern either using electrical current or using the probe's movement up and down. Field Ion microscopes can be used to image the arrangement of atoms at the surface of a sharp metal tip that typically produce beautiful art patterns sold as wall art. An electric field applied to the needle's tip repels ions of helium, neon or argon that spread out and produce the image of the metal structure on a fluorescent screen.

Microscopes rank high in our list as the foundation for modern biology. To a first approximation biology can be reduced to two theories, *Cell theory* and Darwin's *theory of natural selection*. It contributed to our modern understanding of cells including organelles such as the nucleus, particularly based on its modern offshoots such as electron microscopes. Understanding of the cell organelles led to the eventual discovery of chromosomes and then DNA, modern foundation for biology and vaccines among others. X-ray crystallography led to the discovery of DNA double helix structure. Our study and understanding of microbes contributed to medical advances which saved millions of lives. It has been with us for around 400 years, around the same age as newspapers, and ranks very high, at the top 10, of the second half of this book.

Petroleum

The Scottish chemist James Young's 1852 patent, "IMPROVEMENT IN MAKING PARAFFINE-OIL" marks the beginning of the petroleum industry; It describes the process of making *paraffine-oil* (coal oil), used in lighting and lubricating machinery, from coal. A year earlier in 1851, he started the world's first oil refinery, *Bathgate Chemical Works,* to produce paraffine-oil at Bathgate near Edinburgh in the UK. In 1852, Abraham Gesner of Canada invented and named kerosene in his patent, "IMPROVEMENT IN PROCESSES FOR MAKING KEROSENE". He introduces Kerosene in his patent:

"...have invented and discovered a new and useful manufacture or composition of matter, being a new liquid hydrocarbon, which I denominate "Kerosene." I obtain this product from petroleum, maltha or soft mineral pitch, asphaltum, bitumen, or bituminous and asphaltic rocks, and shales by dry distillation, and subsequent treatment with powerful reagents and redistillation."

In 1854, Ignacy Łukasiewicz built the first oil well at Bóbrka, near Krosno in Poland near the Slovakia border. In 1859 Edwin Drake's oil well, first in the US, at *Titusville* in northern Pennsylvania near the New York border started the oil boom in the US and the world. *Standard Oil* was started by John Rockefeller in 1870; In the early 20th century, it refined nearly 75% of all US crude and marketed over 80% of domestic kerosene. Modern oil giants such as *ExxonMobil, Marathon* and *Chevron* trace their roots to Standard oil. Rockefeller is the richest American ever with a wealth of 313 Billion dollars, adjusted for the US 2020 GDP of 20.93 Trillion. All petroleum fuels are a complex mixture of *hydrocarbons*:

- *Kerosene* is a mixture of hydrocarbons usually with a chain length of 10-16 carbon atoms per molecule, average being 12 with 35% alkanes (paraffins), 60% cyclic alkanes (naphthenes) and 15% aromatics.
- *Petrol* (Gasoline) is a mix of hydrocarbons with a chain length of between 5-12 carbon atoms. Virgin gasoline usually contains: 50% alkanes, 40% cyclic alkanes and 10% aromatics. Blended gasolines are mixtures of virgin gasoline, catalytically cracked gasoline, and thermally reformed gasolines, and may contain up to 30% alkenes (olefins).
- *Diesel* is made of hydrocarbons containing 12 or more carbon atoms with the average being 15 with 30% alkanes, 45% cyclic alkanes and 25% aromatics.

Diesel fuel is denser than petrol and contains about 15% more energy by volume (roughly 36.9 MJ/litre compared to 33.7 MJ/litre). Airplanes use Kerosene, with its energy density between petrol and diesel, low freezing point, higher octane rating providing more power and is cheaper than gasoline. Diesel has a *flashpoint* [indication of flammability] of 52°C and petrol has a flashpoint of < -21°C; That is petrol is highly flammable at room temperature.

Different hydrocarbon chain lengths all have progressively higher boiling points the longer the chain, so they can be separated by the *fractional distillation* process. During the process, crude oil is heated in a distillation column, and the different hydrocarbon chains are extracted as a vapour according to their vaporisation temperatures and then recondensed. A standard *U.S. barrel* contains 159 litres (42 gallons) of crude oil which yields about 167 litres (44 gallons) of petroleum products, 43% gasoline, 22% diesel and 9% jet fuel.

Knocking in spark ignition internal combustion engines ("gasoline engine") occurs when the petrol/air mixture explodes even before it is ignited resulting in suboptimal energy release. *Octane rating* is used to measure this tendency of a compound to 'knock', higher the rating lower the 'knock'. In this scale, *2,2,4-trimethylpentane* (Isooctane) is assigned a rating of 100 (good) and n-heptane is assigned a rating of 0 (bad). All other compounds such as *Cyclopentane* (141) and *Benzene* (99) are assigned a value on this scale. Gasoline at the pump has a rating that is a weighted average of compounds in the mix.

World produced 95.2 Million *barrels per day* (bpd) in 2019. The US is the world's largest producer of crude oil with 17 Million bpd followed by Saudi Arabia and Russia with 11.8 Million and 11.5 Million respectively. The US became the world's biggest oil producer in 2017 through the *shale fracking revolution*. There are 1,734 billion barrels worth of proven reserves in the world now or 50 years worth of oil at the rate of 95 Million barrels per day. The US consumes 20.5 Million bpd or 10 liters per person per day. China consumes 14.1 Million bpd (2 liters per person) followed by India at 5.3 Million bpd (0.75 liters per person). That is, US per capita consumption is 13 times that of India. The US led the world in natural gas production in 2019 with 920.9 billion cubic metres followed by Russia and Iran at 679.0 billion and 244.2 billion respectively. Jamnagar Refinery in India, Owned by *Reliance Industries*, with a crude processing capacity of 1.24 million barrels per day (bpd) is the world's biggest refinery in the world.

Petroleum ranks very high as the fuel for automobiles, airplanes and railways. It played a crucial part in the plastics revolution and ranks higher than plastic. It is one of the most important parts of the economy with more than 6 Billion dollars worth of crude oil alone produced every day; This ranks petroleum high, at the top 10, in the second half of the book.

Plastic

We produce more than *380 Million tonnes* of plastics in a year, the weight of more than 7,300 *Titanic*, the classic steam behemoth that perished on its first voyage in 1912; Put another way, it is almost the combined weight of all the *7.8 billion* people on earth. Plastics are very long organic polymers containing carbon chains which are *plastic* during manufacture and can be molded into various shapes easily using custom molds at relatively low temperature, hence the name *plastic*. In this sense, it is similar to *cast iron* but cast iron requires much higher temperatures to mold than plastics (see the chapter on *Iron metallurgy*); It is also much heavier than plastics as anyone who lifted a cast iron skillet would know (see the chapter on *Pot*). Plastics are very likely the most common item around us after iron and steel; It is used for packaging containers, clothes, tires, chairs, electronics, car parts, airplanes, medical instruments, sports equipment, toys, bags and water pipes among others. *Rubber* is a naturally occuring plastic made up of *isoprene* molecular chains; *Vulcanization of rubber,* invented by Charles Goodyear in the US in 1839, can be considered the start of the plastic age (see the chapter on *Rubber*). Nitrocellulose [called *Parkesine*],

invented by Alexander Parkes in 1856 in England, was the first man-made polymer made from *Cellulose*, found in plant cell walls, in 1856. *Bakelite* is the first synthetic [fully man made] plastic, invented by Leo Baekeland in the US in 1909. Given that plastics are any soft organic polymer, important varieties were invented at various times by different people and continue to be made to this day. Here is a quick summary of most commonly used plastics:

- German chemist Eugen Baumann discovered *Poly Vinyl Chloride* [PVC] in 1873, used for wire insulation and water pipes. However it became usable only after Waldo Semon at *B F Goodrich* in the US invented in 1926 the plasticization of PVC, originally a rigid, brittle polymer.
- German chemist Karl Zeigler invented the process for *High Density PolyEthylene* [HDPE] at scale in 1953 and Italian chemist Guilio Natta invented *Polypropylene* [PP] in 1954 using Zeigler's catalysts; Polypropylene and HDPE are the most commonly molded plastics with uses in furniture, toys and other household items. *Low Density PolyEthylene* (LDPE), lighter than HDPE as the name implies and used in packaging foams surrounding electronic appliances and plastic bags, was invented by *Imperial Chemical Industries* at Northwich near Manchester in England in 1933.
- Nathaniel Wyeth of Dupont in the US invented the *PolyEthylene Terephthalate* [PET], in 1973 now heavily used as containers for soft drinks and juices due to its high strength relative to the weight; PET is called *polyester* in textiles and is heavily used in fabrics as a [wrinkle free] replacement for natural fibres such as cotton.
- *Nylons*, a group of polyamide polymers of *Nylon stockings* fame, were invented first by *Dupont* in 1935 and are used as a synthetic replacement for natural silk. *Acrylic fiber*, a polyacrylonitrile polymer, was invented by *IG Farben* in Germany and Dupont in the US in 1942; Acrylic is the third most common fabric, after polyester and cotton, used in textiles as a synthetic replacement for natural wool from sheeps. In fact now more than 60% of our clothing material is plastics, dominated by these three synthetic fibers. Polyester alone accounted for more than 50% of the clothing fibers in 2018 with cotton a remote second at 25%.

Plastics provide a cheap alternative to metal, ceramic and glass items which made them affordable for most of humanity for the first time. Most of the kitchen containers are plastic now. Modern medicine has greatly benefited from plastics, for instance the disposable plastic syringe; Vaccines for viruses such as the seasonal flu, measles or polio are stored in individually wrapped, single-use plastic syringes. Plastics provide an affordable alternative to our natural fibers such as cotton, silk and wool; As plastic synthesis is more easily scalable to billions of people instead of growing more sheeps or silkworms, they now provide affordable clothing to the majority of the people in the world.

Plastics have very large organic chains with strong carbon bonds which make them very durable. This also contributes to it being not biodegradable leading to *plastic pollution*. Glass is also not biodegradable but they degrade faster through natural processes. Glass is easily recyclable with glass shards directly used as a raw material in glass making. There is evidence that *Bisphenol A* (BPA) produced from the hydrolysis of some plastics such as *polycarbonate* may interfere with human reproduction. Plastics also burn to produce toxic glasses unlike glass. *MIT school of engineering* says:

"When plastic is burned, it releases dangerous chemicals such as hydrochloric acid, sulfur dioxide, dioxins, furans and heavy metals, as well as particulates. These emissions are known to cause respiratory ailments and stress human immune systems, and they're potentially carcinogenic."

Plastics rank very high in human inventions for providing us with an ubiquitous, cheap and lightweight raw material for our household items and clothes. It contributed to a higher standard of living by making common items such as electronics and appliances affordable. In less than a century it has become one of the most important items in modern civilization. Its importance is comparable to glass and it is replacing glass as a safer alternative. Its use in fabrics makes it more important in modern society. However glass has been with us for more than 2000 years, making a bigger impact in our history.

Rubber

"All branches of the rubber business, as you find it in this country today, took their real rise from Goodyear's discoveries of the process of vulcanization and of the hard rubber process originated with his brother, Nelson Goodyear. At first the making of shoes engaged the attention of the rubber men. Then came the making of rubber belting, rubber clothing, blankets, and pontoons, which were much in demand in Mexico and the Far Southwest. ... The boot and shoe industry increased so rapidly along with other branches of the industry that from an output in 1860 of the value of $795,000 the yearly output in 1870 had grown to $8,000,000.

It would take columns of THE TIMES merely to catalogue the articles made wholly or in part of rubber and to relate the uses to which this material is put. Nearly all of the rubber output of this country is used at home, although there is a steadily increasing export trade. The aggregate capital used is in excess of $100,000,000 and the value of the product annually exceeds $200,000,000. Several hundred thousand men are employed in this business."

This is from an article on Rubber in The *New York Times* of Sept 23, 1906, "The ROMANCE of RUBBER". 200 Million in 1906 is around 6 Billion in 2021 dollars and this was before automobiles took off. Every year around 2 Billion tires are sold in the world. As a major component of tires, rubber plays an important part in the world. Although technically a plastic, natural rubber was widespread more than 100 years before petroleum based plastics became mainstream in the mid-20th century and deserve to be ranked on its own. After experimenting with rubber for years, including in debtor's prison, Charles Goodyear in 1839 invented the *vulcanization of rubber* by accident when rubber and sulfur over a hot stove *vulcanized* it; *Rubber* is a naturally occuring plastic made up of *isoprene* molecular chains; In natural rubber, isoprene chains can slide over each other similar to graphite layers; Vulcanization process interconnects isoprene chains via sulfur atoms making them strong and elastic (*"rubbery"*); It was the start of the rubber (and the plastics) age. Joseph Priestley, discoverer of o*xygen*, found out that rubber makes the pencil marks rub off and named it "rubber" in 1770, marking the first use of a plastic for wider humanity (outside of native Americans). Rubber is produced from the latex of rubber trees (*Hevea brasiliensis*) native to South America and Congo rubber plant (*Landolphia owariensis*), the latter with a bone chilling, horrific history in Congo at the hands of Belgians and their king Leopold II, one of the worst villains in human history in a century with plentiful supply of them.

Natural rubber, also known as *caoutchouc* and *India rubber,* consists of polymers of isoprene molecules; Isoprene (*2-Methylbuta-1,3-diene*) itself is a relatively simple five carbon molecule with two double bonds. These long, linear chains in rubber latex, a viscous white milk like liquid, are cross linked together by sulfur during vulcanization leading to much stronger and elastic ("rubbery") material. It takes around 7 years from planting for the trees to start producing rubber latex. In 1909, Fritz Hofmann at Bayer in Germany produced the synthetic "natural" rubber, *polyisoprene* rubber. In 1929, Walter Bock and Eduard Tschunkur at IG Farben in Germany developed the most important synthetic rubber, *styrene butadiene rubber* (SBR) from *butadiene* (a 4-carbon molecule with two double bonds, simpler than isoprene with no methyl group) and *styrene* (a benzene ring with a single double bond). It played a crucial role in world war II after natural rubber supplies from Asia were disrupted by Japan. Earlier, the Russian chemist Sergei Lebedev polymerized butadiene in 1910 as *polybutadiene*, the second most important synthetic rubber. In 1929, Arnold Collins at Dupont in the US developed *Neoprene*, a polychloroprene rubber used to make wetsuits. *Ethylene Propylene Diene Monomer* (EPDM) rubber, invented in 1967 at *B.F. Goodrich*, is another important synthetic rubber used in washers and door seals in cars and refrigerators.

Thailand leads the world in rubber production with 4.9 Million tons in 2019 followed by Indonesia and Vietnam at 3.3 Million and 1.2 Million respectively with a world total of around 13 Million. Synthetic rubber production is comparable with SBR production worldwide is around 8 Million tons followed by polybutadiene at around 4.5 Million. China is the largest producer as well as the consumer of synthetic rubber with the majority of it going into tires. Although synthetic rubbers are widely available now it is not directly substitutable for natural rubber. Commercial vehicle tires such as that of heavy-duty trucks and buses require more natural rubber to deal with the increased physical strain of heavier load with 75% of rubber used being natural rubber and 25% synthetic rubber. In passenger car tires it is only 45% natural rubber and 55% synthetic rubber. *Bridgestone*, founded in 1931 by Shojiro Ishibashi at Kurume near Nagasaki in Japan, is the biggest tire manufacturer in the world with a revenue of 27 Billion dollars in 2019. *Michelin* (of *Michelin stars* fame), founded in 1889 by brothers André and Édouard Michelin at Clermont-Ferrand in France near Geneva, is a close second with a revenue of 26.5 Billion dollars. *Goodyear*, founded in 1898 by Frank Seiberling at Akron near Pittsburgh in the US, and *Continental*, founded in 1871 at Hanover, Germany, follow at a distance with around 15 Billion and 13 Billion dollars respectively.

Rubber, both natural and synthetic, ranks high as a critical invention enabling all the automobiles and bicycles in the world in the form of wheel tires. Rubber is important in the manufacture of footwear such as shoes and flip flops, providing people with more affordable footwear. It has been with us for almost 200 years playing an important role in human history, both positive and negative, including the horrific tragedy in Congo.

Lock

"The archaeological evidence for locking mechanisms fits well with the contemporary description of a lock of the Neo-Assyrian period: in 714, Sargon II (721–705 BCE) had the lock of the Haldi temple at Musasir removed and taken to Assyria as booty. In his inscriptions the four components of the lock are described individually and in great detail, as they are made out of gold and fashioned as works of art. Andreas Fuchs recently succeeded in identifying these components and reconstructing the lock (Fig. 2): The locking mechanism consists of a heavy transverse bar, the *askuttu*. In order to lock the gate a smaller holding bar, the *sikkuru*, is pushed through the appropriate hole in the transverse bar. The holding bar is in turn kept in place with the help of one or several bolt-pins, the *sikkatu* or, in plural, *sikkate*. In order to open the lock, the bolt-pins have to be removed from the holding bar with the help of a key called *namzaqu*. The mechanism of this lock closely resembles that of the 'Egyptian

lock', also known as the Greco-Roman balanos lock. In accordance with the naming of this lock type, which takes its name from the bolt-pin, Greek βαλανοδ 'acorn', Fuchs called the Assyrian lock type *sikkatu* lock, after the same component."

Karen Radner, An Austrian *Assyriologist* and Professor of Ancient History at the *University of Munich* and *University College London* describes the oldest known locks, *sikkatu* locks, in beautiful detail in her book chapter, "GATEKEEPERS AND LOCK MASTERS: THE CONTROL OF ACCESS IN ASSYRIAN PALACES". These locks and parts of locks have been found in the palaces of Nineveh, Dur-Sharrukin (Khorsabad) in Assyria (near Mosul in the current northern Iraq bordering Syria) dated to around 2700 years ago. From Radner's description we can clearly see that it is a form of *pin tumbler lock*; This works by aligning pins of different heights which have to align perfectly for the lock to open. The key has grooves with different heights to align the pins correctly. Radner also mentions that pin-tumbler locks were in wide use as Egyptian locks and Greco-Roman *balanos* ("acorn") locks. Interestingly this is one of the most common locks still in use today, via a re-invention in the 19th century United States. Modern form of pin tumbler locks was re-invented and patented by the father and son duo of Linus Yale Sr and Linus Yale Jr in 1848 and 1861 respectively in New York; This is the lock commonly found in the US. Another common variation of this lock is the *lever-tumbler lock* invented by Robert Barron in England in 1778. Instead of pins, this aligns the levers just the right way to let the bolt pass through when the correct key is used. It is the common model used in Europe, India and South America. In 1784, Joseph Bramah of England designed yet another variation on tumbler locks, using sliders instead of pins, a round lock mechanism operated by a distinctive looking tubular key. He started *Bramah* in the same year, one of the oldest extant lock companies in the world now. The story has it that he put a lock in his shop window and offered a reward of 200 guineas to anyone who could open it; In 1851, after more than half a century, at the Great Exhibition in London, A.C Hobbs an American locksmith spent 52 hours spread over 16 days and claimed the prize. One of the most innovative locks was the *Chubb detector lock* that can detect if someone tried to open the lock unsuccessfully. Jeremia Chubb invented this lever tumbler lock with a twist in 1818 as part of British Government Challenge to invent an unpickable lock. If one of the levers were to be lifted high during a picking attempt with a different key it would trigger a mechanism to prevent the opening till it was reset by the original key. This would also indicate to the owner that someone tried to pick the lock as the original key needs to go in the reverse direction for the reset mechanism.

Swedish *Assa Abloy* founded in 1994 is the world's biggest lock manufacturer in the world with 10.2 Billion in revenue mainly through their acquisitions of old, well known companies. It acquired well known companies such as *Yale* (later *Yale & Towne*) locks founded by the inventors of pin tumbler locks in 1868 and *Chubb* locks founded by the inventor of the Chubb detector lock in 1818. Nowadays *RFID* based key cards are replacing physical keys in *smart locks* in places such as hotels and office buildings (see the chapter on *RFID*). Smart locks, that include car locks, are electronically controlled by microprocessors and require electric current and an electric motor for operating the bolt.

Locks are used to lock mailboxes, desk drawers, suitcases, cars, bicycles in addition to houses and offices. However *Smithsonian Magazine* reports that much talked about chastity belts and keys are (thankfully) only a myth. The *U.S. Postal Service* produces over 250,000 locks a year. Millions are used in prisons with over *10.35 million people* incarcerated throughout the world as of 2015 with the most being in the United States at *2.2 million*.

Locks get their high rank as one of the most commonly used day to day inventions. We lock our houses and our cars everyday without even thinking about it. It is a critical component in one of our basic necessities, shelter. It is one of the critical inventions for the modern economy; If all of us were to sit at home or hire security guards to go to work, the modern economy would come to a standstill. It has also been with us for almost 3,000 years making a high impact in human history and deserves to be a high ranking invention.

Battery

As I write this in Feb 2021, Market capitalization of *Tesla*, an electric car maker in the US is 750 Billion dollars, or a *whopping 3.5%* of the US economy in 2020 of 20.93 Trillion dollars. It is more than 16 times the market capitalization of Ford which started in 1903, exactly 100 years before Tesla started or three times that of *Toyota*, world's biggest car maker from Japan which produced more than 10 million vehicles in 2017 as compared to Tesla which produced slightly more than 0.1 million vehicles in 2017, a factor of 100. Even after discounting the stock market's *irrational exuberance*, this shows electric cars with huge batteries in them are thought of as the future of automobiles at least by investors if not by common people. Looking at this investment trend, other automobile manufacturers such as *GM* and *Volvo* have declared that they will stop producing internal combustion engine vehicles by 2030-2035. Alessandro Volta at *University of Pavia* near Milan in Italy invented the first battery in 1800, called the *Voltaic pile*. It has alternate pairs of zinc and copper discs separated by *brine* (concentrated salt water) soaked cloth as the electrolyte; One cell [one disc pair with brine separator] could produce *0.76 volts*. The *Daniel cell*, a much more useful variation on the voltaic pile with liquid electrolyte, was invented by John Daniel in London, England in 1836; It also contains zinc and copper electrodes but in *zinc sulfate* and *copper sulfate* solutions respectively. It produced exactly 1.0 volts, *by definition,* as it was used to define the voltage unit in 1881. (However with the modern accuracy of the SI system it is now known to produce slightly more voltage, *1.1 volts*). This cell powered the 19th century electrical inventions including telegraph and telephone.

A *dry cell*, with no liquid electrolyte making it more portable, was developed in 1886 by German Carl Gassner at Mainz, of Guternberg's press fame, in Germany and he patented it in 1887 as "GALVANIC BATTERY". It has zinc anode and carbon cathode with *Ammonium Chloride* as the dry electrolyte; It also requires *Manganese (IV) Oxide* as a *depolarizer* to complete the chemical reaction by consuming hydrogen formed at the electrolyte. It evolved from the *Leclanché cell* invented by Georges Leclanché of France at Brussels in Belgium in 1866 that had the same components except for ammonium chloride as a solution. Modern alkaline batteries replace Ammonium Chloride with *Potassium Hydroxide*, an alkali, providing higher energy density and longer shelf life. In 1859, Gaston Planté at Paris, France invented the *lead-acid battery*, the first rechargeable battery now heavily used in automobiles to start the engine and power accessories; It powers more than a billion automobiles around the world. Lead-acid battery contains a *lead oxide* positive plate and *lead* as the negative plate with *concentrated sulphuric acid* as the electrolyte in charged state; In discharged state both plates become *Lead (II) Sulfate* with very *diluted sulphuric acid* as the electrolyte. In 1899, the *Nickel Cadmium* (Ni-Cd) battery was invented by Waldemar Jungner in Sweden. Thomas Edison also invented and patented the nickel cadmium battery in 1901 as "REVERSIBLE GALVANIC BATTERY". This battery contains *Nickel(III) Oxide-Hydroxide* as the positive plate and *Cadmium* as the negative plate and an alkaline electrolyte such as *Potassium Hydroxide*. *Nickel Metal Hydride* (Ni-MH) batteries are a variant of this with a hydrogen absorbing alloy, *Lanthanum-Nickel alloy* for instance, replacing cadmium as negative plate; This replaces cadmium, a poisonous heavy metal, with a safer alternative while providing much higher power density.

Energy density of batteries is measured using two measures; *Specific Energy Density* in Watt-Hours/Kg measures density against weight and *Volumetric Energy Density* in Watt-Hours/Litre measures density against volume. Lead-acid scores lower in both at 30-50 Wh/Kg and 50-80 Wh/L; Ni-Cd scores 50-80 Wh/Kg and 100-180 Wh/L; Ni-MH scores 60-120 Wh/Kg and 100-300 Wh/L; Li-ion performs the best at 100-200 Wh/Kg and 250-400 Wh/L. However the cost also increases in the same direction with lead-acid being the cheapest; lead-acid batteries are also easily recyclable and 99% of them are recycled. Heavy metals such as Lead and Cadmium are toxic for people (see the chapter on *Paint*).

Lithium ion batteries have the highest energy densities, the reason for its use in devices such as smartphones and electric cars. Akira Yoshino at *Fujisawa* near Tokyo in Japan working for *Asahi Kasei*, a Japanese chemicals giant founded in 1931, invented the lithium ion battery in 1985 building on the work of Stanley Whittingham at *Clinton* near New York City working for *Exxon Research* and John Goodenough at the *University of Oxford*. It was commercialized by *Sony* in 1991. Lithium ion batteries have *intercalated* Lithium compounds such as *Lithium Cobalt Oxide* as anode with *Graphite* anode; Lithium ions are interspersed ("intercalated") between *Cobalt Oxide* layers in the discharged state.

Batteries rank very high as the main electricity provider in the 19th century for high ranking inventions such as telegraph and telephone. Cells played a crucial role in electrolysis and discovery of many chemical elements. Batteries are used in automobiles to start the engine, for the spark plug and to power accessories and many household devices such as toys and remote controls. Lithium ion batteries are used in smartphones and electric cars. Batteries are not ranked as high as electricity as the mainstream electricity is produced by generators using *electromagnetic induction* and not from batteries (see the chapter on *Electricity*); Batteries are not ranked as high as telegraph although batteries clearly provided the electricity for them; The reason is telegraph only requires electricity and not necessarily stored electric power from batteries; It is the difference between *information* and *computer memory*; Computer memory stores the information; However that doesn't mean information necessarily depends on having computer memory (a book will do just fine). Even if batteries weren't not invented, telegraphs would have been invented with electricity from electromagnetic induction and very likely in the same 19th century. The *telegraph*, the idea of sending pure information without physical carriers such as trains or horses, is a more fundamental concept with huge implications for humanity and ranks higher. Batteries are not ranked higher than internal combustion engines, at least not in this edition of the book, as automobiles based on internal combustion engines have proven themselves over the last century and electric vehicles so far haven't lived up to the hype yet. Batteries will likely replace internal combustion engines in automobiles in a few decades, increasing its ranking.

Electric motor

"Sir :—I have lately succeeded in producing motion in a little machine by a power, which, I believe, has never before been applied in mechanics—by magnetic attraction and repulsion.

Not much importance, however, is attached to the invention, since the article, in its present state, can only be considered a philosophical toy; although, in the progress of discovery and invention, it is not impossible that the same principle, or some modification of it on a more extended scale, may hereafter be applied to some useful purpose. But without reference to its practical utility, and only viewed as a new effect produced by one of the most mysterious agents of nature, **you will not, perhaps, think the following account of it unworthy of a place in the Journal of Science."**

In 1831, Joseph Henry of Albany, New York in the US wrote an article named *"On a Reciprocating Motion Produced by Attraction and Repulsion"* with this understatement [emphasis added] to *Silliman's American Journal of Science* describing what would become one of the most important inventions in history. Now it is found in appliances all

around us, washing machines, refrigerator compressors, fans, elevator and escalator, mixer and juicer, water pumps, electric cars, electric trains and trolleys and power tools; Modern cars have multiple of them including one for starting the car, 40 or more in all; Computer hard drives have them. Converting motion into electricity as in hydro-electricity and converting electricity into motion, particularly rotating motion, have been two of the most important energy conversions for the modern world. Hans Christian Ørsted at the *University of Copenhagen* in Denmark discovered *electromagnetism*, the foundation of electric motors in 1820. Michael Faraday at the *Royal Institution* at London in England discovered the generation of current in a moving magnetic field in 1831 and its inverse effect, the foundation of the electric motor, much earlier in 1821. William Sturgeon at London in England invented the first direct current (DC) electric motor and the all important *commutator* that allows the change of direction of ("commute") the current so that the motor rotates in the same direction in 1832. In 1834, Thomas Davenport, a blacksmith, in the US developed an electric DC motor. His patent, the first on electric motors, "IMPROVEMENT IN PROPELLING MACHINERY BY MAGNETISM AND ELECTRO-MAGNETISM", describes the invention:

"For instance, the north pole No. 2 having now become a south pole by reason of its wire being brought in contact with the conductors of the zinc plate, and No. 4 having in like manner become a north pole, its wire having changed its position from the zinc plate to the copperplate, the poles of the galvanic magnets are, of course, now repelled by the poles that before attracted them; and in this manner the operation is continued, producing a rotary motion in the shaft, which motion is conveyed to machinery for the purpose of propelling the same."

The first alternating current (AC) commutator-free induction motors, the most commonly used electric motor in household appliances as we use AC power at home, was invented by Nikola Tesla in the US in 1887 and his patent, appropriately named, "ELECTRO-MAGNETIC MOTOR" explains the invention:

"In a system for the electrical transmission of power, the combination of the following instrumentalities, to wit: a motor composed of a disk or its equivalent mounted within a ring or annular field-magnet, which is provided with magnetizing-coils connected in diametrically-opposite pairs or groups to independent terminals, a generator having induced coils or groups of coils equal in number to the pairs or groups of motor-coils, and circuits connecting the terminals of said coils to the terminals of the motor, respectively, and in such order that the rotation of the generator and the consequent production of alternating currents in the respective circuits produces a progressive shifting of the poles of the motor, as hereinbefore described."

Three phase motors run on three power lines with each 120° out of phase providing three times the power for the increase of one extra power line; This is typically required in industrial motors such as lathes and grain mills. The world's first electrically (that is, with an electric motor) operated public transport, a streetcar by Werner von Siemens, was inaugurated on May 12, 1881 at Berlin, Germany. Brushless DC motors were invented by T.G. Wilson and P.H. Trickey at *Durham* near Washington DC in the US in 1962; In this "smart" DC motor, changing the direction of current is accomplished electronically (*microprocessor*, see the chapter on *Transistor*) instead of a physical brush or commutator so that the rotor continues to rotate in the same direction; It is more efficient due to reduced friction as it has no physical commutator contacts and is used in computer hard drives and airplanes among others.

Electric motors rank high as one of the most ubiquitous and useful inventions. There are more than a billion cars in the world accounting for ten Billion motors conservatively. Assuming an average of three electric motors per household - a refrigerator, a washing machine and a kitchen juicer or mixer - counting households in the US, India, China and Euro zone alone, there are another 2 Billion motors in our homes. With the replacement of internal combustion engines with electric cars the number of electric motors will grow to another Billion in the next two-three decades, increasing its importance and ranking it high.

Photography

"The children were at play last Thursday beside a large pink pagoda on the edge of the town of Trangbang, 40 miles from Saigon. Some 150 yards away, North Vietnamese soldiers were entrenched in the town's marketplace. Two single-engine, propeller-driven Skyraiders of the South Vietnamese air force began dropping bombs on the enemy, and some of the bombs fell near the pagoda. The children ran down a road."

These words from The *New York Times* of June 11, 1972 doesn't quite convey the horror of the *Vietnam war* as much as the picture associated with the article, *"napalm girl"*, a 9-year old running naked down a road, screaming in pain after a *napalm* attack. (If you have never seen this picture, it is definitely worth taking a look before continuing). This picture does more to bring out the horrors of the Vietnam war for

common people than thousands of newspaper articles and books written about the war; A picture is indeed worth a thousand words. Pictures at the *Library of congress* of Nazi concentration camp with emaciated jews brings out the horror of nazis much more than any of the books in the same library. Joseph Nicéphore Niépce at *Chalon-sur-Saône* near Lyon in France took the first primitive photograph, *View from the Window at Le Gras*, in 1827 using Bitumen of Judea after more than 8 hours of exposure to the scene. Louis Daguerre, "father of photography" who worked with Niépce, invented the *daguerreotype process*, the start of modern photography based on silver salts in 1839. His *Boulevard du Temple* photograph is arguably the first high quality photograph and the first photograph to include people. *Daguerreotype* is a plate coated with *silver halide* which is exposed to light via the camera to form the *latent image*. Latent image was developed by mercury fumes and fixed using *Sodium Thiosulphate*. This washes off the extra silver halide, making the picture permanent. Many of the early pictures of famous people such as Abraham Lincoln were made with Daguerreotype. Each Daguerreotype produced one unique photo directly and can't produce copies. There is a website *daguerreobase.org* run by *Netherlands Fotomuseum* containing beautiful collections of daguerreotypes from the mid 19th century and the photos are beautiful, high quality and makes one spend hours on them (I definitely did).

William Henry Fox Talbot of England created the first negative based print photography, *calotype*, in 1841. First color photography was proposed by James Maxwell, of *electromagnetic equations* fame, using three images with red, green and blue filters and combining them to create a composite photo and was taken by his assistant Thomas Sutten in 1861. Sergey Prokudin-Gorsky of Russia produced stunning color images of Russia [for instance, that of Ėmir Bukharskiĭ. Bukhara] using this technique in the early 20th century that is available at the *Library of Congress*. George EastMan's introduction of *Kodak cameras* with flexible roll film instead of plates made it ubiquitous; He started *Eastman Kodak* in Rochester, New York in 1888 to handle photography processing for normal people with the slogan *"You Press the Button, We Do the Rest"*. Kodak introduced *Kodachrome film*, the first widely used color film, in 1935. Eastman did to photography what Henry Ford did for cars, making the cameras a mass market commodity.

First digital camera was invented by Steven Sasson in the US working at Kodak in 1975. Digital cameras convert an image into electrical signals using a *charged coupled device* (CCD) with each pixel made of a capacitor or more common CMOS sensors with each pixel consisting of a *photodiode* for light detection and a *transistor* for amplification. These electrical signals are then processed and converted into a digital image. Nowadays cameras are mostly built into cell phones with CMOS sensors. Kodak was reluctant to capitalize on it as it would have cannibalized their film business and paid the price for it in its bankruptcy of 2013, a backhanded compliment to why innovations matter for even a mature, well established company and even more importantly, for a country.

Photography, as the *"television for time"*, ranks very high in our list of inventions as its contribution to modern life is crucial. Newspapers carry news with photos which make it much more vivid and impactful [For instance, the well known "napalm girl" photo from the *Vietnam war* mentioned above]. It serves to capture the priceless memories of our life. It helps us understand history better, for instance the Nazi concentration camps. Photography has been very important for government functions as well, identification documents such as *driver's license* or *passports* require photographs to work and they are used as evidence in courts. Photography is also the pioneer of movies, TV and videos that provide dominant forms of entertainment in modern life. In fact, Eastman's films were crucial in making the movies by the Lumière brothers and others (see the chapter on *Cinema*). Photography has been a productive invention similar to telegraph which in its turn contributed to many high ranking secondary inventions such as the telephone. However telegraph's impact in communications speed has been far more important to the development of modern civilization and ranks much higher. As can be seen from the daguerreotypes of the 1850s, It deserves its high ranking for making people immortal, at least in history.

Pasteurization

LONDON, May 22. — It is just a year since Nathan Straus, the New York Philanthropist, concluded his campaign of education in this country in favor of the pasteurization of milk, and in the year that has gone by since then there has been a great change in the attitude of both the medical profession and the public authorities toward this great question.

Dr. Joseph Priestley, the Medical Officer of Health for Lambeth, one of the poorest and most populous boroughs of London, is a man who has figures to prove the value of pasteurization. Prompted by the success attained by Mr. Straus in reducing the infant death rate of New York and other American cities, he induced the Lambeth Borough Council to provide him with funds to start a small experimental pasteurized milk depot in the Marsh Ward, which is one of the black spots on the poverty map of London.

... The result was astonishing. When Dr. Priestley began the distribution the infant mortality in the whole borough was 134 per 1,000. In the Marsh Ward and Bishop's Ward, the parts which were served by the depot, it was 187 and 272 per 1,000, respectively. During the first six months the death rate among the infants fed on milk from the depot was only 58, and Dr. Priestley

points out that the depot-fed infants were selected from the worst lives in the district.

... Dr. Priestley has no patience with doctors who declare that scurvy and rickets are caused by pasteurized milk. "It is all the veriest rot." he said emphatically. ... "Some day we will look with as much horror and disgust on a man who drinks raw milk as we would to-day on a man who ate his beef raw."

The *New York Times* of May 23, 1909 reported in an article with the subtitle, "Former Skepticism as to Merits of Pasteurized Milk Is Disappearing. SOME AMAZING RESULTS Demonstrations In Germany and In Parts of London Show Great Reduction In Infantile Mortality.", the impact of a simple invention in saving the lives of babies around the world, in the US, UK and Germany. Nathan Straus of New York City, one of the owners of *Macy's* (a well-known department store in the US), mentioned here played an important role in the pasteurization movement in the US and Europe at the beginning of the 20th century.

Pasteurization is the process of heating milk to a high enough temperature for a long enough time to kill illness-causing germs. It is in some sense a common sense measure of boiling the milk before giving it to children and an invention, particularly for one with life altering consequences for millions, can't get any simpler. FDA defines pasteurization with a table of temperature and processing time duration, 30 minutes at 145° F (63° C) or 1 second at 191° F (89° C) and so on in its *"Sec. 1240.61 Mandatory pasteurization for all milk and milk products in final package form intended for direct human consumption."* According to CDC [emphasis added]:

"Raw milk can carry harmful bacteria and other germs that can make you very sick or kill you. While it is possible to get foodborne illnesses from many different foods, **raw milk is one of the riskiest of all**.

Some people who chose raw milk thinking they would improve their health instead found themselves (or their loved ones) sick in a hospital for several weeks due to infections caused by germs in raw milk. Getting sick from raw milk can mean many days of diarrhea, stomach cramping, and vomiting. **Some people who drank raw milk have developed severe or even life-threatening diseases, including Guillain-Barré syndrome, which can cause paralysis, and hemolytic uremic syndrome, which can result in kidney failure, stroke, and even death."**

Louis Pasteur of France, *"father of germ theory"* who makes multiple appearances in the book, was looking for a way to save wines from spoiling due to fermentation and the resultant acidity in 1864. The prevailing theory in the 1850s was that fermentation was a chemical reaction that had no biological cause. Pasteur's work discovered that microorganisms cause fermentation's chemical reactions to happen in the first place. Armed with this knowledge, he invented *pasteurization* to preserve wines and patented it. However it took a long time for countries to adopt this simple invention due to the notion of "natural, pure milk" being despoiled by heating it and the thinking, "if it is good enough for generations before us then it should be good enough for us". As late as 1938, in the US, disease outbreaks caused by milk were about one-quarter of all food outbreaks. The problem was made worse by the growth of big cities. Milk transported from the countryside to the city would spend a good deal of time at higher temperatures, making it a target for bad bacteria. It took activists such as Straus for this invention to take hold in society. In Sweden and Denmark, pasteurization started in the 1880s. In the US, the first law requiring the pasteurization of milk was passed in Chicago in 1908.

Pasteurization played an important role in saving the lives of children, particularly infants. It ranks high for this life giving ability. It again emphasizes the recurring theme, that simple, incremental inventions matter, *a lot*. Straus alone is estimated to have saved the lives of 240,000 people. It helped reduce infant mortality over the last century and saved hundreds of millions of lives in the last 100 years, ranking it high in our list of inventions.

Television

"London, Friday, May 27. - A successful television test arranged by the Daily Mail was carried out last night by John L. Baird, British inventor of the process, over the 438 miles of telephone line between London and Glasgow. ... Two ordinary post office lines were used. ... The images of the speakers and their movements in London were distinctly seen in Glasgow."

The *New York Times* reported on May 27, 1927 about one of the earliest demonstrations of television over a very long distance at that time of 704 km (438 miles) by British inventor John Baird. It is clear from the report that the moving images "were distinctly seen in Glasgow", with that *television* a reality. In an interview with The *New York Times* in April 1926, John Baird describes his invention:

"Practical Television, therefore, boils down to the very rapid transmission of light dots and a synchronizing mechanism. Suppose we want to transmit a moving picture of an object, say two inches square. We have to

transmit at least ten complete pictures of it every second, and by the most conservative estimate this requires the transmission of about 25,000 light dots a second."

As Baird says above, television transmits a scene as a sequence of continuous images in quick succession; Each image is formed from the scene by continuously scanning the image into scan lines containing light dots [*pixels*, in modern parlance]. Each light dot falls on a photoconductive cell such as a *selenium cell* which converts the light into a continuously varying current. This electric current is amplified and transmitted via a wire as in telegraph or radio as in wireless telegraph. At the receiving side this process is reversed; The electric current from the wire is amplified and fed into a *neon lamp* which produces a continuously varying light dot when seen through the same scan line mechanism used at the transmission and produces the exact image replica. Baird's television scan lines were produced by a spinning mechanical *Nipkow disk,* the mechanical image scanning device invented by Paul Nipkow at Berlin in Germany in 1884, with holes in a spiral pattern. Each hole made a curved scan line, part of the circle the hole was traversing, one at a time from top to bottom through the narrow view port. Number of holes controlled the number of scan lines which required very big disks for transmitting real images (if the description sounds confusing, it may help to look at a picture of the Nipkow disk on the Internet). In Summary, TV requires:

- A way of scanning the scene continuously using scan lines - Nipkow disk, Video camera
- A way to convert the scan line light dots to varying current signal - selenium cell, video camera
- A way to transmit the current signal - telegraph lines or radio
- A way to convert the current back to varying light - neon lamp, Cathode ray tube (CRT)
- A way to redraw the scan lines - Nipkow disk, CRT

The Nipkow disk is an ingenious device used to convert the scene into [curved] scan lines continuously. This was critical in Baird and others' mechanical television. The era of electronic television started when video camera tubes replaced Nipkow disks. *Iconoscope* [Emitron], invented by Russian born American Vladimir Zworykin, was one of the earliest video cameras that converted an image falling on it into a sequence of scan line currents using the cathode ray tube. This was developed into the super-emitron in Europe to amplify the image current for a better quality signal by splitting the image capture part and the cathode ray scanning part with an amplification in between them. First transistor TV *TV8-301* was launched by Sony in May 1960; It was an 8-inch portable television set. First experimental BBC programme was broadcast in September 1929 using Baird's system. In 1939, NBC and CBS started broadcasting regularly scheduled TV shows in the US using electronic television in New York City. Earlier CBS started an experimental mechanical TV station in 1931 in New York City that stopped in 1933. In Brazil, the first TV broadcast in South America started in September 1950.

People spend a lot of their time in front of TVs all over the world. According to the 2019 *American Time Use Survey*, Americans spend 2.81 hours watching TV and 0.27 hours (16 minutes) reading. According to the *New York Times*, Average American watches more than 5 hours of TV per day working out to 30% of their waking lives. An invention which takes up 20-30% of people's productive [not spend sleeping] life should count very high in the list for better or worse. Live sports television is one of the biggest entertainments of all time; The 2018 football (soccer) world cup was watched by more than 3.5 Billion people, almost half the people in the world; *Super bowl* (American "football" championship game) 30 second advertising spots cost around 5.5 million dollars. Television is one of the most powerful public communication media as can be seen from the power of *Fox news* in the US in shaping public opinion. Its importance is comparable to radio and newspapers in public broadcast reach and impact. Radio and newspapers have been with us for much longer and reach more people in the world. (I grew up in India without a TV till 2000). Radio is a more foundational invention with non-broadcast uses, ranking it much higher than television.

Refrigerator

Thomas Midgley Jr was an important character in Refrigeration. The *Smithsonian Institution* quotes author Bill Bryson as saying that he had "an instinct for the regrettable that was almost uncanny."

- One thing he invented was *tetraethyl lead* and leaded gasoline for improving engine efficiency (anti-knocking agent, see the chapter on *Petroleum*); *Lead* in gasoline, or in general for that matter, was later known to be a serious poison, particularly for children's nervous systems.

- Second one was *chlorofluorocarbon* (CFC), the first non-toxic and non-flammable refrigerant, in refrigerators instead of *Ammonia*; It was later known to cause serious ozone layer depletion and was banned by the 1987 *Montreal protocol*.

A modern refrigerator uses a *refrigerant*, the working fluid, to cool the inside of the fridge. It removes the heat from the fridge when it goes from liquid to gaseous state. This gas is compressed by the *compressor* [using electricity, work done on the gas] and turns into a liquid by losing heat to the outside air in condenser coils. Carl Von Linde of Germany, the "father of refrigeration", invented this mechanism called *vapor-compression refrigeration*, used in current fridges for liquefaction of the gases in 1876. He started the commercial refrigeration business with Ammonia as the working fluid. The company he started, appropriately named *Linde*, is still a "leading global industrial gases and engineering company with 2020 sales of $27 billion (€24 billion)". Earlier, William Cullen at the *University of Edinburgh* in the UK demonstrated that boiling *diethyl ether*, that is turning a liquid into a gas, can remove the heat from surroundings in 1756. (This is the same *diethyl ether* that became the first anesthetic making its second appearance, see the chapter on *Anesthetic*).

As Ammonia is corrosive, CFC replaced Ammonia in home refrigerators. In the 1980s, CFCs were known to cause ozone layer depletion and hydrofluorocarbons such as *tetrafluoroethane* [R-134a/HFC-134a] were started to be used as the new refrigerant. However, HFC-134a is a potent greenhouse gas with 1,430 times the potential of Carbon Dioxide. The US *Environmental Protection Agency* (EPA) mandated that HFC-134a in turn be replaced with hydrocarbons such as *isobutane* [R-600a] and *propane* [R-290] with much lower greenhouse warming potential starting Jan, 2021. European Union also introduced similar *'f-gas' regulation* in 2014 to phase out HFCs by 2030. Ammonia is still used in industrial scale refrigeration systems due to its low cost and exceptionally high efficiency. Household refrigerators started in the US in 1915 as additional units for an existing icebox. Fully independent fridges started to become commercially available in 1925. Mechanically refrigerated trucks started operation in 1927 by *Golden Gate motor transportation company* in California. Air conditioners work using the same principle as the refrigerators, vapor-compression refrigeration using a working fluid. First air conditioner was invented by Willis Carrier in 1902. In 1915, he founded *Carrier* to commercialize the air conditioning systems; It developed the modern air conditioning system in 1933 using "using a belt-driven condensing unit and associated blower, mechanical controls, and evaporator coil". Carrier remains the leading company in air conditioning with 18.6 Billion in revenue in 2019.

Refrigeration revolutionized the economy by being able to move perishable goods over long distances in refrigerated containers replacing ice boxes completely. Refrigeration is critical for many medicines including vaccines such as the one for coronavirus. In 2015, an estimated 164 million domestic refrigerators and freezers were in operation in the United States according to the EPA with more than 10 Million new units produced every year. Almost 100% of US homes have a refrigerator with 23% having two or more. Germany also has almost 100% coverage in her homes and it is very likely to be the case in other developed countries. Even developing countries such as India have more than 30% refrigerator coverage in their households. Ice cream market, which owes its existence to refrigeration, alone will be worth around 100 Billion dollars by the end of this decade (2020s). It has eliminated quite a bit of manual labor such as salting and pickling for meat and vegetables by keeping them fresh without any preprocessing.

Refrigerator ranks high in the list of modern inventions as the mainstay of modern household kitchen and industry; This is one of the inventions that touch our lives daily. In 2012, *Royal Society* named refrigeration as the most important invention ever in food and drinks, outscoring even fishing nets and ploughs. One of the reasons famines such as the *Great Irish Famine* of 1846 that killed more than a million people have become uncommon, apart from fertilizers (see the chapter on *Fertilizer*), may be the invention of refrigeration as it enables us to bring food from faraway places using refrigerated trucks and refrigerated shipping containers. Air conditioning is important in tropical regions such as Africa, India, UAE, Singapore or in the southern United States. With the proliferation of data centers, cloud computing and even crypto coin mining (an unfortunate and unproductive development in computing and finance), air conditioning plays a key role in information technology securing its high rank.

Airplane

Steven Pinker quotes Louis CK, an American comedian, in his book *Enlightenment now* which succinctly and humorously explains the importance of airplane as an invention that is worth quoting at length, to shake off the "anaesthetic of familiarity" as Richard Dawkins would say [emphasis added]:

"Flying is the worst because people come back from flights and they tell you...a horror story...They're like: "It was the worst day of my life. First of all, we didn't board for twenty minutes, and then we get on the plane and they made us sit there on the runway..." Oh really, what happened next? Did you fly through the air incredibly, like a bird? Did you partake in the miracle of human flight you non-contributing zero?! You're flying! It's amazing! **Everybody on every plane should just constantly be going: "Oh my God! Wow!" You're flying! You're sitting in a chair, in the sky! You're like a Greek myth**

Right now. ... Air travel's too slow. New York to California In six hours. That used to take 30 years, To do that, And a bunch of you would die On the way there. You'd get shot in the neck With an arrow and you'd go- And fall down. And the other passengers Would just bury you And put a stick there with your Hat on it and keep walking."

The *Wright Flyer* built by Wright brothers was the first successful heavier-than-air powered airplane that flew on Dec 17, 1903 at *Kitty Hawk* near Washington DC, the US capital city. Wright Flyer is now displayed at the *Smithsonian National Air and Space Museum* in Washington DC. First commercial flight in the world was the *St. Petersburg-Tampa Airboat Line* that flew across *Tampa Bay* at the *Gulf of Mexico* in the US near Havana in Cuba on January 1, 1914. Oldest extant airline is *Koninklijke Luchtvaart Maatschappij* (KLM, "Royal Aviation Company") which started in 1919. William Boeing started *Boeing,* one of the most successful commercial airlines manufacturers in the world, in 1916 and *United Airlines* is an outgrowth of Boeing. The *Airbus consortium* was founded by European countries in 1970 to compete with American aircraft manufacturers such as Boeing. In 2019 *Airbus* became the biggest manufacturer with 863 aircraft deliveries and revenue of € 70.5 billion (83.3 billion US dollars) mainly after self goals by Boeing such as the *737 Max* that has killed 346 people in its short, five year life.

Airplanes are literally lifted by air molecules and [obviously] won't work without an atmosphere of air, in space for instance. A 747 in flight diverts its [takeoff] weight, about 400 tons [weight of around 100 elephants], in air every second. An MIT website on *Theory of Flight* says [emphasis added]:

"In order for an aircraft to rise into the air, **a force must be created that equals or exceeds the force of gravity**. This force is called lift. In heavier-than-air craft, lift is created by the flow of air over an airfoil. The shape of an airfoil causes air to flow faster on top than on bottom. The fast flowing air decreases the surrounding air pressure. **Because the air pressure is greater below the airfoil than above, a resulting lift force is created."**

A Boeing 747 burns jet fuel at the rate of *20 litres [5 gallons] per minute*. A flight from New York to London with a distance of 5,570 km [3,461 miles] burns 70,000 litres [17,000 gallons] of fuel; A flight from San Francisco to Singapore with a distance of 13,574 km [8,434 miles] uses up 170,000 litres [42,000 gallons] of fuel in a single flight. This is the reason intercontinental aircrafts such as *Boeing 747* or *Airbus A380* require huge fuel tank capacities; Boeing 747 fuel tank capacity is 240,000 litres [63,500 gallons]; Airbus fuel tank capacity is 310,000 litres [82,000 gallons]. For contrast, an average car has a fuel tank capacity of 60 litres [15 gallons], making these behemoths' fuel capacities as big as that of 5,500 cars combined. Surprisingly, air travel is more energy efficient than automobiles. A *Wall street Journal* article notes [emphasis added]:

"With cars, you measure miles per gallon in how far the vehicle can travel on one gallon of gas. With airlines, it's how far one seat (occupied or not) can travel on one gallon of jet fuel. And U.S. major airlines average about 64 mpg, according to calculations using Department of Transportation data for 2009. **For each gallon of jet fuel, airlines could, on average, fly one seat 64 miles.** That's better than your SUV or hybrid car, unless you pack lots of people into the car."

It has had a huge impact in our modern life making our planet a global village. It has made people more productive by reducing travel times even to remote corners of the world. According to *ICAO*, 4.3 Billion people, almost 30% of the people in the world assuming a roundtrip per person, flew in 2018. Airplanes also play a critical role as military technology. Compared to its predecessors ship and railways, airplanes have only been with us for a century and both ship and railways rank higher, in the top 50, as they have played a much bigger role in human history; Airplane ranks relatively high, in the top 20, in the second half of the book.

Paint

"ROME, Dec 12. - The government to-night received a dispatch from Florence containing the information that the 'Mona Lisa' popularly known as 'La Joconde' and 'La Gioconda', the celebrated picture that was stolen from the Louvre, Paris, in August, 1911, had been recovered in the city. Prof. L. Credaro, Minister of Public Instruction and Fine Arts, announced the fact in the lobby of Italian Parliament. There is no doubt as to the picture's authenticity, for it has been identified by Dr. Conrado Ricci, General Director of Fine Arts at the Ministry of Public Instruction, who is one of the most competent art critics in Italy. It is now in the Prefecture of Police at Florence. The man who stole it is under arrest."

The frontpage of The *New York Times* of Dec 13, 1913 reported excitingly about the recovery of *Mona Lisa,* the Renaissance masterpiece created by great artist Leonardo Da Vinci at Florence, Italy between 1503 and 1519; It was stolen from *Louvre Museum* in Paris two years earlier. It was valued for insurance at *868 Million* in 2021 dollars making it the most expensive painting in the world. Paint as a human

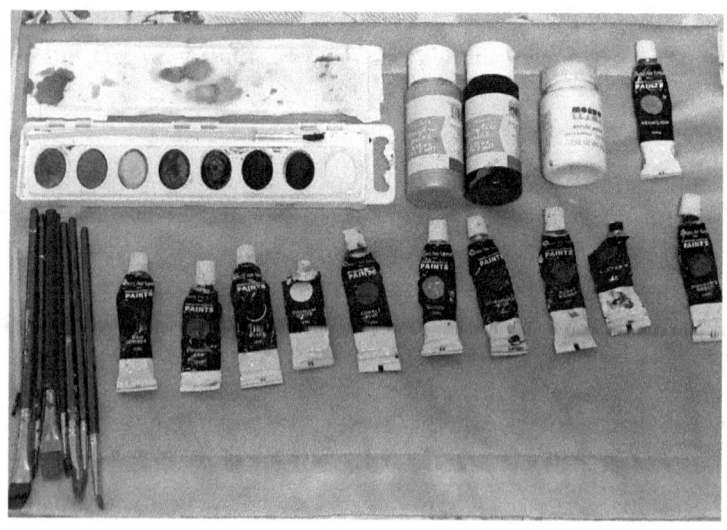

invention was critical for humanity's cultural heritage in the world as well as playing a more mundane but important role as household and automobile paints. From prehistoric times, naturally occurring *Red Ochre* and charcoal have been used in cave paintings such as the ones from *Altamira Cave*, overlooking the *Bay of Biscay*, in northern Spain dated to as early as 36,000 years ago. *Egyptian blue* is found in egyptian paintings such as the beautiful tomb painting of *Nebamun* from around 3,350 years ago now at the *British Museum*. It is made by mixing sand, lime, and copper ore and heating them together. *Lead white*, lead(II) carbonate, is one of the oldest white pigments known for around 2,300 years and heavily used in medieval paintings. It was made by sealing strips of lead in earthenware pots with vinegar and covering with manure. *Ultramarine* is a deep blue color pigment which was originally made by grinding *lapis lazuli*, a mineral from Afghanistan, into a powder; It was the most expensive pigment used by Renaissance painters. The first modern synthetic pigment, Prussian blue - *iron(III) hexacyanoferrate(II)*, was invented around 1706 by Johann Diesbach (by mistake); It replaced the very expensive ultramarine and made blue affordable in paintings. *Blueprints* derive their name from Prussian blue used in the early reproduction process of technical drawings. *Whitewash* is a cheap white paint, made from slaked lime, *calcium hydroxide*, used for exteriors and as interior painting in developing countries. Whitewash forms the white layer by absorbing carbon dioxide in the air to form the calcite, *calcium carbonate*, layer. *Acrylic resin*, the binder holding the pigments in place in acrylic paints that dries very quickly compared to oil paints with oils such as linseed oil used as the binder, was invented by Otto Röhm at *University of Tübingen* near Stuttgart in Germany in 1901.

An analysis of the colors in *Monalisa* at the appropriately named website *monalisa.org* turns up colors like so:

"Probe #10: Dress

Layer 1: Red ground of earth pigments, **natural ochres, lead white**, minium, bone black, carbon black, **smalt**.

Layer 2: Grey layer priming of lead white, calcite, carbon black, and granules of red ochres.

Layer 3: Layer of carbon black, lead white, and umber.
Layer 4: Dark grey layer of carbon black, umber, and lead white.
Layer 5: Thin grey layer of calcite, and carbon black...."

Lead-based paints were banned for residential use in 1978 in the US and UK. Apart from lead white as a pigment, lead was added to paint to speed drying and resist corrosion. Due to the toxicity of lead, now lead white is replaced with *Titanium Dioxide* in white paints. Lead is especially damaging to children. Paint containing *lead(II) chromate* ("chrome yellow") is an extremely toxic carcinogen as it contains the double whammy of heavy elements, lead and chromium.

Sherwin-Williams with an annual revenue of more than 18 Billion dollars (2020) is the biggest paint manufacturer in the world; It was founded in 1866 by Henry Sherwin and Edward Williams at Cleveland on *Lake Erie* overlooking Detroit in the US. *PPG industries* with a revenue of more than 13 Billion dollars (2020) is second; It was founded as the *Pittsburgh plate glass company* in 1883 by John Ford and John Pitcairne Jr at *Creighton* near Philadelphia in the US. *AkzoNobel* at Amsterdam, Netherlands that produces the famous *Dulux paints* is third with a revenue of 10 Billion dollars. *Winsor & Newton*, founded in 1832 by William Winsor and Henry Newton at London, England, is one of the well known art paintings manufacturers in the world.

There were around 40 Billion litres of paint produced in the world in 2019. More than two billion houses in the world are painted in one form or another. More importantly, paints were crucial for our shared cultural heritage in great paintings such as the *Nebamun painting* or the *Monalisa*. It has been part of human history for at least 36,000 years, making it older than the wheel, agriculture and many other foundational inventions; It is ranked relatively lower as it is part of our cultural history, a relative luxury, as compared to the foundational inventions addressing our basic necessities. That said, its long history still increases its human impact and ranks it relatively high, in the top 20, of the second half of the book.

Tea and coffee

"Only half a century ago it was commonly believed that the 'Camellia theifera' or tea plant, was confined to a belt of land within the Chinese Empire, lying between twenty-five and thirty-three degrees of north latitude. It chanced, however, that in the year 1823 an enterprising Scottish trader, named Robert Bruce, conveyed a cargo of miscellaneous articles up the rivers of Assam to the town of Rungpore, at that time the most important collection of huts in those parts. This Mr. Bruce being something of a botanist, discovered to his great surprise, that the lofty trees beneath whose grateful shade he was one day reposing were of the same genus, and even species, as the lowly tea-producing shrub of China. Many of these trees exceeded forty feet in height, while not a few attained to sixty feet, though they seldom measured more than a yard in circumference. In answer to his inquiries, he learned that a decoction of tea leaves had long been a favorite beverage of the Singphos, the tribe inhabiting the district, and that the plant was indegenous to all the portion of Assam which is included between twenty-seven and twenty-eight degree north."

An article in the *New York Times* of Oct 1, 1876 titled "Indian Tea" describes the discovery of Indian tea in India's northeastern state of

Assam bordering China and Burma (Myanmar) by a British trader. Tea is indigenous to parts of China, Indian northeast, Tibet and Myanmar. Food, to the extent it was invented, is covered in the chapter on *Agriculture*; However tea and coffee are not exactly food as they typically have no nutritional value (*Carbohydrate, Protein, Fat* or *Vitamins*) and consumed only for their effect as nervous system *stimulants;* For this reason, they rank on their own as an independent invention. Earliest evidence for tea, dated to around 2,150 years ago, was discovered in the tomb of ancient *Han* emperor Jing Di near *Xi'an*, a city of 12 Million near Chengdu and Wuhan, in western China. In the Tang dynasty period (618-906 CE) tea was popular among northern Chinese and people to the west such as *Uighurs*. First book entirely on tea, *Ch'a Ching* ("The Classic of Tea"), was written by Lu Yu in the late 8th century. In 1606, the first consignment of tea to Europe was shipped from China to Holland via *Java* (Indonesia), a dutch colony at the time. Tea soon became a fashionable drink among the Dutch, and spread to other countries in continental western Europe. The first dated reference to tea in England is from an advert in a London newspaper, Mercurius Politicus, from September 1658. It announced that 'China Drink, called by the Chinese, Tcha, by other Nations Tay alias Tee' was on sale at a coffee house in London. *Twinings* was founded by Thomas Twining, of England, who opened Britain's first known tea room, at Strand, London, in 1707 that still operates today after 300 years. In 1707, *Twinings Gunpowder Green Tea* sold for $2,200 per kg (in 2021 dollars) or 100 times the current tea price.

Coffee traces its origin back centuries to the ancient coffee forests on the *Ethiopian plateau*. By the 15th century, coffee was being grown in the Yemeni district of Arabia and by the 16th century it was known in Persia, Egypt, Syria and Turkey. Around this time, Coffee was available in public coffee houses (*"qahveh khaneh"*) in cities across the middle east. By the mid-17th century, there were over 300 coffee houses in London alone. *Alois Dallmayr*, a coffee house chain in Germany, traces its roots to 1700. Tea was the favored drink in the New World until 1773, when the colonists revolted against a heavy british tax on tea; Known as the *Boston Tea Party*, a very important event in American Revolution, it changed the American drinking preference to coffee.

One cup of Tea has around 25 mg (around 1 oz) of *Caffeine* and One cup of Coffee has around 100 mg (around 4 oz). The recommended daily caffeine intake is less than 400mg. Caffeine is a *purine*, a class of molecules that includes *Adenine* and *Guanine* in DNA and RNA; It resembles *adenosine* that's naturally present in our brain. Caffeine works as a *stimulant enabler* rather than being a stimulant itself. Caffeine fits into adenosine receptors in the brain blocking them off. Brain's neurotransmitters such as *dopamine* work more effectively when the adenosine receptors are blocked, and the surplus adenosine in the brain cues the adrenal glands to secrete adrenaline. There is clear scientific evidence that an intake of at least three cups of tea per day reduced the risk of coronary heart disease mainly due to its antioxidants such as *catechins* in green tea and *theaflavins* and *thearubigins* in black tea that undergoes oxidation during its fermentation from green tea. There is also a semi-fermented tea, *Oolong*, that is mid-way between green and black tea.

According to the *Food and Agriculture Organisation* (FAO), in 2013, China was the largest tea producer with 1.9 Million tonnes (around 40% of the world total), while India was the second largest producer at 1.2 Million tonnes in 2013 (20%) followed by Kenya at 0.4 Million tonnes (5%) and Sri lanka at 0.3 Million tonnes. Brazil is the biggest coffee producer with 4.1 Million tons (around 40% of the world) followed by Vietnam and Colombia with 1.7 Million tons (20%) and 0.9 Million tons (10%) respectively. One of the biggest coffee chains, *Starbucks*, founded in Seattle in the US in 1971, had revenue of 26.5 Billion dollars in 2019.

Tea is the world's most consumed drink, after water. Tea is consumed by more than two-thirds of the population in the world. Tea has a global market of 200 Billion dollars a year. The global coffee market is valued at 466 billion dollars. Both drinks have a long history of more than 500-2,000 years, to go with the current impact and importance. As drinks enjoyed by around five Billion people in the world and importantly with positive benefits unlike alcohol, an anti-invention, tea and coffee rank high as one of humanity's greatest inventions.

Cinema

"Thomas A. Edison began thinking about the development of motion pictures in 1888 after studying the successful motion-sequence still photographic experiments of Eadweard Muybridge and Étienne-Jules Marey. ... By June of 1891, Dickson *[Edison's assistant]* produced a series of successful experimental motion pictures that were shown to visiting groups at the Edison laboratory in New Jersey. ... Over the next two years Dickson worked to perfect the two basic machines required for successful motion pictures: **a device to record moving images, which he and Edison called the Kinetograph; and a machine to view the results, which they called the Kinetoscope.** A major problem that slowed Dickson's work in the beginning was the nonexistence in the commercial marketplace of another essential invention—motion picture film stock. After **Eastman Kodak began supplying quantities of reliable film stock** in the fall of 1893, the road to commercial development of the movies was opened."

The *Library of Congress* in Washington DC in the US has documented the history of cinema for us given that it was a relatively recent invention and the library itself was involved in the early process of developing a *"phonograph for the eye"* as the protector of Intellectual property. Thomas Alva Edison and his employee William Dickson in the US developed the *kinetoscope*, a device for a single person to view moving images through a peephole. Edison's *Kinetoscopic Record of a Sneeze*, received in the Library of Congress on January 9, 1894 is the earliest surviving copyrighted motion picture. Lumière brothers, Auguste and Louis Lumière of France invented cinema in 1895 with a crucial breakthrough, projecting moving images onto a large screen for shared viewing. Lumière brothers' device *Cinématographe* projected ten short films with a total showtime of 15 minutes with a ticket price of one *franc* [equivalent to around five US dollars now]. Later Edison switched to public projection of movies as well with his *Vitascope* projector. *The Great Train Robbery, produced* by Thomas Edison in 1903 with a length of 13 minutes was a path breaker with a story plot that would define cinema proper; It was a commercial success and it is now preserved in the Library of Congress (and available to watch online). As films had no copyright in those times, Edison submitted a copy to the Library of Congress to protect his intellectual property. These early movies were shown in *vaudevilles*, a theatre showing a variety of entertainment such as live music, dance and comedy. Earlier in 1888, Louis Le Prince made a short film, *Roundhay Garden Scene*, of a few seconds using a camera and Kodak's film although it was never projected for public viewing; It is now preserved in the *Science Museum* at London and was made into a film in 1933 by the museum. From these beginnings in a few years Cinema grew to a feature-length film in 1906, interestingly outside of major film centers in the US and Europe, in Australia. UNESCO's memory of the world register's entry for the 60 minutes film *The Story of the Kelly Gang* says: "The Story of the Kelly Gang, directed by Charles Tait in 1906, is the first full-length narrative feature film produced anywhere in the world."

First feature length color film was a british documentary *With Our King and Queen Through India* produced using *kinemacolor*; This uses black and white film with red and green filters for filming and projection, reminiscent of Maxwell's color photography technique (see the chapter on *Photography*). *Technicolor* was a series of color processes that produced color in the projection reel itself using chemical processing and could be played with a normal projector. This produced classics such as *The Ten Commandments* (1923) and *The Wizard of Oz* (1939). First feature length movie with sound ["talkie"] was *The Jazz Singer* in 1927 produced in the US. Now India is the biggest movie producer and consumer in the world with around 2,000 cinemas a year and more than 2 Billion cinema tickets sold. Tom Sherak, a former president of *Academy of Motion Picture Arts and Sciences*, the organization which awards Oscars said in an interview about the role of the cinema in the society [emphasis added]:

"**Film is a reflection of society, both present and past.** I think the film and it's innovations sometimes has to catch up to society but sometimes it leads society too. Movies are stories, movies are people who come out with ideas about something they want to say, something they want to tell someone. Movies are a form of communication and that communication, **those stories, come from societies- not just where society is presently and what it's doing now- but where society has been.** It's been that way for as long as movies have been around!"

Cinema is the most dominant form of entertainment in the world with it surpassing more than 100 Billion dollars in 2019. In comparison, books make up less than half of this number. It has been our predominant form of storytelling with huge cultural impact. Cinema serves as the record of our recent history and an instructive medium. Most pleasurable way to understand 1930s America is to watch movies from that era. First few minutes of *Saving Private Ryan* makes one understand the messy affairs of war better than reading about it in a book. As a prerequisite for cinema, photography ranks higher. Television also ranks higher as it serves an additional purpose as a public broadcast medium apart from entertainment.

(Image credit:Warner Bros.)

Phonograph

"The invention consists in arranging a plate, diaphragm, or other flexible body capable of being vibrated by the human voice or other sounds, in conjunction with a material capable of registering the movements of such vibrating body by embossing or indenting or altering such material, in such a manner that such register-marks will be sufficient to cause a second vibrating plate or body to be set in motion by them, and thus reproduce the motions of the first vibrating body."

One day before Christmas in 1877, Thomas Edison filed his patent for "IMPROVEMENT IN PHONOGRAPH OR SPEAKING MACHINES". The phonograph for the first time enabled sound to transcend space and time, doing what Gutenberg's printing press did to writing for sound, a much more natural medium for people including the illiterate. Edison used a primitive recording medium, *tinfoil wrapped around a cylinder*, with indentations on tinfoil serving as the recording. Alexander Graham Bell, of telephone fame, and others working with him at *Volta laboratory*, which would later become *Columbia records*, introduced *wax coated cylinders* as the recording medium with engravings on wax serving as the recording that highly improved the quality. Edison immediately after

learning this set to work producing his all wax cylinder phonograph and patented the "PHONOGRAPH RECORDER" in 1888. Emilie Berliner at Washington DC patented "GRAMOPHONE", a form of phonograph, that used *flat discs with spiral grooves* in 1887; It could hold around 2-3 minutes of music. The phonograph disc record was the primary medium used for music reproduction throughout the 20th century. Berliner experimented with a variety of materials for records, including glass, celluloid, and hard rubber and eventually used *Duranoid rubber*. The main advantage of the disc as compared to wax cylinder was that duplicate copies could be mass produced from a single master. *Long Playing* (LP) Vinyl records that transformed the industry and remained the standard for much of the 20th century were introduced by Columbia records in 1948; This 12 inch PVC disc had a playtime of 45 minutes, compared with 8 minutes for the two sides of the conventional record. There is an editorial in the The *New York Times* of Nov 7, 1877 on phonograph [emphasis added]:

"...Instead of libraries filled with combustible books, we shall have vast storehouses of bottled authors, and though students in college may be required to learn the use of books, just as they now learn dead languages, they will not be expected to make any practical use of the study. Blessed will be the lot of the small boy of the future. He will never have to learn his letters or to wrestle with his spelling-book, and **if he does not revere the name of the inventor of the phonograph, he will be utterly destitute of all gratitude.**"

I think this very much applies to the school children in the current (2020) *coronavirus pandemic* taking online school classes. This also implies that the phonograph was well known and anticipated even before Edison filed his patent on Christmas Eve of that year.

Philips, founded in 1891 by father-son duo Frederik and Gerard Philips at Eindhoven in the Netherlands near Cologne in Germany, introduced *cassette tape* in September 1963. Cassette tapes are made of a *polyester* plastic film with a magnetic coating of *iron or cobalt oxides*; Signal is encoded in the magnetic domains of the oxide particles. *Sony*, founded in 1945 by Masaru Ibuka at Tokyo in Japan, introduced the *Compact Disc* (CD) in 1982. CD is a digital recording device that works optically; In its tracks, a flat reflective area represents a binary one, while a non-reflective bump represents a binary zero and a laser is used to read the disc (see the chapter on *Laser*). Nowadays music has become completely digital, stored in computers (including smartphones).

Original microphone came as part of the telephone invention (see the chapter on *Telephone*). *Carbon microphone* was independently invented around 1878 by Edison, Berliner and David Hughes of London in England. It works by using a *diaphragm electrode*, a fixed electrode with carbon granules between them; Varying carbon electrical resistance due to sound waves encodes the sound signal as varying electric current. The first *dynamic* (moving coil) microphone, the common microphone used by public speakers on stage, was patented in 1874 by Ernst Siemens of Germany but it caught on only in 1931, with the release of Western Electric *618A Electrodynamic Transmitter*. In dynamic microphones, a coil of wire surrounds a magnet and is connected to a diaphragm which vibrates in response to incoming sound waves. When sound waves hit the diaphragm, the coil oscillates back and forth past the magnet, generating a current which creates the audio signal. *Audio-Technica* website explains:

"A dynamic microphone operates like a speaker in reverse. The diaphragm is moved by changing sound pressure. This moves the coil, which causes current to flow as lines of flux from the magnet are cut. So, instead of putting electrical energy into the coil (as in a speaker) you get energy out of it. In fact, many intercom systems use small speakers with lightweight cones as both a speaker and a microphone, by simply switching the same transducer from one end of the amplifier to the other! A speaker doesn't make a great microphone, but it's good enough for that application."

Phonograph ranks very high as the father of all forms of music recording today from records, cassettes, CDs to digital devices. Its cultural significance in human history is very important and it spawned the huge music industry. It ranks just below cinema, as cinema includes both video and sound, for its rich historical impact.

Rocket

"That's one small step for man, one giant leap for mankind."

Neil Armstrong said this on first reaching the moon on July 20, 1969, which clearly shows the importance of this achievement for humanity. We for the first time ventured out of our piece of rock in the universe to another piece nearby in the history of humanity. The invention which enabled this landing, rocket, richly deserves its rank in this list. Wernher von Braun designed the *Saturn V* rocket which took man from his home, Earth, to our nearest neighbor in space, moon, in 1969; It is a journey with significance comparable to Columbus' arrival to the New World in 1492. In 3020 AD many of the inventions in this list may be forgotten, similar to us forgetting cassettes, pony express or pretty soon even disk drives. However man's first journey to the moon will still be remembered and taught to children in Mars classrooms and beyond. Von Braun, a German Engineer, developed the dreaded

German *V-2 rockets* for the Nazis. After World War II, he played a pivotal role in rockets for the US space programs. Sergei Korolov played the role of von Braun for the USSR which launched *Sputnik*, the first artificial satellite in 1957 using the *Sputnik rocket*, a modified R-7 missile (see the chapter on *Satellite*). Russian rocket scientist Konstantin Tsiolkovsky published the "rocket equation" in 1903; It describes a rocket, a device that self-accelerates using thrust by expelling part of its mass with high velocity and moving itself forward due to the *conservation of momentum*. Robert Goddard of the US achieved the first successful flight with a liquid-propellant rocket in 1926, fueled by liquid oxygen and gasoline. There is a well known editorial from the *New York Times* criticizing Goddard's achievements in 1920, in a backhanded compliment, This serves as the contemporary record of Goddard's inventions.

"That Professor Goddard does not know the relation of action to reaction, and of the need to have something better than a vacuum against which to react ... Of course he only seems to lack the knowledge ladled out daily in high schools."

On 12 April 1961, a *Vostok-K* rocket was used to launch *Vostok 1*, the first manned spaceflight, which made Yuri Gagarin the first person to fly in space. The Saturn V rocket used in the moon landing was 111 meters (363 feet) tall, about the height of a 36-story-tall building, and 18 meters (60 feet) taller than the *Statue of Liberty*. Fully fueled for liftoff, the Saturn V weighed 2.8 million kilograms (6.2 million pounds), the weight of about 400 (adult) elephants. The Saturn V could launch about 43,500 kilograms, about seven elephants, to the moon (the other 300 odd elephants are fuel!). The first Saturn V mission, *Apollo 4*, was launched in 1967. On July 20, 1969, *Apollo 11* was the first mission to land astronauts Neil Armstrong and Edwin Aldrin on the moon.

Rockets are a foundational invention as it helps us reach beyond our home planet. Space travel by private companies will likely be realized in the lifetimes of people reading this book. Mining asteroids and other objects for minerals may become common in a few centuries as well. It may be called the ships of the 3rd millennium, exploring space instead of sea. In fact Von Braun himself said [emphasis added]:

> "It filled me with a romantic urge. Interplanetary travel! Here was a task worth dedicating one's life to. Not just stare through a telescope at the Moon and the planets but to soar through the heavens and actually explore the mysterious universe. **I knew how Columbus had felt.**"

Moon landing and human space flights can be compared to the discovery of new world's in the 15th century. It changed our destiny earlier [for the worse in case of native Americans]. Space flights have the future potential for us to expand to other planets and satellites in the solar system like Moon and Mars, particularly if climate change were to severely impact our motherland later in this century or in the early 22nd century. It can even become literally a matter of survival (I hope it won't come to that for the sake of our grandchildren and their children). It is one of the biggest life insurance policies for humanity. However climate change may turn out to have a milder impact due to our proactive actions or our understanding of the process improves as we reach closer to the second half of the century. We haven't reached Mars yet and the Moon remains unoccupied more than 50 years after our landing. The *Mayflower* (for non US readers: an iconic ship in American history carrying early settlers to the US fleeing religious persecution called *pilgrims*) reached the new world in 1620, more than 125 years after Columbus' first visit to the new world. That means we still have time to go, at least till the end of the century before we start to see the real impact. Once we see the impact, Rockets may get a ranking very close to ships [that is in the top 15] in this list. In the meantime they get their ranking for a more practical achievement, sending artificial satellites such as GPS satellites to space. As a critical invention required for satellites, they are ranked higher than satellites and GPS.

Satellite

"SOVIET FIRES EARTH SATELLITE INTO SPACE; IT IS CIRCLING THE GLOBE AT 18,000 M.P.H; SPHERE TRACKED IN 4 CROSSINGS OVER US"

The *New York Times* headline on Saturday, October 5, 1957 screamed about the launch of *Sputnik*, the first man-made satellite for our earth. Sputnik [literally "satellite" in Russian], the artificial moon was 901 km (560 miles) above the earth and had a diameter of 56 cm (22 inches) and a weight of 83 kg (184 pounds) - roughly ⅓ of a person's height and about his weight; It circled the earth once every hour and thirty five minutes, more than 15 times a day. *Sputnik 2* was launched on Nov 3, 1957 one month after Sputnik 1. Third one and the first from the US launched on Jan 31, 1958. Fast forward 63 years, less than one human lifetime, we have more than 6,000 satellites [including 3,200 defunct] orbiting the earth. The US owns 1,308 satellites as of April

2020, China has 356 followed by Russia at 167. As of Feb 2021, *SpaceX*, founded in 2002 at Los Angeles in the US, owned more than 1,000 satellites as part of its *starlink* satellite broadband system.

One of the greatest among them is the *Hubble space telescope*, a satellite working to understand our knowledge of the whole universe; It helped us estimate the age of the universe itself by observing the rate of expansion and better value for *Hubble's constant*. Upon its launch in April 24, 1990, it had a flaw in primary mirror blurring its vision threatening the abandonment of billion dollar telescope; However NASA found a work around, *Corrective Optics Space Telescope Axial Replacement* (COSTAR), effectively a corrective eyeglass for its primary mirror and ever since it has performed its mission very well; It has produced the most beautiful pictures of our universe. The *James Webb telescope*, its successor, is expected to be launched in Nov, 2021. Another major group of satellites is *GPS satellites* (see the chapter on GPS).

The largest one, *The International space station* (ISS), weighs 419,725 kg (925,335 pounds), more than 5000 times heavier than Sputnik and has the volume of a *Boeing 747* plane inside. ISS is a habitable satellite that is a collaboration of the US, Russia, Japan, Europe and Canada. It was launched on 20 Nov, 1998 and costs around 100-150 Billion over its lifetime of 30 years including development, launch and operating costs. NASA alone currently spends $3 billion to $4 billion a year, or more than $8 million a day; However for comparison that is just *0.004%* of the United States defense budget of more than $740 billion.

First telecommunications satellite *Telstar 1* was built and owned by AT&T connecting the US, UK and France on July 10, 1962 and it relayed the first live transatlantic television feed. First commercial satellite providing live television broadcast, *Early bird* [IntelSat 1], launched on April 6, 1965. IntelSat transmitted the first moon landing around the world in 1969. Communications satellites are placed in a *geosynchronous orbit* to make it stationary with respect to earth; They are at 35,786 kilometres (22,236 miles), above Earth's equator and follow the direction of Earth's rotation; This is almost 3,000 times the altitude of an *Airbus A380* at 13 kilometers [43,097 feet], the commercial jet that can fly at the highest altitude. (Typical intercontinental flights fly at a lower altitude of around 10 kilometers or 30,000 ft, somewhat above the height of Mount Everest at 8.85 km). However it is still pretty close to earth at just 10% of the distance to the moon of 384,402 km (238,856 miles).

NASA launched the *Television Infrared Observation Satellite* (TIROS-1), the world's first successful weather satellite on April 1, 1960. Operating for three months, the satellite transmitted thousands of images of cloud patterns to ground stations. *Geostationary Operational Environmental Satellites* (GOES) are the new weather monitoring satellite series; This monitors continental US, atlantic and pacific oceans particularly for tropical cyclones. The satellite is also designed to observe fog, fires, dust storms, tornadoes, volcanic eruptions and lightning. The latest in the series are GOES-16 covering East at 75.2 W [at the US east coast] and GOES-17 covering West at 137.2 W [at the *pacific ocean* between Hawaii and continental US] and providing coverage for the western hemisphere. *Himawari-8* at 140.7 E [east coast of Japan] operated by *Japanese space agency* and *Meteosat-11* at 0.0 [Greenwich Meridian] operated by *European Organisation for the Exploitation of Meteorological Satellites* (EUMETSAT) cover the eastern hemisphere of the earth. (The live weather imagery from these satellites is available to watch online to the public and the link is in the *Bibliography*).

For a puny species on a rock in the solar system in the milky way galaxy one among billions of galaxies containing billions of stars each to even estimate the age of the vast universe is no mean feat. Satellites now provide navigation via GPS, communications including sports live telecast, weather monitoring and forecast, defense among others. GPS based navigation alone has transformed modern life. Rockets are the prerequisite for satellites and play an even bigger role in space exploration and rank higher.

Mobile phone

"According to one estimate, every cell phone adds $3,000 to the annual GDP of a developing country."

Steven Pinker says this in his 2018 book *Enlightenment Now*. Martin Cooper of *Motorola*, founded in Chicago in the US in 1928 by brothers Paul and Joseph Galvin, invented the first mobile phone in 1973 and patented "RADIO TELEPHONE SYSTEM" in 1975. The first mobile phone weighed *1.1 kg* (2.5 lb) and was *25 cm* (10 inches) long. Motorola commercialized this as *DynaTec 8000x* in 1983. It still weighed almost a kilogram (1.75 lb), stood ⅓ of a meter (13 inch) high, took 10 hours to recharge and cost 10,500 in 2021 dollars adjusted for inflation [$3,995 in 1984]. AT&T [as *Ameritech*] launched the first mobile network using the first mobile phone in 1983. Now mobile phones weigh around *100-150 grams* (4-6 oz) [*Google Pixel 4a* at 143g, *Apple iphone 12* at 164g]; *Nokia*, founded in *Tampere* near Finnish capital Helsinki and *Gulf of Bothnia* that separates Finland and Sweden in 1865, launched the legendary *3310* in 2000; It sold 126 million units worldwide and weighed only 133g despite its *"brick phone"* image. World's best selling phone ever, *Nokia 1110* that launched in 2003, weighed only 86g and cost around 20 dollars in India in 2008. In contrast, when Nokia launched one of its first mobile phones *Mobira Cityman 900* in 1987, it weighed 760g,

9 times the weight, and cost around 9,000 dollars in 2020 dollars, 450 times the price. *Transistors* and *lithium ion batteries* played critical roles in 100g mobile phones becoming a reality.

Mobile phone *cell networks* emerged from the infrastructure for car phones. The first "Generation" [1G], was the first commercially automated, analog cellular network. First 1G network was launched in Japan by *Nippon Telegraph and Telephone* (NTT) in Dec 1979. *Nordic Mobile Telephone* (NMT) was released in the Nordic countries in Oct 1981. These 1G car networks evolved into mobile phone networks once phones became available. *Advanced Mobile Phone System* (AMPS), the first 1G network in the US, launched in 1983 as a mobile phone network with Motorola's DynaTec 8000x mentioned above.

First *Global System for Mobile Communications* (GSM) network, 2G digital network, was launched in Finland in 1991 by *Radiolinja* with the network built by *Nokia* and *Siemens*. GSM also introduced the *Subscriber Identity Module* (SIM) card. 3G Networks were introduced by *NTT Docomo* in Oct, 2001; It provided upto 40 times the internet speed compared to 2G. 3G consists of *Universal Mobile Telecommunications System* (UMTS), evolution of GSM protocol with *W-CDMA* as the air interface. A competing technology existed in 2G and 3G networks called CDMA that would later merge with the GSM based standards in 4G.

In 2004, NTT Docomo of Japan proposed *Long Term Evolution* (LTE) as the international standard that was the next evolution (4G) of GSM and UMTS and provided speeds upto 10 times faster than 3G. In Dec, 2009, the first commercial LTE deployment was in the Scandinavian capitals Stockholm and Oslo by the Swedish-Finnish network operator *TeliaSonera*. With 4G, mobile phones left their wired counterparts and *circuit switching* and switched completely to more efficient packet-switching that created the internet (see the chapter on *Internet*). It supported a speed upto 300Mbps. However, phone calls that didn't use *Voice over LTE* continued to fall back to the old UMTS [3G] standard for some time. First 5G network was launched in South Korea in April 2019. 5G speed can reach upto 10Gbps, that is an entire DVD (4.7 GB) download over 5G network takes just four seconds, a 20 fold increase from the LTE. 5G is capable of supporting up to 1 Million devices in a square kilometer (important for crowded venues such as stadiums).

Blackberry devices by *Research In Motion* starting in 1999 introduced the smartphone features such as corporate email and mobile web without the phone call itself. The iphone, the first widely successful smartphone, was launched in June, 2007 by Apple. First Android phone (HTC dream) was released in Sep, 2008. Now 85% of the worldwide smartphones are powered by the Android phone operating system owned by Google.

Its importance as a modern invention can't be overstated. In fact, many developing countries including India and in Africa have leapfrogged the normal landline telephones completely to skip to mobile phones. Modern phones combine phone, maps, internet, news and videos into one device. There are 5.2 Billion mobile subscribers in the world or two out of three people (including children) own a mobile connection; Over half of these connections are smartphones. Now mobile radio spectrum itself is considered a precious national resource with 5G spectrum auction raising $80.9 billion in 2021 in the US alone, underscoring their importance. It ranks relatively lower as compared to its enabling technologies of radio, telephone and computer for three reasons. It has been with us only for around 30 years, with smart phones for only 15 years, both much shorter time frames, and hence shorter impact on human history, than most inventions ranked higher. "Smart" aspects of phones are to be considered more as part of computers than a phone and is already ranked higher (see the chapter on *Computer*). Both radio and telegraph, with telephone being a form of telegraph, are more foundational inventions that represented bigger breakthroughs in human history and communications (see the chapter on *Telegraph*). They made much bigger societal changes in the 19th and early 20th century than what we, as people living through the mobile phone era, think the mobile phone is contributing to. In the long run, it will be seen as an incremental invention that it is, building on top of the more fundamental inventions of radio, telegraph and computers. However, remember incremental inventions matter, *a lot*.

Soap

"The harder the water, the more valuable are synthetic detergents for the family wash, David R. Byerly, Procter & Gamble chemist, pointed out yesterday at a luncheon meeting in the Pierre Hotel to introduce "Tide," the company's new product.

Demonstrating the differences between soaps and synthetic detergents, Mr. Byerly showed how soap forms an insoluble curd which reduces its effectiveness. The scum causes streaks on dishes and grayness in clothes washed repeatedly in such cloudy water.

Synthetic detergents dissolve completely in either hard or soft water, leaving it clear and effective. But the first detergents, introduced as early as 1934, did not have the power to do a thorough cleaning job on heavily soiled clothing, such as overalls, in the family wash.

As a result of constant research, chemists have now developed a synthetic detergent that has the "muscle power" to remove heavy dirt from clothing. Mr. Byerly declared. Tide, he said, is effective for hard or soft water, dishwashing or heavy duty laundering."

The *New York Times* of Oct 15, 1948 introduced the first successful synthetic detergent, *Tide*, in the article, "NEW WASH DETERGENT IS INTRODUCED HERE"; It was invented by David Byerly, the chemist in this story, at *Procter and Gamble* after more than a decade of "constant research". A soap is a large molecule with a *hydrophilic* and *hydrophobic* (or more appropriately *lipophilic*) parts; Lipophilic part binds with dirt such as oil and grease and hydrophilic part binds with water (that is, the molecule partially dissolves in water) helping to remove the dirt when water with part of the soap molecule in tow flows out; The soap molecule moving with water pulls the dirt along for the ride removing it from surfaces such as clothes or dishes. *Sodium stearate* is the most common fatty acid salt in today's soaps. Common sources of the starting material, *stearic acid*, are vegetable triglycerides obtained from coconut and palm oils and animal triglycerides from tallow. The names stearic and stearate are derived from *stéar*, the Greek word for tallow. For those unfamiliar with *tallow*, it is rendered beef fat. The rendering process cooks the raw fat, and turns it into a hot liquid (or solid at room temperature) fat. The highest quality beef tallow comes from the fat around the kidneys. Sodium stearate ($C_{17}H_{35}COO^-Na^+$) has both a lipophilic part, the long hydrocarbon chain ($C_{17}H_{35}$), with Carboxylate group (COO^-) at the end the hydrophilic part that dissolves and ionizes in water.

A detergent, such as *Tide* mentioned above, is a synthetically made cleaning agent (*surfactant* - surface acting agent) with a sulfonate group ($SOOO^-$) replacing the carboxylate group of soap as the hydrophilic part. These substances are usually *alkylbenzene sulfonates*, a family of compounds that are similar to soap but are more soluble in hard water, because the sulfonate group is less likely than the carboxylate group to bind to calcium and other ions found in hard water. Alkylbenzene sulfonates have two long alkyl groups and one sulfonate group attached to the benzene ring.

Soap is a very effective tool, better than hand sanitizers, in our fight against microbes with lipid membranes as the membranes share a strong similarity to soap molecules, with hydrophobic tails and hydrophilic heads. When we wash our hands with soap and water, we surround any microbes on the skin with soap molecules. The hydrophobic tails of the soap molecules wedge themselves into the lipid envelopes of viruses, prying them apart.

William and James Lever started *Lever brothers* at Warrington midway between Liverpool and Manchester in England in 1884 to manufacture soap. *Lifebuoy*, introduced by Lever Brothers in 1895, is one of the most popular soaps of all time with it being the popular soap in the US from 1920s to 1950s and still very popular and one of the most affordable in India; It was originally a soap made from *phenol* with the characteristic medicinal smell. *Unilever*, the giant formed after the merger of Lever brothers with *Margarine Unie*, a dutch company in 1929, had a revenue of around 60 Billion dollars in 2020. *Proctor and Gamble* was founded in 1837 by candlemaker William Procter and soapmaker James Gamble at Cincinnati near Chicago in the US. In 1946, *Tide*, the first successful synthetic detergent went on sale and has been the leading laundry detergent in the US since then. Proctor and Gamble had a revenue of 71 Billion dollars in 2020. In 1806, soap and candle maker William Colgate founded *William Colgate & Company* in New York City. In 1873, the company introduced its first *Colgate Toothpaste* that remains popular to this day. Toothpastes also contain a form of soap, *Sodium laureth sulfate* (SLS) apart from the main ingredient, abrasives such as Aluminium Hydroxide or Calcium Carbonate. In 1928, *Palmolive*, founded in 1864 by Benedict Johnson at Milwaukee near Chicago in the US, acquired Colgate to create *Colgate-Palmolive*. Colgate-Palmolive had a revenue of 16 Billion dollars in 2020.

Soap ranks high for providing our cleanliness, an important part of our health and in our fight against microbes. It played an important role in our recent increase in life expectancy. Every year, more than 10 Billion soaps alone are sold in the world (assuming a conservative estimate of 10 bars per household per year and a billion households in the world). It has been with us for more than 200 years in its modern form alone, ranking it high.

Pain killer

"I will greatly multiply thy sorrow and thy conception; in sorrow thou shalt bring forth children"
Genesis 3:16, The King James Bible

Any woman who went through childbirth without an *epidural injection* [a combination of *local anesthetic* and *opioid painkillers*] can appreciate the benefit of modern painkillers. Opium poppy (*Papaver somniferum*,"sleep-bringing poppy") has been used as a painkiller (*analgesic*) from time immemorial. *Laudanum*, opium in alcohol, was a common pain killer from the 17th century till early 20th century. *Morphine*, the active ingredient in opium, was isolated from poppy by German Friedrich Sertürner at *Paderborn* near Cologne in 1804 and was initially sold as a painkiller; Later it was found that Morphine is more addictive than opium or alcohol. Heroin (*Diacetyl Morphine*) was synthesized from Morphine in 1874 and in 1898 marketed by *Bayer*,

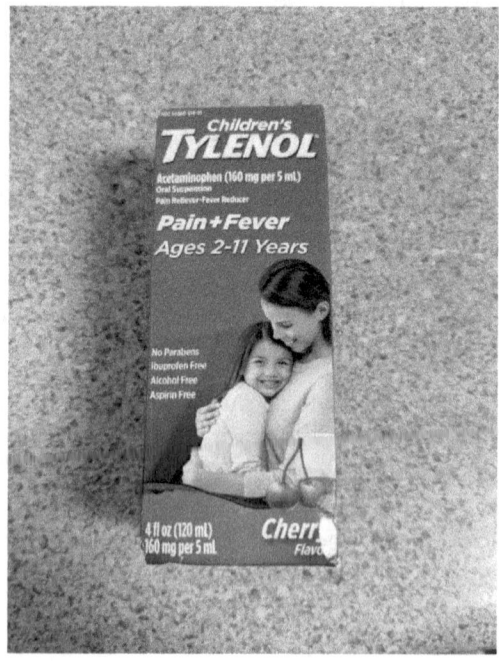

founded in 1863 by Friedrich Bayer at Barmen near Cologne in Germany, as an analgesic. Aspirin (*acetyl salicylic acid*) was made by Bayer in 1899 and sold as an analgesic. Ibuprofen (*isobutylphenyl propionic acid*) was made by *Boots*, founded in 1849 by John Boot in Nottingham near Birmingham in the UK, in 1961. *Paracetamol* (Acetaminophen) has an interesting history. It was first made by German chemist Joseph von Mering at Bayer in 1893 and he thought it could have toxic effects based on his trials and it was discarded. Paracetamol was rediscovered in 1949 again when it was found that it actually had no toxic effect. In the 1950s it came to market as an analgesic. *Naloxone*, an important drug to reverse opioid overdose, was invented by Jack Fishman and Mozes Lewenstein in New York in 1961. *Epidural injection* of analgesic has been a boon for mothers during childbirth. According to *American pregnancy association*:

> "Epidural medications fall into a class of drugs called local anesthetics, such as bupivacaine, chloroprocaine, or lidocaine. They are often delivered in combination with opioids or narcotics such as fentanyl and sufentanil in order to decrease the required dose of local anesthetic."

Robert Hingson and James Southworth at the *Marine hospital* in New York developed the technique of *continuous caudal* analgesic and anesthesia using an indwelling needle in 1941 and used it for the first childbirth in 1942. In the US, now as high as 79% of the mothers use epidural injection in some states. *Sufentanil*, used in epidural injection, is a very potent drug with approximately 5 to 10 times as potent as its parent drug, *fentanyl*, and 500 times as potent as morphine.

There are two main types of over the counter painkillers. *Acetaminophen* and non-steroidal anti-inflammatory drugs (*NSAID*). Paracetamol belongs to the former and Aspirin, Ibuprofen and Naproxen belong to the latter group. Paracetamol's mechanism of action is unknown. Aspirin and other NSAIDs work by inhibiting an enzyme that produces *prostaglandins--hormone-like messenger molecules* that trigger many processes in the body, including inflammation. Prescription medications include opioid and non-opioid medications. Opioid medications are derived from opium plants. These include Morphine,

Oxycodone, Codeine and Hydrocodone. Heroin is no longer used as an analgesic as it is very strongly addictive. Prescription non-opioids include *Diclofenac* invented in 1965 by *Novartis* in Basel, Switzerland that traces its roots to two companies Geigy and CIBA founded in 1857 and 1859 respectively in Basel; Diclofenac is one of the most commonly prescribed drugs now.

Opiate refers to drugs derived directly from the opium plant such as Heroin. *Opioid* (*narcotic*) refers to drugs which mimic opiates including synthetic drugs such as *Hydrocodone*; However almost all of these start with opium plant products as the starting raw material. It is not clear why opium poppies evolved their morphine and codeine producing genes, but it's likely because the chemicals ward off herbivores, according to biologist Ian Graham who decoded the opium poppy genome in 2018. Opioids work by activating opioid receptors known as G protein-coupled receptors (*GPCR*) in nerve cells. (GPCR is an important drug target for many modern drugs). *Naloxone* is an opioid antagonist and works by attaching to opioid receptors in nerve cells and blocking off opioids. Painkillers also have a darker side due to their addictivity, particularly for opioid medications. In 2018, 67,367 drug overdose deaths occurred in the United States alone. The *New York Times* reports that [emphasis added]: "Overdoses have passed car crashes and gun violence to become the **leading cause of death for Americans under 55.**"

Painkillers rank very high in our inventions with its amazing ability to mitigate avoidable suffering during childbirth, after surgery, terminal illnesses or even in very common cases such as tooth pain. It also has life saving consequences for mothers who are very weak to go through the labor pain. However it also has the serious downside of addiction and death and has played a role of an anti-invention. On balance, it has been a boon to humanity to mitigate human suffering and ranks high in the list of inventions.

Telescope

Telescopes put humanity in its place, literally; No other scientific discovery other than Darwin's *theory of evolution* could be considered as providing a bigger ego bust to humanity. For the first time we could literally see that earth is not at the center of the universe. There are other planets and they have moons; There are billions of stars. Ole Christensen Rømer, a Danish Astronomer in 1676 measured the *speed of light* using a telescope for the first time demonstrating that light has finite speed. The speed of light eventually led James Maxwell to deduce the existence of electromagnetic waves via his *electromagnetic equations*. This in turn paved the way for Hertz producing these waves which led to the invention of radio, wireless telegraphy, TV broadcast and mobile phones among others. Dutch eyeglasses maker Hans Lippershay applied for a patent for a telescope in 1608 and actually was rejected because the patent office considered it as a widespread invention with competing claims at that time. When Galileo in Italy heard

of it in 1609 he immediately built one and made pioneering discoveries in astronomy. These earliest telescopes used two lenses and light refracting through them to make images and they were called *refracting telescopes*. First *reflector telescope*, with a mirror collecting the light and focusing the image to another smaller mirror was made by Isaac Newton in England in 1668. One of the first reflector telescopes Newton made is still preserved at the *Royal Society* in London. Most modern Astronomical telescopes are of the reflector kind including the *Hubble space telescope* with 2.4 m (94 inch) mirror and the soon to be launched *James Webb space telescope* with a huge 6.5 meters (252 inch) across mirror. The James Webb space telescope is scheduled to be launched in November 2021 and it will collect infrared light from faraway galaxies, red-shifted due to the *doppler effect* instead of visible light. The Hubble space telescope helped us estimate one of the most profound discoveries of humanity, the age of the universe at *13.8 Billion years old*. In fact, Hubble himself used the telescopes [and stars called *Cepheid variables*, standard light candles in Astronomy] to discover that the universe is expanding and there are more galaxies outside our own milky way, dealing yet another blow to humanity's ego.

It is much easier to make a large, near-perfect mirror than to make a large, near-perfect lens. Also, since mirrors are one-sided, they are easier than lenses to clean and polish. Refracting telescopes are also longer and bulkier for the same power compared to reflecting telescopes. *Yerkes Observatory* houses the largest refracting telescope; It has a 102 cm (40 inch) diameter lens with 19m (63 ft) tube that weighs 20 tons. It was built in 1897 and is no longer used in Astronomical research reflecting the fact that reflectors have come to dominate in the recent centuries. *Lick observatory* near San Francisco Bay Area at Mount Hamilton in California in the US has a 91 cm (36-inch) refracting telescope that was the world's largest when it opened in 1888.

Both telescopes at the *Keck Observatory* near the summit of Mauna Kea in the U.S. state of Hawaii have 10 m (33 ft) aperture primary mirrors. *Gran Telescopio Canarias* (GTC) located at the *Roque de los Muchachos* Observatory on the island of La Palma, in the

Canaries, Spain is the largest reflecting telescope with a diameter ten times the largest refracting telescope (Yerkes). GTC website describes:

"We can compare its power of vision to 4 million human eyes and, with it, we could distinguish car headlights some 20,000 km away, or at the distance that separates Spain from Australia.... The GTC has a primary mirror of about 10.4 m in diameter, composed of 36 segments of about 450 kg each....The limit of the polishing error may not exceed 15 nm, or 3,000 times smaller than a human hair, while any irregularity cannot exceed 90 nm. ... In comparison, if the mirror would be as big a Iberian Peninsula, the highest "mountain" irregularity would be only a few centimeters tall!"

There are telescopes that work at frequencies higher than visible and infrared spectrum as well. X-ray space telescopes such as *Chandra observatory*, launched in July, 1999, use a *grazing angle telescope* called Wolter telescope as normal incidence similar to a reflecting telescope would simply absorb the x-rays. *Gamma ray space telescopes* such as the *Compton Gamma ray observatory* launched in April, 1991, use instruments that use *scintillators* that produce visible light when hit by (invisible) gamma rays.

Telescopes deserve their high rank for their role in our understanding of the universe and our place in it. It helped us understand that earth is one of the many planets which orbits the sun instead of the other way around; It helped us measure the speed of light with its profound impact through radio waves, compute the age of the universe and the fact the universe is expanding. Telescope doesn't rank as high as Microscope which has had more impact on life saving discoveries such as germ theory.

(Image credit: NASA and ESA)

PART 4

Alcohol

"Worldwide, more than a quarter (26.5%) of all 15–19-year-olds are current drinkers, amounting to **155 million adolescents.** . Prevalence rates of current drinking are highest among 15–19-year-olds in the WHO European Region (43.8%), followed by the Region of the Americas (38.2%) and the Western Pacific Region (37.9%) ... Mortality resulting from alcohol consumption is higher than that caused by diseases such as **tuberculosis, HIV/AIDS and diabetes.** "

According to the World Health Organization (WHO) *Global status report on alcohol and health 2018* [emphasis added]. Alcohol (*Ethanol*, C_2H_5OH) is another high ranking anti-invention. It causes 3 million preventable deaths (5.3% of all deaths) in the world; It is as if we have a hidden *coronavirus epidemic* every year. 21.6% of all deaths [more than 1 in 5] in Russia is attributable to Alcohol. Its use is much more widespread than tobacco with more than 50% of the US population reporting alcohol usage. In 2017, 51.7% of the people in the US used Alcohol with a shocking 19.4% of 16-17 year olds. Data from other

years corroborate this, about 24.6 percent [almost 1 in 4] of 14- to 15-year-olds reported having at least one drink in 2019. For comparison, Tobacco use is around 14% in the US. If Alcohol were to be invented in the 20th century it would have been classified as a controlled narcotic substance.

Ethanol's close relative on the lower molecular weight side with one less methyl (CH_3) group, *Methanol* (CH_3OH) is highly toxic. A potentially lethal dose of methanol is 30-240 ml or 1g per kg of body weight. Permanent visual damage may occur with 30 ml (1 fl oz) of methanol. The body metabolizes methanol to *formic acid*, a toxic compound that can cause blindness as a result of permanent injury to the optic nerve. Ethanol's relative from the higher side with one additional methyl group, *Isopropyl alcohol* (C_3H_7OH), is also very toxic. It is used in hand lotions, hence the name *rubbing alcohol*. Isopropyl alcohol is about twice as toxic as ethanol with a fatal dose of only 160-250 ml [8 oz] or 4-8 g/kg. In fact, people who don't have access to ethanol or have high ethanol tolerance have turned to drinking it for a fast "high". Sandwiched between these two toxic alcohols, ethanol is necessarily toxic and it would have been banned for consumption if discovered recently. The lethal dose of alcohol is 5-8g/kg (3g/kg for children) – for a 60 kg person, 300g of alcohol can kill, which is equal to 20 standard US drinks (about 1 litre of spirits or 4 bottles of wine), not much higher than isopropyl alcohol.

In the United States, a standard drink is defined as any beverage containing 0.6 *fluid ounces* or 14g of pure alcohol (also known as an alcoholic drink-equivalent), found in:

- 12 ounces of beer with about 5 percent alcohol content (by volume)
- 5 ounces of wine with about 12 percent alcohol content
- 1.5 ounces of distilled spirits with about 40 percent alcohol content

The *Substance Abuse and Mental Health Services Administration* (SAMHSA) classifies binge drinking as 4 or more alcoholic drinks for females or 5 or more alcoholic drinks for males within a two-hour time period. This is around 0.16% Blood Alcohol Content (BAC), twice the legal limit of 0.08% BAC. Around 0.20% [5-6 drinks] makes one pass out and around 0.30%-0.40% [7-10 drinks] leads to coma or death. Excessive alcohol use is responsible for 95,000 deaths in the United States each year, including 1 in 10 total deaths among working-age adults according to *Center for Disease Control* (CDC) in the US. In 2010, excessive alcohol use cost the US economy $249 billion or about $1,000 per adult in the US. Another alarming statistic in the US was more than *15% of surgeons had alcohol abuse with around 14% male surgeons and 26% of female surgeons.* According to the *National Institute of Health* (NIH), Alcohol is a significant factor in the deaths of people younger than age 21 in the United States each year.

There is a theory that our ability to digest ethanol may have evolved for us to eat rotten fruits. Earliest evidence of Alcoholic drinks is from *Jiahu* in China around 9,000-8,600 years ago. Alcohol was discovered and consumed by ancient societies in all continents for more than 2,000 years. Now Russia and her neighbours are among the top consumers of alcohol per capita. According to the WHO, in 2016, the average *Belarusian* drinks 15 litres of alcohol per year. In India, the southern state of *Kerala* - the state the first European explorer to India, *Vasco de Gama,* reached in 1498 (see the chapter on *Ship*) - with 34.7 Million people and a GDP of $104.6 Billion spends $1.6 Billion on Alcohol (1.5% of GDP). For comparison, it spends $2.9 Billion on Education (2.8% of GDP) and $1 Billion on Health (1% of GDP), both considered at the higher end for indian states.

For the opposite reason the vaccine or fertilizers get a high ranking, Alcohol gets a high ranking for its enormous harmful impact on human life. Two prominent historical inventions, Alcohol and Religion have been with us for a long time and share intertwined, symbiotic history. It has been so powerful in our society that when the US experimented with prohibition in the 1920s it led to widespread illicit alcohol black market and rise in crimes. It has impacted human history for around 9,000 years and in all continents with human civilizations (except Antarctica) which increases its [anti-] invention ranking.

Tobacco

Tobacco industry is huge and justifies the name *"Big Tobacco"*. In 2019, the legal global tobacco market was around US $818 Billion, almost ⅓ of India's GDP of $2,869 Billion in 2019. For comparison, the movie industry including the famous *Hollywood* and *Bollywood* accounted for $42.5 Billion, or just 5% of tobacco revenue. Even the giant oil and gas industry only makes around two times the revenue of tobacco. This would be even bigger if we include the illegal tobacco market estimated to be around 11% of the total market. The largest global tobacco category is the *combustible cigarettes*. Tobacco industry sells 10,000,000 cigarettes every minute. With over 5,200 billion (5.2 x 10^{12}, that is 5 followed by 12 zeros) cigarettes consumed annually, it alone is valued at US$705 billion. Total economic cost of smoking is more than $300 billion a year in the US alone.

Tobacco (*Nicotiana tabacum*), the plant, is a new world plant native to North and South America. Its leaves are used to produce tobacco.

Archeological studies suggest the use of tobacco around the first century BCE, when Maya people of Central America used tobacco leaves for smoking in sacred and religious ceremonies. It then spread to large parts of the Americas including *Carribean* by the time of arrival of European explorers. Early Portuguese and Spanish explorers spread it to Europe and other parts of the world. Tobacco cultivation and export formed an important part of early American history and economy. It was used as a form of money in American colonies. In 1542, tobacco started to be grown commercially in North America, when Spaniards established the first cigar factory in *Cuba*. John Rolfe in 1612 planted the first commercially successful tobacco crop in the US. By the early 17th century there was a big market in Europe for tobacco and from there it spread to all the European colonies in Asia and Africa such as India and Indonesia. Now the major tobacco-growing and consuming countries are China, US, the Former Soviet States, India, and Brazil. Over 80% of the world's *1.3 billion tobacco users* live in low to middle income countries. In the US, 13.7% of all adults (34.2 million people) - 15.6% of men, 12.0% of women - are smokers as of 2018. In India, nearly 29% of all adults (267 million people) are users of tobacco as of 2017. In China, 50.5% of men smoke with 307.6 million adults in all.

An average cigarette yields about 2 mg of *Nicotine*, the main active ingredient in tobacco. Tobacco also contains 4,000 minor chemical ingredients that play an active role. Nicotine consists of a *Pyridine* (Benzene ring with a Nitrogen atom instead of Carbon) ring and a *Pyrrolidine* (5 carbon ring with a Nitrogen atom instead of Carbon) ring. It binds to *Acetylcholine*, neurotransmitter, receptors and induces the release of *dopamine*, the neurotransmitter responsible for the feeling of pleasure. As with many drugs, dopamine reinforces the brain to repeat the same behavior, in this case using tobacco, forming the addiction. It may be harder to quit smoking than to stop using cocaine or opiates like heroin. In 2012, researchers reviewed 28 different studies of people who were trying to quit using the substance they were addicted to and found that about 18% were able to quit drinking, and more than 40% were able to quit opiates or cocaine, but only 8% were able to quit smoking.

In 2019, *Philip Morris*, founded in 1847 at London, England, was the largest tobacco company with a revenue of 78 billion US dollars followed by *British American Tobacco*, founded in 1902 at London, England, at 33 Billion. In 2019, Philip Morris manufactured around 706 Billion cigarettes and British American Tobacco manufactured 668 billion cigarettes. *Lorillard Tobacco*, founded in 1760 in New York and now part of British American Tobacco, was one of the oldest companies in the world. *W.D. & H.O. Wills* was founded in 1786 and is now part of *Imperial tobacco*, another tobacco giant in Bristol, England.

Smoking forms of tobacco used are pipe, cigar, cigarette, *hookah* (water pipe), *beedi* and *kretek*. Cigar is a rolled bundle of dried and fermented tobacco leaves. Cigarette is a narrow cylinder containing tobacco rolled into thin paper for smoking. *Beedi* is an indian leaf-rolled cigarette made of coarse, uncured tobacco, tied with a string at one end. *Kretek*, similar to beedi, is an indonesian leaf-rolled cigarette made of tobacco and clove among others. First cigarettes were made in France in the middle of the 19th century from native American and spanish leaf-rolled precursors and became popular in the early 20th century. Beedis and kreteks were made from leftover tobacco leaves in the early 20th century. The most prevalent form of tobacco use in India is smokeless tobacco and commonly used as betel quid (*paan*) with betel leaf and chewable tobacco. *Snuff* (roasted and finely powdered for inhalation through the nose) is another common form of smokeless tobacco.

Tobacco ranks high as one of the most consequential anti-inventions of the world. WHO calls smoking "the single biggest preventable cause of death". Over 19% of the world's adult population, or one in five people, are smokers, making it one of the most widely used addictive substances. It kills half its users and smokers die 10 years earlier than nonsmokers, making it one of the deadliest (anti-) inventions. Tobacco also had a long time to impact human history particularly after its introduction to the wider world around 500 years ago, predating the industrial revolution and its inventions.

Bridge

"I have built a bridge that will last forever."

A plaque at the center of the *Alcántara bridge* in Western Spain near the Portugal border contains these words. It was built in 106 CE, 1900 years back with stone and characteristic Roman arches; (Alcántara literally means *"The Arch"* in Arabic and is unlikely to be the original name of the bridge). True to the claim at the bridge, It is still in use today and is a masterpiece of Roman Engineering. Another Roman Engineering marvel, *Pont du Gard* was mainly an *aqueduct* ["bridge for water"] that was built in 1st century CE. The Iron bridge in *CoalBrookdale* in England near Birmingham, designed by Thomas Pritchard and built in 1781, is the first iron bridge in the world; First steel bridge was the *Eads Bridge* across the Mississippi river connecting St.Louis, Missouri and St.Louis, Illinois near Chicago in the US built in 1874. First recorded man made bridge is *Arkadiko bridge* in Greece built around 1200 BCE, 3200 years ago. *Brooklyn Bridge* in New York City, designed by German-American John Roebling, was the longest suspension bridge in the world when it was built in 1883. China now has the world's longest bridge, the 164.8 kilometres (102-mile) *Danyang Kunshan Grand Bridge*,

a high-speed rail viaduct running parallel to the *Yangtze River* near Shanghai.

Beam bridges are the simplest type of bridges; It consists of a sequence of short spans supported at either end by piers. It typically requires more support beams and is built low to save on the cost of the piers making it impassable to boats except at designated places. (One could say it is the *"brutalist"* architecture for bridges). The world's longest beam bridge is *Lake Pontchartrain Causeway* in southern Louisiana in the United States, at 38.35 km (23.83 miles), with individual spans of 17 meters (56 ft). The six-lane elevated *Bang Na Expressway* that runs for 54 km (33.5 miles) through Bangkok is one of the longest bridges and longest elevated expressway [overpass] in the world; It is a kind of beam bridge. *Truss bridges* are a variation of beam bridges with additional support, constructed with *triangular load bearing lattice structures made of steel* supported by piers. Truss structure is typically used for railway bridges to provide additional support. The *Bogibeel bridge*, a combined road and rail bridge of 4.94 km over the *Brahmaputra River* in the northeastern state of Assam, India, that opened in Dec, 2018 is a truss bridge.

Arch bridges are the oldest bridges including *Alcántara bridge* and *Roman aqueducts*. It uses the pillars and characteristic *arched abutments* to support the bridge. Arched structure allows the water to pass through it while providing strong support for the bridge by passing the load across its curve effectively to the piers. It allows for longer spans than beam bridges. *Sydney Harbour Bridge* in Australia is a *through-arch bridge*, a variation on the Arch bridge, with inverted arch, in which cables hanging from the arch support the bridge. *Cantilever bridges* are supported at one end by piers and at the other end is the bridge-bearing *cantilever arm*. Two cantilever arms from either side support the suspended span of the bridge. It is easier to construct as they can be built from both ends, typically on land; In fact, other types of bridges start out as cantilevered during construction. They typically don't support long spans with the longest, *Pont de Québec* in Canada, only 549 meters long. *Bascule bridges* such as the famous *Tower Bridge* in

London can be moved up (opened up) to provide clearance for boat traffic.

Suspension bridges are supported by cables on either end and make up the most *beautiful bridges* in the world such as *Golden Gate Bridge* in San Francisco or *Tower Bridge* in London. *Cable stayed bridge*, a variation on suspension bridge but operating under different principles requires strong load bearing beams acting as anchors for cables; It *doesn't require strong land anchors for cables* and can be built around soft ground. The beautiful Golden Gate bridge at San Francisco in the US was both the longest and tallest bridge in the world when it opened in 1937; It was designed by Joseph Strauss and has a height of 227.4 m (746 ft) and a length of 2,737 m (8981 ft). The *New York Times* wrote on May 30, 1937 in its editorial "Golden Gate Bridge" just three days after the opening:

"A century ago RICHARD DANA wrote of the golden gate: 'if ever California becomes a prosperous country, this bay will be the center of its prosperity.' The new bridge is the very highway of the prosperity he foresaw. Beneath it ply vessels that carry petroleum, fruits, grains and lumber. Over it streams an endless current of trucks and automobiles from sixty towns and cities that cluster on the peninsula of San Francisco and along the bay. A great state has fulfilled DANA's prophecy with one of the greatest engineering structures of man."

Bridges are both economically and culturally important. The *Bogibeel bridge* reduced the travel time to cross the Brahmaputra river by 705 kilometers (438 miles). In some cases it made it possible in the first place, as in the Brooklyn bridge connecting Manhattan and Brooklyn in New York City through land for the first time. The *George Washington bridge* connecting New York City with Fort Lee, New Jersey was used by 103 million people in 2016, or the equivalent of almost ⅓ of the US population. Bridges have also been aesthetically and culturally important with Golden Gate Bridge serving as the icon of the San Francisco Bay area in the US and Tower Bridge on Thames river becoming the famous symbol of London, England.

Dam

First Dam in the world, called *Jawa Dam,* was constructed in Modern day Jordan near the capital Amman *5,000* years ago. Dams have been important for *hydroelectric power, irrigation, water supply* and *flood control*. The *Three Gorges Dam* across *Yangtze River* near Wuhan has been the world's largest power station in terms of installed capacity (*22,500 MW*) since 2012. For comparison, the largest hydroelectric dam in the US, *Grand Coulee Dam* in Columbia River near Seattle close to the Canada border produces *6,809 MW*, less than ⅓ of Three Gorges Dam. The Three Gorges Dam is a hydroelectric gravity dam that spans the Yangtze River by the town of *Sandouping*, downstream of the Three Gorges. It has a dam volume of *27.2 million cubic meters*, enough to fill almost 11,000 olympic sized swimming pools. It is 181 metres high and spans 2.34 kilometers, almost double the *main span* (distance between towers) of the *Golden Gate bridge*. The Three Gorges Dam is the world's biggest man-made producer of electricity from renewable energy. On the flip side, it also displaced 1.13 million people.

There is now 4.4 Peta Watt-hour (PWh) of the total 25.9 PWh electricity produced in the world, or around 17%, is hydro electricity. To make sense of this abstract number, the US consumes 4 PWh of electricity in a year, 12,000 kWh per person or 30 kWh per day. In 2020, hydroelectricity accounted for about 7.3% of total U.S. electricity

generation and 37% of the total renewable electricity generation. First dam specifically designed for hydro electricity became operational in 1893 in Austin, Texas across *Colorado river* (a different and smaller river from the one *Hoover Dam* is on in the south west of the United States); The dam was constructed at a cost of $1,000,000 (29.2 Million in 2021 dollars) and the electric generators cost $600,000 (17.5 Million in 2021); It was 1143 ft long, 60 ft deep and 60 ft across at its base. Just after 7 years, It was washed away in a flood in April, 1900. *Kallanai* (literally "stone dam"), constructed in the second century CE on the *Kaveri River* near Chennai, is the fourth oldest dam in the world and the oldest in India. The *Kariba Dam* near Lusaka, the capital of Zambia, that impounds the *Zambezi River* between Zimbabwe and Zambia is the largest man-made reservoir in the world, impounding 181 km^3 of water; The dam, completed in 1960, now provides 35% of the basin's hydroelectric capacity at 1,830 MW. *Tarbela Dam* in Pakistan is the largest earth filled dam in the world and one of the largest hydroelectric dams at 4,888 MW.

China has the largest number of big dams - 23,841 in all - in the world. It is followed by the US and India at 9,263 and 4,407 dams respectively. Japan and Brazil have 3,130 and 1,365 dams respectively. Around 16,000 dams are built for *irrigation*, the most common purpose for dams, followed by *hydro-electricity* at around 6,000 dams. Dams built for irrigation are dominant in Asia with around 12,000 of them, while in Europe it is mostly built for hydro electric power and in North America mainly for flood control. Japan has the oldest dams among the top five countries with the most dams discussed above, at an average age of 111 years. India has the youngest dams among the top five, at an average age of 42 years. The UK has the oldest dams overall with an average age of 106 years and a median age of 111 years. Iran and Turkey have some of the youngest dams in the world.

Arch dam, similar in principle to *Arch bridge* (see the chapter on *Bridge*), is curved upstream so as to transmit the major part of the water load to the abutments. A gravity dam such as the Grand Coulee dam relies on only its weight and internal strength from the foundation for stability. A *buttress dam* consists of a watertight part supported at intervals on the downstream side by a series of buttresses. An *embankment dam* with earth or rock fills, such as *Oroville dam* near San

Francisco, the tallest dam in the US at 235 meters (770 ft), is constructed of excavated natural materials with only a concrete core.

Columbia river system is the most dammed river in the world with more than 400 dams and 60 major dams including Grand Coulee Dam, the biggest hydroelectric dam in the US (6,809 MW); It was built in 1942 with a height of 168m and a length of 1,592 m. *John Day dam* and *Chief Joseph dam* in the system produce 2,160 MW and 2,069 MW respectively. Hydroelectric power plants located on the river and its tributaries account for *29 GW* of capacity and contributed 44% of the total hydroelectric generation in the US in 2012. The *Amazon river* is surprisingly the least dammed in the world given it is one of the longest.

Biggest dam failure was *Typhoon Nina–Banqiao dam* failure, catastrophic dam failure in August 1975 in western Henan province near Wuhan in China, caused by a *Typhoon* (tropical cyclone). Floods led to the cascading failures of dams, including two large dams, *Banqiao* and *Shimantan*, and 62 smaller reservoirs in the upper Huai River basin, Henan Province. The ensuing floods caused more than 150,000 casualties. *Johnstown flood* near Philadelphia in the US after the failure of *South Fork Dam* killed more than 2,200 people in 1889. The *Mullaperiyar Dam* near Chennai in the state of *Kerala*, the oldest modern dam in India, with 53.6 m in height and a reservoir capacity of 443 million m^3 was built in 1895 by the British government to provide irrigation and began to generate power in 1959. This *125 years old dam is at high risk* and a dam failure would be catastrophic: nearly 3.5 million people will be affected. Some of the old dams are being demolished now due to the risk of failure. The *Glines Canyon Dam*, also known as *Upper Elwha Dam*, built in 1927 was demolished in 2014; It was the largest dam removal ever in the world.

Dams rank high for hydro electric projects required for electricity. It is critically important in flood control (North America) and irrigation (Asia). However it also displaces a lot of people such as more than a million for the *Three Gorges Dam*. Dam failures can cause serious casualties such as the Banqiao dam failure and old dams such as Mullaperiyar dam pose serious risks to the millions of people living near them. Its huge human impact both positive and negative, ranks it high in the list of inventions.

Canal

"A journey from the Suez Canal in Egypt to Rotterdam, in the Netherlands — Europe's largest port — typically takes about 11 days. Venturing south around Africa's Cape of Good Hope adds at least 26 more days, according to Refinitiv, the financial data company. The additional fuel charges for the journey generally run more than $30,000 per day, depending on the vessel, or more than $800,000 total for the longer trip.

The shutdown of the canal is affecting as much as 15 percent of the world's container shipping capacity, according to Moody's Investor Service, leading to delays at ports around the globe. Tankers carrying 9.8 million barrels of crude, about a tenth of a day's global consumption, are now waiting to enter the canal, estimates Kpler, a firm that tracks petroleum shipping.

...More than 200 ships are now stuck at either end of the Suez Canal, with no clarity on when they will be able to continue their journeys. Some 80 additional ships are scheduled to arrive over the next three days, Mr. Singh said."

The world watched with anxiety and curiosity as a container ship, *Ever Given* ran aground in the Suez Canal, blocking the channel completely on March 23, 2021 and efforts were being made to tug the behemoth out of the channel using tug boats, david and goliath style. The *MV Ever Given* is one of the world's largest container ships, 400m-long (1,312ft) and weighing 200,000 tonnes with space for 20,000 containers carrying goods across the sea. The *New York Times* explains the effect of the Suez canal being blocked as above. The canal saves almost a Million dollars in fuel per ship and a month in travel time in travel from Asia to Europe. The Suez Canal carries almost 10 Billion dollars worth of traffic every day, underscoring the critical importance of the canal to the global economy. That equates to $400 Million worth and 3.3 million tonnes of cargo an hour, or $6.7 Million a minute. The Suez Canal (In Arabic: *Qanat as-Suways*) was built in 1869 by The Suez Canal Company (*Compagnie Universelle du Canal Maritime de Suez*) established by the former French diplomat Ferdinand Marie de Lesseps. It took 10 years to complete and 3,000 workers lost their lives. The Suez Canal stretches 193 km (120 miles) from *Port Said* on the Mediterranean Sea in Egypt at the northern end to the *Gulf of Suez* in the Red Sea at the southern end. It doesn't require any locks as the water level on both sides are almost the same. The canal separates the African continent from Asia, and it provides the shortest maritime route between Europe and the lands lying around the Indian and western Pacific oceans. The canal is extensively used by modern ships, as it is the fastest crossing from the Atlantic Ocean to the Indian Ocean. It takes 12-16 hours for a ship to cross the canal. The 2015 expansion of the canal added a parallel lane to most of the canal. (Unfortunately, *Ever Given* got stuck near the Gulf of Suez end of the canal with only a single line).

One of the most complex engineering projects ever undertaken by humanity was the *Panama canal*, an 82 km artificial waterway connecting the Atlantic and Pacific Ocean through *Panama*, a country pretty much carved out of *Colombia* by the US just to build this canal, in Central America. France tried to build the canal in 1881-89 with the same de Lesseps of Suez canal fame and failed disastrously. They lost the lives of 20,000 workers and 287 Million dollars (8 Billion dollars in 2021). When the US finally pulled off the feat, It cost 5,500 workers mainly due to infectious diseases and 375 million in 1914, a whopping 9.8 Billion in 2021 dollars. The canal saves a total of about 12,600 km (7,800 miles) for a ship from New York to San Francisco.

The *Grand Canal* in China started in the 5th century BCE and fully connected around the sixth century CE is one of the longest and oldest canals in the world. It connects the *Yellow* (Huang he) and *Yangtze* rivers and was critically important for the transportation of grains to the capital, Beijing. The *Erie Canal*, built by New York state Governor Dewitt Clinton in 1825, to create a navigable route from the New York City and the Atlantic Ocean in the east to Great lakes near the Canada-United States border (Chicago Area) in the west; The Erie Canal played an important role in the development of the US; It was built at a cost of 7 Million dollars (around 200 Million dollars in 2021). The *New York Times* of Jan 14, 1863 writes about the importance of the canal in the article, "The Enlargement of the Erie Canal a National Necessity":

"It costs $20 a ton for one hundred miles on an ordinary road; $2 on a railroad, and 20 cents on the ocean, for the same distance. The cost on the great lakes exceeds that on the ocean, and in the Erie Canal that on the lakes; but there is no doubt on the average, fright can be brought from Chicago to New York, for example, for less than one-half of what it can be transported between the same points by railroad. ... Indian corn is never brought to market on railroads; nor is wheat all the way from the districts of its production."

Canals rank high due to their critical importance in the modern economy, particularly the *Suez canal* and the *Panama canal*. They have been historically very important and contributed to the development of trade and economy by providing connectivity and shortening travel times. Canals have been important for irrigation and flood control throughout history. Canals have also been with us for 2,500 years increasing their impact.

(Image credit: Deutsche Welle)

Bicycle

"I think [the bicycle] has done more to emancipate women than any one thing in the world."

Susan Anthony, one of the important leaders of the American *women's suffrage movement* said this in 1896. Bicycles are so ubiquitous and relatively low tech that they seem to have been with us forever. In fact the first modern bicycle is only around 135 years old, invented in 1885 by the british inventor John Kemp Starley at Coventry near Birmingham, called the *Rover 'safety' bicycle*. This bicycle, on display at the *Science Museum* in London, looks very much like a modern bicycle and has hardly changed since then. Interestingly, it is newer than railways or even automobiles [internal combustion engines] and not much older than the airplane itself (see the chapters on *Railways, Automobile* and *Airplane*). First bicycle, *"velocipede"*, was made in *Karlsruhe,* Germany, a very productive city for inventions along the Rhine river bordering France near Stuttgart, of Automobiles fame, by Karl von Drais in 1817; It had two wheels, and qualifies as a bicycle, but no pedal to drive them; Riders simply pushed them along with their feet

as small children sometimes do with their bicycles. In 1866, Pierre Lallement of Paris, France, then residing at New Haven, home of *Yale University* and mid-way between Boston and New York, in US, took out an US patent for the "improvement in velocipedes" that describes for the first time, pedals attached to the front wheel (called *treadles* in his very readable and concise patent worth reading in full, the link is in *Bibliography*). Parts of the bicycles were made with wood such as wooden spokes and some parts were made with iron such as iron tires. It was not widely successful, likely due to the condition of the roads at that time; Some riders took to pedestrian pavements and were later banned by cities. It was also very expensive; In 1868, bicycles cost around 160 dollars (*$3,000 in 2021 dollars*), this at a time when the average daily wage of a typical worker such as a carpenter was only 2 dollars; This makes a bicycle as expensive as 3 months of wages, or more than 8,500 dollars, adjusted for 2019 per capita income in the US of 34,103 dollars. The next big breakthrough came with *"a hollow or tubular india-rubber tire for the wheels of cycles and other vehicles which is is inflated with air or gas under pressure"*; John Boyd Dunlop of Belfast, Ireland patented, in yet another eminently readable patent, "Wheel-tire for cycles", the familiar rubber tire inflated with air in 1889, making the ride much smoother and comfortable. As mentioned above, John Starley's *Rover safety bicycle* of 1885 completed the innovation process and fixed bicycles in its current form. In 1878, Albert Pope started *Columbia bicycles* for manufacturing bicycles in the US for the first time at *Hartford*, mid-way between Boston and New York, which is still producing bicycles today. Apparently bicycles were not well received by society at the beginning. Police routinely stopped riders and shooed them off city streets according to *Smithsonian magazine*. In 1881, three cyclists who defied a ban on riding in New York's *Central Park* were jailed, this one apparently for good reason as in 2014 a 58 year old woman died in Central park after being struck by a bicycle.

Gottlieb Daimler, of internal combustion engine fame, in Germany built the first motorcycle, *Daimler Reitwagen,* in 1885. It was literally a bicycle with an internal combustion engine added-on. Two steam powered motor cycles preceded this, *Roper steam velocipede* by Sylvestor Roper of Boston, US in 1867 that looks more like a stunt bicycle, used in exhibitions and Louis Perraeaux's *Michaux-Perreaux steam velocipede* based on a velocipede bicycle; Neither of them were practical nor were commercially available and generally not considered as motorcycles proper. The *Royal Enfield bullet*, by *Royal Enfield* that started in 1899, has been in continuous production from 1932 pretty much in the same form first in England and now in India. *Harley-Davidson*, the most popular motorbikes maker in the US, was founded in 1901 in Milwaukee at *Lake Michigan* near Chicago. *Yamaha*, the well-known motorbikes company from Japan, was founded in 1955 in Iwaka near *Sendai* and Honda at Tokyo in Japan, world's largest motorcycle producer as well as the largest internal combustion engine maker, started producing motorcycles in 1955 as well. Motorcycles have a *fatality rate* 28 times higher than cars per kilometer driven, making it one of the most dangerous vehicles on the road in the US. (However this statistics doesn't apply to clogged city traffic in developing countries that move at a snail's pace).

Railways, automobiles and airplanes have had a much bigger impact than bicycles in human history including in two world wars. However it has been very important as a faster alternative to walking in many developing countries including India, China and many countries in Africa. Its internal combustion engine form, motor bikes are vital in developing countries such as Indonesia, India or Thailand. It has also been an important tool for education, particularly in rural areas that some states in India, such as the state of *TamilNadu* in southern India, have started providing them for free to high school students. Rubber tires invented for bicycles have carried over to many modern forms of transport including automobiles and airplanes. Bicycles also contributed to the development of modern roads. It has been culturally important including in women's emancipation as mentioned at the start of the chapter. In developing countries, two wheelers still dominate the cars by a huge margin and play the role of automobiles in the developed world. In 2016, 50% of households had a bicycle in India and 36% of households had a motorbike. However only 11% owned a car. This increases the importance of the bicycles to humanity overall and ranks it in the top 100.

X-ray

"LEXINGTON, Ky, Oct. 17. - By the aid of the X-ray, Prof. Pence of the State College here has located the bullet in the spine of James B. Ferguson, the race horse starter, which came so near killing him this Summer.

The bullet, which was fired into Ferguson by Capt. May in a fight here twenty years ago, divided in two sections, lodging in the spinal column about an inch and a half apart. The picture shows the lead imbedded in the bone, almost, if not quite, touching the spinal cord. The pressure of these missiles on the spine has caused Mr. Ferguson to become almost paralyzed. He was quite low for a long time in California, and was brought home, it was thought, to die. Several operations were performed, but the bullet had not been located up to this time.

Dr. Archibald Barkley will operate on Mr. Ferguson, cutting out the lead, within the next few days."

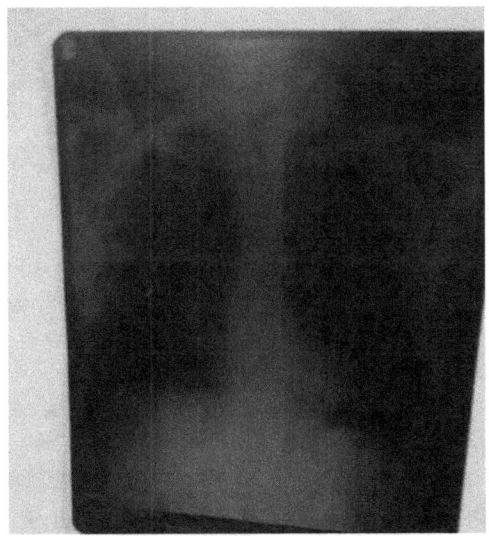

This remarkable story from The *New York Times* of Oct 18, 1900, just four years after X-ray was discovered by Wilhelm Röntgen, shows the life saving power of x-ray imaging as an invention. Being able to see inside our bodies without cutting through it, as was done to the poor Mr. Ferguson above multiple times in search of a bullet before the invention, is a medical miracle that transformed medicine and saved countless lives. Wilhelm Röntgen at the *University of Würzburg*, more than 600 years old University near *Nuremberg*, discovered x-rays in 1895. World's first human x-ray imaging was the hand of Anna Bertha Ludwig, Röntgen's wife, who reportedly exclaimed after seeing the image, *"I have seen my death."* X-rays are electromagnetic waves with much higher frequencies than light visible to the human eye and can penetrate human soft tissues but not bones. A *Columbia University* article on x-ray, "History of Medicine: Dr. Roentgen's Accidental X-Rays" says:

"In today's world, doctors order X-rays to diagnose all sorts of problems: a broken bone, pneumonia, heart failure, and much, much more. Mammography, the standard screening method for breast cancer, uses X-rays. We barely think about it, it's so ubiquitous. But not so long ago, a broken bone, a tumor, or a swallowed object could not be found without cutting a person open.

News of his discovery spread worldwide, and within a year, doctors in Europe and the United States were using X-rays to locate gunshots, bone fractures, kidney stones and swallowed objects....early use of X-rays was widespread and unrestrained, even to the degree that during the 1930's and 1940's, shoe stores offered free X-rays so that customers could see the bones in their feet."

Computed tomography (CT) scan is a computerized x-ray procedure in which a narrow beam of x-rays is aimed at a patient and quickly rotated around the body, producing cross-sectional *"tomographic"* images or slices of the body. Once a number of successive slices are collected by the machine's computer, they can be combined using computer algorithms to form a 3D image that allows for easier identification of problems. Godfrey Hounsfield, an electrical engineer at

EMI in London, England, invented the CT scan in 1971 based on the work of physicist Allan Cormack at *Tufts University* near Boston in the US.

In 1901, Willem Einthoven at *Leiden* in the Netherlands invented the first practical *Electrocardiogram* (ECG), used to detect heart issues. It is a graph of voltage versus time of the electrical activity of the heart using electrodes placed on the skin. *Ultrasound* uses sound waves with frequencies higher than 20 Khz, inaudible to people, to form an image, typically a fetus in *Gynaecology*. The technique is similar to the *echolocation* used by bats, whales and dolphins. Ultrasound started with 1958 *Lancet* (a popular journal in medicine) paper *"The investigation of abdominal masses by pulsed ultrasound"* by Ian Donald at *Glasgow University*, John McVicar at *Royal Maternity Hospital*, and Tom Brown at *Kelvin & Hughes Scientific Instrument Company*, all from the UK; This paper reported the very first ultrasound images of a fetus.

Magnetic resonance imaging (MRI) is an imaging technique that uses a magnetic field and radio waves induced in an antenna by atomic nuclei in human tissues due to the field to create detailed images of the tissues in the human body using computers. Raymond Damadian at *SUNY Downstate Health Sciences University*, invented the first *MR scanner* for cancer detection in 1972. Paul Lauterbur at *Stony Brook University* near New York City in the US developed the technique for generating 3D MRI images in 1973. Peter Mansfield at *Nottingham University* near Manchester in England developed the mathematical technique, *echo planar imaging*, that made MRI fast enough to be practical in 1975.

X-ray, with other medical imaging inventions, ranks high for improving medicine by leaps and bounds, providing better quality of life (think of poor Mr. Ferguson above suffering for twenty years with a bullet) and saving lives. Almost everyone reading this has had an x-ray or some form of medical imaging such as ultrasound making it an ubiquitous and useful invention. According to WHO, worldwide around *3.6 Billion diagnostic procedures* are performed every year. It has a long history of around 120 years, ranking it in the top 100.

Cement

Portland Cement Association, the foundation of American Cement Manufacturers, calls cement *"The foundation of civilization"*, which is not much of an exaggeration for once, coming from partisan industry advocates. Cement literally built the foundations of millions of skyscrapers, homes and factories and also their walls and floors in addition. After water, concrete is the most widely used substance in all of human civilization. Portland cement contains about 60% *lime* (calcium minerals), 25% *silica* (sand) and 5% *alumina* (Aluminium Oxide) with *gypsum* (Calcium Sulfate Dihydrate, $CaSO_4 \cdot 2H_2O$) and *iron oxides* making up the rest of the 10%. The gypsum regulates the hardening time of the cement. Concrete consists of 60%-75% of *aggregates* (sand, gravel and crushed stone), 7-15% of *cement* and 14-21% of *water*. *Reinforced concrete* (or RCC for Reinforced Cement Concrete) is made by casting concrete around steel rods and bars, called *rebars*. Large structures such as highways, bridges and skyscrapers require the additional strength of Reinforced concrete. Around 31% of cement produced every year is used for highways and bridges, 28% for houses and 22% for offices, hotels and hospitals and other public buildings.

Joseph Aspdin at *Leeds* near Manchester in England invented and patented the *portland cement* in 1824 and gave it the name because it has the same color as stone quarried on the *Isle of Portland* at English channel that separates England from Europe. Joseph Monier at Paris in France developed and patented the reinforced concrete for different uses graduating from *horticulture* (Monier was a professional gardener) to bridges over 1850-1880; He used iron bars instead of steel as Bessemer's process for steel was invented only in the 1850s (see the chapter on *Iron Metallurgy*). It was popularized in the US by Ernest Ransome; Ransome's first rebar concrete building was the *Arctic Oil Company Works* warehouse in San Francisco, which was completed in 1884. In 1889, Ransome built the *Alvord Lake Bridge* in Golden Gate Park – the world's first reinforced concrete bridge, now the oldest RCC structure in the world.

Portland cement is a *hydraulic cement* that sets and hardens through a chemical reaction with water. Soon after the aggregates, water, and the cement are combined, the concrete mixture starts to harden. The longer the concrete is kept moist, the stronger and more durable it will become. Concrete surfaces are cured by sprinkling with water to ensure the continued hydration of the cement. Concrete continues to get stronger as it gets older. Most of the hydration and strength gain take place within the first month of concrete's life cycle, but hydration continues at a slower rate for many years. Concrete is both fire and water resistant.

Around *4 Billion tons* of cement is produced every year in the world with China producing more than half of that amount. For comparison, all the plastics produced in the last 60 years weigh the same as just two years of cement. China is the largest producer of cement in the world at 2.2 Billion tons in 2020. India is a distant second at 0.3 Billion tons. With operations in 90 countries, *LafargeHolcim* based in *Jona* near Zurich in Switzerland is the largest cement producer in the world and produced 386 million tonnes with a revenue of 28 Billion dollars in 2019. LafargeHolcim was formed in 2015 after the merger of two giants *Lafarge* and *Holcim*; Lafarge was founded in 1833 by Joseph-Auguste Pavin de Lafarge at *Le Teil* midway between Lyon and Marseille in France; Lafarge was a major provider of materials for the Suez Canal project in the 1860s. Holcim was founded by Adolf Gygi in 1912 at

Holderbank near Zurich in Switzerland. *Anhui Conch*, owned by the government of Anhui, a province near Shanghai and Wuhan and based in Wuhu, a city of 3.6 Million in Anhui, is the second largest with 288 million tonnes in 2019. *China National Building Material* (CNBM), a chinese state owned company based in Beijing is the third largest with 176 million tonnes. China is also home to the largest concrete structure in the world – the *Three Gorges Dam* across the *Yangtze River* with 27.2 Million cubic meters of concrete, around the size of 11,000 olympic sized swimming pools (see the chapter on *Dam*). Tallest skyscraper in the world, *Burj Khalifa*, has 330,000 m^3 of concrete, the size of around 132 olympic sized pools. The largest concrete structure in the US, *Grand Coulee Dam* that is also the largest hydroelectric dam in the US, has around 9.2M m^3 of concrete or 3,700 olympic pools worth. *Boston City Hall* in the US, with an architectural style called *'brutalism'*, is considered one of the *ugliest* concrete structures in the world. (Taking a look at it, it definitely deserves its title!)

On the flip side, apart from the ugliness of brutalist buildings, cement manufacturing is one of the biggest polluters in the world. China used more cement between 2011 and 2013 than the US did in the entire 20th Century. According to The *Guardian*:

"concrete is said to be responsible for 4-8% of the world's CO2. Among materials, only coal, oil and gas are a greater source of greenhouse gases. Half of concrete's CO2 emissions are created during the manufacture of clinker, the most-energy intensive part of the cement-making process."

Cement and Concrete ranks high as it serves one of our basic necessities, shelter. Our dwellings from skyscrapers to one story houses are built using them. National infrastructure such as dams, bridges, highways and runways are built using cement; Concrete pavement is all around us. Concrete has been with us from Roman times, the magnificent *Pantheon*, a 1900 years old Roman temple and now a church, in Rome in Italy being the prime example; Cement has been with us for around 200 years and RCC for 150 years. It had plenty of time and volume - at 4 Billion tons a year - to make an impact in human history. As the invention consumed the most by our civilization it ranks in the top 100.

Elevator

"The great object of my endeavors, has been to construct an elevator for the conveyance of persons from the different stories of hotels, public buildings and even private residences, which shall be free from the extreme and ordinary dangers, of suspension upon chains, ropes, or cords, of any kind, in the safety of which, every additional experience has led me to place less and less reliance.

... I have adopted the principle of a screw or revolving helical column, extending from bottom to top of the building, and to a nut, which is fitted to this screw, the passenger car is attached, the screw is revolved by steam or other kind of motive power. I am well aware that the screw itself, is an old, and much used device for raising weights in general; but, as far as I am informed the principle of the screw, has never been used to take the place of that of suspension in elevators ..."

The 1859 patent by Otis, "ELEVATOR OR HOISTING APPARATUS FOR HOTELS, &c." explains the elevator as above.

However one may get the feeling that it doesn't even remotely look like the elevators we use, with a *helical screw like column and a nut* with the elevator car hanging from it; However the inventor's name sounds about right. The reason is an amazing coincidence with truth stranger than fiction; At around the same time in the 1850s there were two people named Otis, inventing the same thing, an elevator. This patent is by Otis Tufts of Boston now known more for the first steam operated printing press in the US than elevator. In an alternative universe we may have adopted this design as it feels much safer although more expensive. In our universe, the other Otis, Elisha Otis at New York City invented the elevator we actually use now. In 1852, Elisha Otis introduced the commercial safety elevator, a crucial invention that made skyscrapers and modern city life possible. Otis's invention, a ratchet device using a wagon spring to engage a toothed bar when the rope broke, enabled him to start the first elevator company in the world that still remains one of the top elevator makers of the world with more than around 13 Billion US dollars in revenue in 2020. *E. V. Haughwout Building* in New York, US installed the world's first elevator from Otis in its five stories building in 1857. This was powered by a steam engine as electricity was not yet widely available at that time as a power source. In fact there is at least one interesting story of the elevator steam motor being ingeniously used to generate electricity required for electric lights. In an article from June 8, 1880, "HOTEL LIGHTED BY ELECTRICITY":

"On Saturday Evening the main hall, dining room, office, and bar-room of the Continental Hotel were brilliantly illuminated with the Brush electric light, six burners being used, each, it is stated, costing 1 cent an hour. In the dining room, where heretofore 144 gas-burners have been used, there will be in future two electric lights. The electro-dynamic machine is run by the steam-engine which moves the elevator, and the lessee of the hotel believes that the saving in the cost of lighting will be very great. In some years when the city gas was used, the bills ran as high as $21,000 for 12 months. It was noticeable on Saturday that the lights burned steadily, with no perceptible flicker. This light is only adapted for large rooms or halls, and will not be used in the sleeping-rooms at the Continental."

The eight-story *Equitable life building* in New York was the first office building to have an elevator in 1870. In 1880, First electric elevator was built by Werner von Siemens, founder of Siemens, in Germany. In less than 100 years from original invention Otis, the company - the inventor was long dead by then, installed 68 elevators in a single building, the iconic *Empire State Building* in 1931. The *New York Times* wrote on the importance of elevators [emphasis added] on Nov 8, 1903 when they installed modern electric elevators from Otis, in their brand new 25-story, 111 metres (363 ft) high The *New York Times* skyscraper at *Times Square*. (Times Square itself is named after this building):

"In the designing and constructing of the modern office building there is no one feature which deserves so much attention upon the part of those interested as the elevator equipment of the structure. **In all tall buildings the rental value of the upper stories depends entirely upon the efficiency of the elevator service.** There is also a fixed ratio between the number of the elevators and size of the structure. Experience has demonstrated that for every 25,000 square feet of total floor area there should be an elevator."

Modern cityscapes are defined by skyscrapers; Skyscrapers themselves wouldn't have existed without elevators. Buildings couldn't have been more than 5-10 floors without this invention. In 1850 before the invention of escalators, tallest buildings such as the *Augsburg Town Hall* in Augsburg near Munich in Germany were only seven stories tall. In the early 1900s, in the *Singer building*, we already see 41 floors thanks to elevators. Current tallest building *Burj Khalifa* is a whopping 163 stories tall. Modern big city life with its support of millions of people in tiny places like Manhattan or Singapore wouldn't have been possible without skyscrapers made possible by elevators.

Skyscraper

"Productivity isn't everything, but in the long run it is almost everything. A country's ability to improve its standard of living over time depends almost entirely on its ability to raise its output per worker."

Paul Krugman, a Nobel Laureate in Economics, said this about human productivity. All important, history changing human inventions increase human productivity, steam engine, electricity and computers to name a few. Same can be said about the land; Our useful land on earth is limited. If we have to feed billions more people using the same or less land, agricultural land simply has to become more productive, that is produce more grains per acre. This is the importance of the Haber-Bosch process (see the chapter on *Fertilizer*) as it made the same land vastly more productive saving billions from famine and starvation Malthus style. Skyscrapers do the same for land in cities; It makes the same land livable for more people, can support more businesses or support more office workers for a business in the same amount of land. It made our third dimension, vertical, more productive for human purposes. This invention was powered by the mass production of steel using the Bessmer process, electric lighting and elevator all of

which came together in the 19th century. Steel allowed for much greater height, strength and stability without the greater weight of traditional masonry construction.

In 1885, the first skyscraper in the world, the *home insurance building,* designed by William Jenney was built in Chicago in the United States. (Jenney's classmate from *École Centrale Paris* where he studied Architecture, Gustave Eiffel, designed the *Eiffel Tower*). It was ten stories high and 42 m (138 ft) tall when it was built; Later two more floors were added making it 55 m (180 feet) high. The *Empire State Building,* designed by William Lamb and Homer Balcom and built in 1931, was the first building in the world to have more than 100 floors. The building is so large that it was assigned its own ZIP Code, *10118*; It has 102 floors and 380 meters (1250 ft) tall; It also has 73 elevators. It cost about 52 Million dollars (900 Million in 2021) and contains 60,000 tons of steel. Governor Franklin Roosevelt (FDR) of New York spoke for many of us, awestruck by the view from a tall skyscraper, on the day, May 1, 1931, the Empire State Building opened:

"I am still a little awestruck. I have not got my sense of proportion back yet. In looking out from this building I have got an entirely new conception of things in the city of New York. As a simple countryman who has only been down here in New York for twenty-five years I still think in terms of fields and creeks. And when I looked out north and saw Central Park, it reminded me of the sides of my cow pasture at Hyde Park. And when I looked over to the Hudson river and East river, they looked from here just about the size of Wappinger Creek, in Duchess County."

Central park is the huge 846 acre park in Manhattan, the main island in New York City and *Hudson river* near New York is a few miles wide. It would be the world's tallest building till 1972, for 41 years. World's tallest building in 1900, *Park Row Building* in New York City designed by Robert Robertson, was only 31 storey high. In 1950, it was the Empire State Building at 102 storey high. In 2000, *Willis (Sears) Towers*, Chicago had the most floors at 108 but *Petronas Towers* in Malaysia was the tallest in the world at 451.9 m (1,483 ft). There are now 18 buildings with 100 or more floors; 7 in China, 4 in the US, 3 in UAE, 2 in South Korea and one each in Saudi Arabia and Taiwan. A floor takes up around 4-5 meters in typical skyscrapers.

Burj Khalifa in Dubai in the United Arab Emirates, designed by American architect Adrian Smith, is currently the tallest skyscraper with a height of 829.8 metres (2,722 ft) and 163 floors; For contrast, the height of the *Great Pyramid* of Egypt is only 146.7 metres (481 ft), making Burj Khalifa 5.5 times taller and 2.5 times taller than Eiffel tower's height of 324 metres (1,063 ft); It cost 1.5 Billion dollars to build when it opened in 2010. World's second most expensive building complex at $16 Billion and 120 floors, *Abraj-al-bait*, was built in 2012 in Mecca, Saudi Arabia near the most expensive building in the world, *Masjid-al-haram*, Islam's holiest site, at more than $100 Billion. Europe's tallest is the 87-story *Lakhta Center* built in 2019 by *Gazprom* in St. Petersburg, Russia. The 88 storey *Petronas Towers* at *Kuala Lumpur* in Malaysia, designed by César Pelli, an Argentine-American Architect, is the tallest twin towers in the world.

Skyscrapers define the cityscape of modern cities such as New York, Chicago or Singapore. Skyscrapers make the same land in places such as the tiny island of Manhattan, New York (59 km^2, for contrast O'ahu island in Hawaii is 1,545 km^2) more productive and support more people (*26,821 people per km^2*). For comparison, the average population density of the world is 25 people per km^2 and average population densities of India and the US are 450 people per km^2 and 36 people per km^2 respectively. In the 1920s New York surpassed London as the biggest metropolitan area in the world mainly on the strength of its vertical space that accommodated more people. In 2011, New York City had 221 buildings 150m or taller. Cities with 10 Million or more population couldn't be sustained without skyscrapers particularly if they have to work in the same city as well. Just like fertilizers providing food and sewing machines providing clothing, skyscrapers provide shelter for work and rest for hundreds of millions of people making it rank high as a provider of basic necessities.

Dynamite

"If I have 300 ideas in a year and just one turns out to work I am satisfied."

said Alfred Nobel, inventor of Dynamite who had *355 patents* to his name. He is today remembered for the *Dynamite* and the prestigious *Nobel prize*. Dynamite is in some sense very simple; It is *Nitroglycerin* with some river sand (*diatomaceous earth*). Nitroglycerin is very powerful but too unstable and dangerous to be used as an explosive. As we have seen many times in this book such as Watt's *condenser* in the steam engine (see the chapter on *Steam Engine*), incremental innovations matter, *a lot*. Nitroglycerin was known before Nobel; Nobel "simply" stabilized Nitroglycerin by adding river sand as an absorbent in 1867 and it had important consequences for humanity. It replaced *gunpowder* which was very ineffective as an explosive in mining, quarries and other places requiring large scale earth moving with explosives. It accelerated mining and construction of large projects such as the *Panama canal*. The Panama canal would have been impossible in a world in which dynamite wasn't yet invented as it required *7,700 tonnes* (17 Million pounds) of dynamite to move *100 Million cubic meters* (3.5 Billion cubic feet) of earth, or 40,000 olympic size swimming pools worth of dirt. The Panama canal is providing benefits in perpetuity by shortening the travel distance for ships from New York to San Francisco by around 15,000 km (8000 miles). *Hoover Dam* on the Colorado *River* in the southwest

United States used *3,900 tonnes* (8.5 Million pounds) of dynamite. These mega projects would have taken an order of magnitude more time without the dynamite and in many cases would have been impossible. *Science history institute* says [emphasis added]:

> "**One thousand times more powerful than black powder**, dynamite expedited the building of roads, tunnels, canals, and other construction projects worldwide in the second half of the 19th century."

Italian chemist Ascanio Sobrero discovered nitro-glycerine (*trinitroglycerin* or *1,2,3-trinitroxypropane*) while studying at the *University of Turin*, a city near French and Swiss border in Western Italy. Alfred Nobel studied in the same University and learned about nitro-glycerine there. Alfred Nobel's 1868 patent, "IMPROVED EXPLOSIVE COMPOUND" explains the *dynamite*:

> "The nature of the invention consists in forming out of two ingredients long known, viz, tho explosive substance nitro-glycerine, and an inexplosive porous substance, hereafter specified, a composition. which, without losing the great explosive power of nitro-glycerine, is very much altered as to its explosive and other properties, being far more safe and convenient for transportation, storage, and use, than nitroglycerine.
>
> In general terms, my invention consists in mixing with nitro-glycerine a substance which possesses a very great absorbent capacity, and which, at the same time, is free from any quality which will decompose, destroy, or injure the nitro-glycerine, or its explosiveness."

Instantaneous destruction of all the molecules in the explosive is known as *detonation*; Rapid expansion of hot gases causes a violent and destructive *explosion*. Nitroglycerin ($C_3H_5\text{-}(NO_3)_3$) derives its power from three nitrate groups (NO_3) tightly packed to a 3-carbon *propane* (C_3H_8) structure. As explained in the chapter on firearms, its nitrogen atoms are eager to get back to their stable, inert nitrogen bonds in nitrogen (N_2) molecules in the air (see the chapter on *Firearm*). Hydrocarbon skeleton, a well known cooking gas *propane*, provides the fuel as well. Current military dynamites don't contain nitroglycerin and are replaced with *RDX* (Royal Demolition eXplosive), *cyclotrimethylenetrinitramine* as it is safer to handle. Explosives such as *C-4* are mostly (around 90%) RDX with a binder and plasticiser. *ANFO*, Ammonium Nitrate/Fuel Oil, is the most commonly used explosive now for commercial blasting as it is readily available. *Ammonium Nitrate* is produced as fertilizer (see the chapter on *Fertilizer*) and is a commonly available chemical and when mixed with fuel oil such as *diesel* it produces a powerful explosive. Ammonium Nitrate's use as an explosive was discovered in a ghastly way. The *New York Times* explains in an article on 1995 Oklahoma City bombing by a white supremacist that also used ANFO [emphasis added]:

"Engineers discovered that ammonium nitrate fertilizer could be used as an explosive in 1947 after **600 people were killed in an accidental explosion aboard a ship** being loaded with fertilizer in Texas City, Texas.

The fertilizer was in paper bags and apparently blew up after sailors tried to stop a fire in the ship's hold by closing a hatch, creating the compression and heat necessary for an explosion, said Dr. Per Anders Persson, the director of the Research Center for Energetic Materials at the New Mexico Institute of Mining and Technology in Socorro, N.M.The force of the blast was so great that **people were thrown to the ground 10 miles away in Galveston**, Dr. Persson said."

Projects such as the Panama canal, many highways such as the beautiful *California highway one* and tunnels such as the *Channel tunnel* between the UK and France owe their existence to explosives. It gave humanity great power over earth by helping move huge amounts of rock and mountains quickly; It improved the productivity of construction projects. Explosives also play the role of an anti-invention with military explosives in wars and landmines and by terrorists to, well, terrorize society. The *Halifax explosion* in 1917 at *Halifax* port in Canada near Boston killed around 1800 people after explosives laden *SS Mont-Blanc* caught fire. Explosives, for both the positives and negatives, rank in the top 100.

Pesticide

"NEW ORLEANS, Dec. 6 - The United States Public Health Service and cooperating Southeastern states are making 'rapid progress' in their five-year, $6,000,000 program to reduce malaria infection to 'a negligible public health problem' it was announced today before a national convention on tropical medicine.

...Frank Tetzlaff, chief of engineering division of the United States Public Health Service's communicable disease center at Atlanta, Ga., said that the malaria-control program consisted principally of the residual spraying of homes with DDT in endemic malaria areas.

...He stated that the objective of the malaria-control program was to use residual spraying to achieve a reduction in infection over a five-year period beginning with 1947. The Federal Government has appropriated $3,000,000 for this purpose and the cooperating states have matched it with a similar sum."

The *New York Times* of Dec, 7, 1948 in the article, "REPORT PROGRESS IN MALARIA FIGHT", reported on the eradication of Malaria in the US southeast, the most backward region in the US consisting of states such as Mississippi, Alabama and Louisiana, using *dichloro-diphenyl-trichloroethane* (DDT), one of the most important weapons we invented in our fight against Malaria, the biggest killer of people in the world. Mosquitoes kill almost half a million people per year, that is it kills someone every minute. Malaria was prevalent in the United States in the 19th century; Abraham Lincoln contracted Malaria twice in Illinois, once in 1830 and the second time in 1835. A path breaking map on "PROPORTION OF DEATHS FROM MALARIAL DISEASES" produced by the *US census bureau* in 1870 shows some areas with a shocking 14 out of 100 deaths or more from Malaria in the US; *Tallahassee*, capital of the southern state of Florida among them; Trinity river basin near the current mega city of *Houston* in Texas with a population of 7 Million is another. In the 20th century malaria claimed between 150-300 Million lives (second only to smallpox and TB), 2-5% of all deaths.

dichloro-diphenyl-trichloroethane (DDT) was first synthesized in 1874 by Othmar Zeidler of Austria. DDT's insecticidal use was discovered in 1939 by the Swiss scientist Paul Hermann Müller at *Geigy* (now *Novartis*) in Basel, Switzerland. He was awarded the 1948 Nobel prize "for his discovery of the high efficiency of DDT as a contact poison against several arthropods." *Arsenic* based insecticides have long been used in agriculture. *Lead arsenate* was the most commonly used arsenic-based insecticide in the early 20th century. Paris Green (*copper(II) acetoarsenite*) was also very popular as a pesticide. There are more than 1,000 pesticides used around the world now. Most common pesticide groups in use now are organochlorines (DDT, *Endosulfan*), organophosphates (*chlorpyrifos*), carbamates (*Carbaryl*) and pyrethrins/pyrethroids (*Bifenthrin*).

Pesticides are double edged swords. DDT's negative effects such as eggshell thinning in birds such as eagles and pelicans are well known. *Agent Orange* used by the US army in Vietnam for *defoliation* ("removal of leaves") of forests has been known to cause long term health issues in around a million people. Arsenic based compounds are as toxic to people and animals as it is to pests. *Endosulfan*, a commonly used pesticide in India, is highly toxic to people. Disorders of the central nervous system are very common among the children of the area with *Endosulfan* use - cerebral palsy, epilepsy and congenital anomalies like *stag horn limbs*. The *New York Times* says in an article in 2011:

"Endosulfan, a powerful 50-year-old insecticide sometimes called DDT's "cousin," was officially banned last week at an international pesticides meeting in Geneva. Partial exemptions were created for India, however; the chemical may be used on some crops there for up to 10 years.

...In India, endosulfan is ubiquitous and controversial. It is blamed for deforming hundreds of children in the southern state of Kerala whose parents worked on cashew plantations. Pictures of them are common there, reminiscent of "thalidomide babies" in the 1950s and victims of mercury poisoning in Minamata, Japan."

Pesticides rank high as an effective weapon against deadly diseases such as Malaria and for their importance in agriculture. *Irish potato famine* that killed a million people in Ireland between 1845-1852 was caused by potato blight from the fungus *Phytophthora infestans*. As late as 1943, The *Bengal famine* killed 2 to 3 Million people that was caused by *brown spot disease* in rice from the fungus *Cochliobolus miyabeanus*. (However, in both these famines there were confounding political factors which made the death toll horribly worse; A modern example could be 2020 United States with an incompetent adminstration that made the coronavirus pandemic much worse with a death toll of half a million people). Pesticides played a secondary but significant role in eliminating famines that were common across much of the world before the 1950s. For their life giving power both from eliminating diseases such as Malaria and increasing agricultural productivity and for their equally powerful negative effects they rank high in our list, in the top 100.

Safety match

"How could John Walker of Stockton-on-Tees have imagined that by the time the centennial of his discovery arrived more than 6,000,000 matches would be lighted every minute throughout the world? How could he have guessed that a single machine of American make would turn out 177,926,400 matches in a day ... The story of the rise of the match during its comparatively short life to a position of probably the most used convenience in the civilized world is one of compelling interest, a tale of a growth that turned whole forests into tiny white splinters so that man might have the gift of fire at his instant command.

The world's consumption of matches has been placed roughly at **3,228,425,000,000 a year**, with five a day a reasonable estimate per capita of population. A billion a day, it is said, are used in the United States alone. In England the annual consumption of matches is set at two hundred billion a year while the annual American output is three hundred billion. If the matches

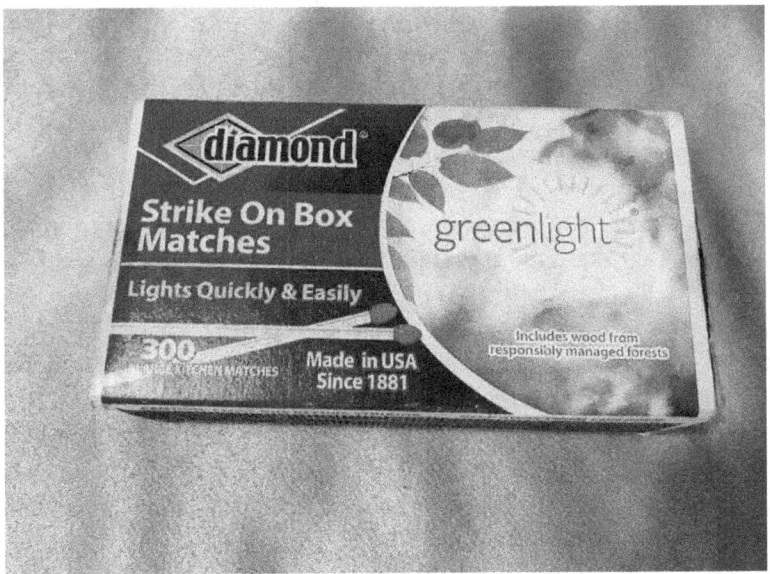

made each year were laid end to end they would reach a distance of 95,538,145 miles or almost 4,000 times around the circumference of the globe."

The *New York Times* of Nov 27, 1927 wrote on the 100th anniversary of safety matches in an article, "OUR LOWLY MATCH HAS ROUNDED OUT A CENTURY" [emphasis added]. First match, an invention for producing fire safely and conveniently on demand, was invented by John Walker at *Stockton-on-Tees* near Manchester in England in 1826. He used *chlorate of potash* ($KClO_3$) and *sulphide of antimony* (Sb_2S_3) mixture attached to the end of a wooden stick dipped first in *molten sulphur* to produce the first friction match. White *phosphorus*, a highly inflammable substance that ignites at 30°C (86°F), replaced Walker's mixture shortly after Charles Sauria at *Jura* near Lyon in France bordering Switzerland introduced the *"strike anywhere"* match in 1831. In 1844, Gustav Pasch of *Karolinska Institute*, which now selects the awardee for *Nobel Prize in Medicine*, at Stockholm in Sweden invented the safety match with *red phosphorus* and a special strike pad. In 1855, Carl and Johan Lundström, two brothers in Sweden founded the *Jönköping match factory*, at *Jönköping* mid-way between Stockholm and Copenhagen, that popularized the safety match around the world. They developed the now familiar matchbox, a slide box with sand paper on its side, that has kept the same look for 170 years now. White Phosphorus caused a serious and very painful condition called *"phossy jaw"* in match workers and it was replaced in the early 20th century with *sesqui-sulphide of phosphorus* (P_4S_3), developed by George Lemoine decades back in 1864.

White (Yellow) phosphorus (*"bringer of light"*) consists of P_4 molecules, four phosphorus atoms in tetrahedral structure similar to methane (CH_4), whereas red phosphorus is amorphous and has a complicated network of bonding. White phosphorus has to be stored in water to prevent natural combustion, but red phosphorus is stable in air. White phosphorus' glow occurs as a result of its vapors slowly being oxidized by the air. White phosphorus is highly toxic. In "phossy jaw", excess phosphorus accumulation caused the match workers' bone tissue, particularly of jaw bone to die and rot away. For this reason, we now use red phosphorus or phosphorus sesquisulfide in matches.

"It's not the order in which things are invented that makes them the most impressive, it's the importance they have to humanity. So my number one is this: fire with a flick of the fingers."

Stephen Fry, a British TV personality and actor, in the program *100 Greatest Gadgets* on BBC Channel 4 said about *lighter* that applies equally well to safety matches. In 1823, Johann Döbereiner at the *University of Jena* near Dresden in Germany invented one of the earliest lighters, *Dobereiner's lamp*, also called *Tinderbox*, which predates matches. This lamp worked by reacting zinc with dilute sulphuric acid in order to produce hydrogen. To use, a valve was lifted, firing the hydrogen towards a porous form of platinum known as *'platinum sponge'*. This then reacted with the atmospheric oxygen, heating the platinum and igniting the hydrogen, producing a steady flame. *Ferrocerium* was invented in 1903 by Carl Auer von Welsbach, of gas mantle fame (see the chapter on *Kerosene lamp*), at Vienna, Austria; It revolutionised the lighter, as it made creating sparks for ignition easy, and was relatively affordable. Ferrocerium, often referred to as 'flint' from flint's previous usage as a producer of sparks, is a synthetic alloy of Iron and rare-earth element *Cerium* (atomic number:58) that produces very hot and bright sparks when struck. George Blaisdell at Bradford near Lake Erie in the US near the Canada border introduced what would become the world's most famous lighter in 1933, *Zippos*. Bic introduced the now ubiquitous disposable lighter in 1973; Now Bic produces more than 2 Billion lighters every year and holds around 50% of the worldwide market. *Butane* is commonly used as the lighter fluid.

Matches rank high as the first mechanism to produce fire affordably, instantly and conveniently on demand. It was one of the first items mass-produced heavily, trillions of them every year around the world. As fire is one of our species' signature inventions, controlling it and inviting it back instantly on demand with a "flick of the finger" counts as one of our important inventions. Matches have been with us for 200 years and lighter for 100 years, having enough time to make a significant impact in human history and ranking them high.

Kerosene lamp

"Miss Sadie Conway, of No 226 Second-street, Brooklyn, E.D., who was so severely burned on Monday evening through the falling and breaking of a kerosene lamp, died yesterday morning after a night of terrible agony. The family were about to move from this house, and in the absence of her parents, Miss Conway was showing a truckman the various parcels to be moved. When the bowl of the lamp she was carrying came off, and, falling to the floor, broke in pieces. The oil ignited and set fire to her skirts. The truckman sprang to her assistance and had almost entirely extinguished the fire, when the girl broke from his grasp and rushed to the street. She ran nearly a block toward Grand-street, screaming for help, but avoiding those who endeavored to hold her. Before she had run the block her clothing was a mass of flame. She then

turned and ran back toward her house, falling in the street just as she reached it. The remnants of the burning clothing were then torn from her, and she was carried into the house."

The *New York Times* of Oct 29, 1879 has this news report of a fatal accident from Kerosene lamp. Nearly two billion people have intermittent or no access to electricity in 2021 according to the BBC. In Nigeria, the biggest country in Africa with a population of *206 Million* around two-thirds of the US population, only 56% people had access to electricity in 2018. In Ethiopia, the second biggest with a population of *110 Million*, only 45% had access to electricity (2018). In Burundi, Chad and Burkina Faso 11%, 12%, 14% respectively had access to electricity (2018). Burundi has more people than Sweden or Switzerland; Chad has around the same population as the Netherlands; Burkina Faso has the population of Norway, Sweden and Denmark combined. As late as 1990, in India only 43% had access to electricity. There were frequent power disruptions due to poor infrastructure for electricity and people relied on kerosene lamps during blackouts including for education and school homework (myself included!). For this reason, the government subsidized kerosene for use in lamps. Kerosene lamps still remain the most accessible form of lighting for almost *930 Million* people in the world or *one in eight people* in the world, mostly in Africa. This lamp is extremely affordable with no need for costly infrastructure in countries which can't afford them. (Once infrastructure is in place electricity is much cheaper). In 1859, 1,800,000 kerosene lamps were sold in the United States and by the end of 1862 the country had between four and five million kerosene lamps out of around six million households. (Based on the 1860 census, the US had 31,443,321 people with an average household size of 5.5). In the US, it is still used in communities such as Amish of Pennsylvania and Ohio who shun modern amenities such as electricity. John Rockefeller became the richest person in the world by selling Kerosene for lamps in the 19th century (see the chapter on *Petroleum*).

Modern kerosene lamp was invented by Ignacy Łukasiewicz in 1853 at Gorlice, Poland near the Slovakia border in the south. As mentioned in the chapter on Petroleum, In 1852, Abraham Gesner at

Halifax at the Atlantic coast, Canada (the same *Halifax* that was devastated by explosives in 1917) gave the name "*kerosene*" to this petroleum oil. The well-known *Petromax* lamp, a kerosene lamp with the characteristic *mantle* instead of a wick and gives off powerful light equivalent to 300 candles, was created in 1910 by Max Graetz of Germany, named after his nickname *"Petroleum Max"* at the lamp factory *Ehrich & Graetz* at Berlin in Germany founded by his father Albert Graetz in 1866. In 1914, William Coleman introduced *Coleman lanterns*, at Wichita near Oklahoma in the central US, that were as popular in the US as Petromax was around the world, becoming an "essential item" for the troops in World War I. In 1920, Tilley in England made the famous *Tilley lamp*, the "Petromax of England". Tilley has been making (non-kerosene) lamps from 1818 and started making kerosene lamps in 1915 for british troops in World War I and railroad companies. Now local manufacturers in India and Africa make mantle kerosene lamps based on these proven, 100 year old models and in India it is still called *Petromax*. Carl Auer von Welsbach of Austria invented the gas mantle, initially for gas lamps, that was adapted for use in these powerful kerosene lamps in 1885. To produce a mantle, *guncotton* is soaked in a mixture of *Actinophor*, a chemical mixture of 60% *magnesium oxide*, 20% *lanthanum oxide* and 20% *yttrium oxide*, and then heated, the cotton eventually burns away, leaving a solid fishing net like structure, which glows brightly when heated.

It deserves its place as the light source powering one-eighth of humanity. Kerosene lamps are widely used for lighting in rural areas of Asia and Africa where electrical distribution is either not available, or too costly for widespread use. It made light affordable to vast numbers for the first time in the 19th century; It was a huge improvement over earlier ones burning vegetable or animal fat oils (and spared whales in the process). At least two billion people in rural parts of Asia and Africa use it, that is there are at least 2 billion kerosene lamps in the world. It has also been with us for around 170 years, playing a significant part in human history, ranking it high in the list of inventions.

Vacuum tube

As with many inventions in this book, it starts with Thomas Edison of Menlo Park in the US and his incandescent bulb (see the chapter on *Light bulb*). In 1884, he filed one of his more than thousand patents with an innocuous looking title, "ELECTRICAL INDICATOR". This patent would prove to be at the foundation of the many revolutions in the following twentieth century.

"I have discovered that if a conducting substance is interposed anywhere in the vacuous space Within the globe of an incandescent electric lamp, and Said conducting substance is connected outside of the lamp with one terminal, preferably the positive one, of the incandescent conductor, a portion of the current will, when the lamp is in operation, pass through the shunt-circuit thus formed, which shunt includes a portion of the vacuous space within the lamp, This current I have found to be proportional to the degree of incandescence of the conductor or candle-power of the lamp."

This describes the phenomenon at the heart of vacuum tubes, *thermionic emission*, emission of electrons by some metals when

heated. Vacuum tubes (*thermionic valves* or simply *valves*) built the first radio, first TV and the first computer. Vacuum tube is conceptually very simple. It can send the electricity in one direction (from cathode to anode by thermionic emission) but not in the opposite direction. Although it sounds like a trivial property it has world altering consequences. For one it allows us to convert alternating current coming from electricity mains to direct current required by most electronic devices. That is the least of its magic. Adding an extra electrode in between cathode and anode transforms this simple device; It allows a control current (typically an information signal encoded as current) to induce a much larger current following the same form in the device, effectively amplifying the signal as the large current has the same waveform, that is the same information content, as the original signal. A transistor operates the same way but more efficiently and takes up less space. If we had waited for a transistor to appear it would have delayed these critical inventions, including radio and TV, by at least half a century. In addition, we may not even have realized the importance of the transistors immediately without working applications already surrounding us. John Fleming of London, England working for the *Marconi wireless telegraph company* of America, New Jersey invented the vacuum tube in 1904, the version without the extra electrode, *diode*. Lee de Forest of New York invented the important variation with the extra electrode, called *triode,* that does the amplification. A *Nature* article written in 1922 describes the thermionic valves and one of its first applications:

"This remarkable invention can be described briefly as a highly exhausted glass bulb, in which is mounted a tungsten or tantalum filament heated by a battery giving about 6 volts. Electrons are emitted by the heated filament. The filament is surrounded by a grid or gauze cylinder, which is insulated and kept at the negative potential of the filament, while a plate of metal mounted inside the bulb is kept at a high potential of from fifty to several hundred volts by means of a battery or some other source of continuous current. The bulb is highly exhausted, and while the grid is kept at a normal negative potential, steady current passes from the filament to the plate or anode, but as soon as the grid is made slightly positive or negative, the current passing between the filament and anode by virtue of the electronic conductivity is increased or decreased.

...One of the most important applications of the valve is the amplification of telephone currents in long distance telephone trunk lines. Here, owing to the length of the cable and to the electrical constants involved, speech becomes greatly attenuated, and thermionic relays or repeaters are introduced about every thirty miles which amplify the speech to its original degree of loudness. In addition, cable of much smaller diameter and weight can be employed, as currents producing almost inaudible sounds can be amplified to any degree of strength."

One application of vacuum tubes is still widespread as the *magnetron* in microwave ovens, a billion of them. It was invented by Percy Spencer in the US working at *Raytheon corp* of Boston in 1947. In the magnetron, thermo-ionically emitted electrons in a magnetic field produce microwave oscillations in resonant cavities that are used to cook the food (see the chapter on *Stove*).

Although vacuum tubes were replaced by transistors later in the 20th century, it played a crucial role in human history. Vacuum tube ranks very high as the "umbilical cord" technology for important electronic inventions. Amplification was critical for long distance telephones, radio, TV and computers. The main idea of vacuum tube itself, that of using a control current to amplify or stop the main current became far more successful and lives on in transistors and with it all the modern digital devices to represent binary 0 and 1. All the 20th century communication revolutions including telephone and wireless telegraph (radio) owe their success to this device and the concept, ranking it high enough to be in the top 100, even as an obsolete invention taken over by transistors.

Washing Machine

"With all the assurances we are being given that the New Deal is bettering humanity and that our President is Great Humanitarian, it is a shock to learn that household washers will be put into the 'luxury' class if the Washington economists have their way.

Of all the blessings which have come to women in the past fifty years, the washer just about tops the list. Why does our Administration force the woman who needs a washer or the woman who owns a worn-out washer to revert to back-breaking toil, or perhaps feed her family less to pay the bills for outside service? Do the laundries of America get the best of the deal? I think they do."

One William Shaw of Chicago wrote this angry letter, "Plea for the Washing Machine", to The *New York Times* on Aug 23, 1941, in the middle of World War II [emphasis added]. Hans Rosling, the swedish

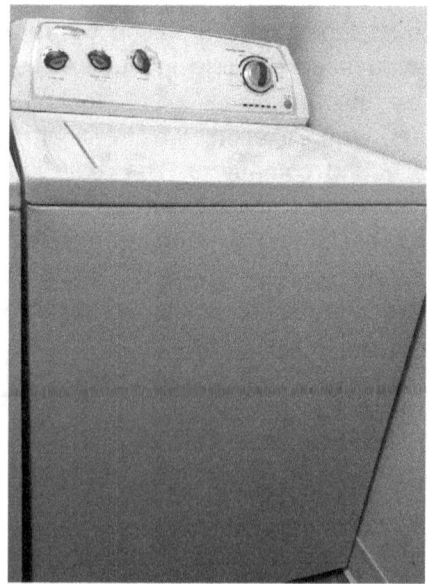

doctor and public health expert named in the *The World's 100 Most Influential People* by the *Times* in 2012, called Washing machine *the greatest invention of the industrial revolution*. (As the rankings in the book indicate, that is somewhat of an exaggeration but the point remains). Earliest washing machines such as the first drum type machine patented by James King at *Suckasunny* near New York in the US in 1851 were hand cranked. The *Science Museum* has a hand turned washing machine made by *Thomas Bradford & Company* of Manchester and London, England around 1880-1890. Initial machines came with a *wringer* for drying clothes. The wringer consisted of two wooden cylinders used to squeeze the water out controlled by a hand crank. The wringer required labor to feed through each article of clothing; It also was notorious for crushing fingers. Later the wringer was replaced by a *dryer*, using the process of spinning the drum of clothes at high speed in a perforated metal container with centrifugal force removing the water from clothes. Alva Fisher of the *Hurley Machine Company* of Chicago, Illinois invented the first electric-powered washing machine in 1908 and it was sold by the brand name *Thor*, marking an inflection point in the history of washing machines. *Museum of Applied Arts & Sciences* in Sydney, Australia curator Margaret Simpson says about the history of the washing machine:

"Throughout the 1800s and into the first half of the 1900s **washing clothes was a laborious and time consuming weekly chore which took a full day to complete, usually on a Monday**. This machine was certainly an improvement on the hard work of scrubbing clothes against a washboard in a tub but it still required a considerable amount of labour. This was to fill and empty the machine twice by hand, to wash, squeeze out and then rinse and remove the heavy wet washing, wring it out again and then hang it on the line.

In an effort to produce an effective washing machine that imitated hand washing, by the 1870s some 2,000 patents had been issued in the USA alone. As well as the machines themselves, there were also many wringers and mangles to squeeze out the washing water, not to mention boilers and coppers. However, the invention of the first electric washing machine, patented in 1910, was the true turning point in washing machine design, leading to washing machines eventually becoming real labour-saving devices. However, most women in Australia had to wait until the 1950s or 60s for their first electric washing machine, when there was greater prosperity and the ready availability of electricity, to release them from laundry drudgery."

Whirlpool and *Electrolux* are the top home appliances (a category that includes washing machines) companies in the world. In 2017, there were around *100 Million washing machines* sold in the world. *Whirlpool* was started by Louis and Emory Upton, an interesting and rare combination of niece and uncle pair, in Nov 1911 to sell electric washing machines. It started as *Upton Machine Company* at *Saint Joseph* across Chicago near *Lake Michigan* in the US. They changed the name to *Nineteen Hundred* after merging with the *Nineteen Hundred washer company* of New York in 1929. They changed the name again in 1950 to Whirlpool. *Electrolux* was founded by Sven Carlstedt as *Elektromekaniska AB* in 1910 in Sweden and Axel Wenner-Gren through a series of mergers made it Electrolux in 1917-18. In 1924, 29% of households already had an electric washing machine in the US, exceeding India's percentage a century later. In India, only 13% of households have a washing machine (2018) and 64% have it in Brazil (2017). In contrast, 96.1 households have washing machines in Germany (2020). New Zealand had 97.7% coverage already in 1998.

Washing machines rank high as one of the most important labor saving devices in the household. Our life is lived in seconds, hours and days. Washing machine returned our *"washday"* and hence 1/7 th of our life back to us or around 10 years for a woman living to be 100 years, excluding childhood and old age. (It is interesting that the washing machine adds the same amount to our life as the amount a smoker loses to smoking). Women's emancipation and washing machines go together (correlated) In human history and washing machines serve as a proxy for the state of women's rights in every country from India to New Zealand, the first country to give women the right to vote in 1893.

GPS

"This summer a small jet aircraft flew from Iowa to Paris, navigating only by the signals from the experimental Air Force Navstar satellites already in orbit. Not only did this system permit transatlantic navigation, but upon landing, the pilot taxied to his assigned parking space using only the satellite signals and came to a stop within less than a wingspan of the designated parking spot.

Production of the Navstars designed for operational use has been funded. The first of these will be launched in 1986 and a full operational complement of 18 will be in orbit before the end of 1989. The Global Positioning System made possible by Navstar, although designed for military use, **will be a major boon to commercial users who will be able to determine their position accurately in terms of latitude, longitude, altitude and speed.**"

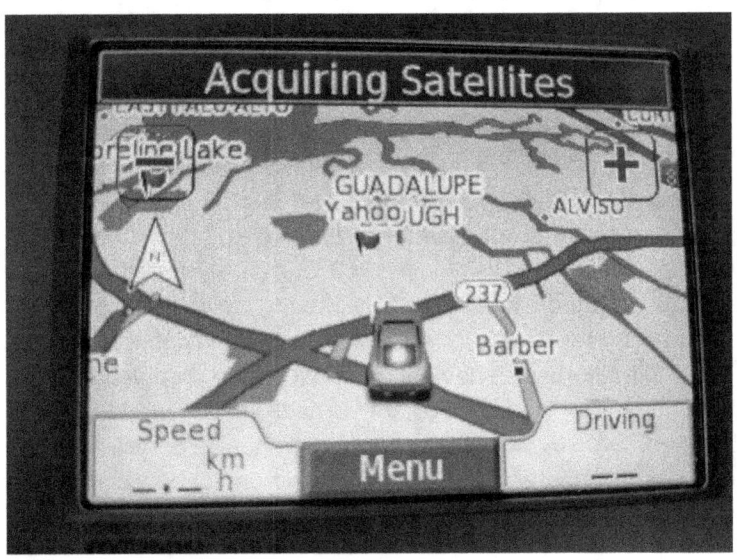

One of the earliest articles [emphasis added] in The *New York Times* in Aug, 1983 about GPS (then called *Navstar*) clearly anticipates its importance a few decades ahead and describes one of its earliest uses for navigation. It is easy to trace the history of GPS with contemporary newspaper articles as it happened very recently. 1,400 sailors lost their lives in the *Scilly naval disaster* of 1707 as mentioned in the chapter on Maps. In this disaster four warships carrying these sailors crashed into the Isles of Scilly near England and sank as they were unable to determine their location accurately and deviated from their path back to England. This disaster led to the invention of the *Chronometer* which enabled one to calculate the longitude at sea accurately as long as one knows the time (very accurately) at a reference longitude like *Greenwich*. The *Global Positioning System* (GPS) vastly improves on this invention by enabling one to calculate the *latitude, longitude* and *altitude* at any point on earth using location and time signals received from satellites in space. That is GPS receiver computes its position x, y, z from known distances x_i, y_i, z_i of satellite i [in a shared coordinate system called *WGS 84*] and the distance travelled, that is speed of light **c** times the time difference t_i using simple (Euclidean) *Spherical Geometry*. The GPS project called *NAVSTAR* by the US military in 1973 started the GPS satellite network and it was fully operational by 1993. The US government made its signals fully available to the public on May 1, 2000 under President Bill Clinton starting the explosion in GPS receivers and maps using them. In 1988, *Magellan Nav 1000* by *Magellan* in California in the US was the first commercial *GPS receiver* in the world. The first mobile phone with a GPS receiver, the *Benefon Esc!* by *Benefon*, Finland, became available in 1999. In 2001, companies such as *Garmin* of *Lenexa* smack in the middle of the US near Chicago founded in 1989 and *TomTom* of Amsterdam, Netherlands founded in 1991 began offering navigation devices that provided turn-by-turn directions. Now it makes navigation very trivial particularly with the use of navigation services like *Google Maps*. Now, GPS receivers are in a few billion devices in the world as part of smartphones. On Nov 11, 2020, India's regional GPS system (*IRNSS*) that operates in India and around 1,500 km from its boundary became operational. Apart from the US, Europe (*Galileo*), China (*BeiDou*) and

Russia (*GLONASS*) have their own GPS satellite systems as well. A *New York Times* article from March 29, 2001, "HOW IT WORKS; Getting There With Help From Above" explains:

"With G.P.S., a receiver on the ground or in the air can calculate its position using time signals from the satellites. The calculation itself is based on a kind of triangulation -- a high-school math technique used to locate an object based on its distance from three points. So signals from three satellites are necessary, although in practice a signal from a fourth satellite is used to improve the accuracy of the other three signals. To calculate these distances, the system uses another basic high-school math equation: distance is equal to the speed of travel multiplied by the time.

In addition to the time, a signal from a G.P.S. satellite also includes information about the satellite's exact location, which is known, tracked and kept accurate by ground control stations. The time signal is also very accurate, because each satellite contains several atomic clocks. These rely on the natural, and very regular, oscillation frequencies of atoms to keep time. The end result is that a G.P.S. receiver -- perhaps an in-dash navigation system in a car -- can produce highly accurate coordinates of latitude, longitude and altitude."

GPS gets the honor of the youngest system to be in the book. This is the United States' second biggest gift to the world after the *Internet* in the 20th century. Although it has been only with us, civilians for two decades it has made a phenomenal impact with all of us relying on it in our day to day lives. Satellites and Rockets were a prerequisite for GPS and rank above it. The *Galileo system of Europe* is used by an estimated two billion users (in Apr 2021). This means US GPS should be used by half the people in the world, around 3-4 Billion people. *Global Navigation System Satellite* (*GNSS,* system agnostic term for GPS) receiver shipments are forecast to grow from 1.8 B units in 2019 to 2.7 B in 2029, each costing less than $5, making it ubiquitous and ranking it very high.

Radar

"The shooting down of Iran Airlines Flight 655 by the Aegis Cruiser U.S.S. Vincennes in which 290 people died was so horrific that a repetition must be excluded from practical possibility if the United States is to maintain its force in the Persian Gulf. This means taking a hard look at the judgment of Capt. Will Rogers and the rules of engagement under which he acted.

...The airbus was "squawking" Mode Three A signals on its civilian transponder. Had the Vincennes, its sister ships or the command vessel Coronado checked these signals with air traffic control at either Bandar Abbas or Dubai, they would have identified the flight number of this regularly scheduled airliner.

...The Mode Two signals, which may not have existed but the Vincennes claims it heard, were interpreted by the ship as an Iranian military code that reportedly had been cracked. Since Iran's troop transports squawk Mode Two signals to identify themselves to friendly forces, and since the code from the airbus differed from one positively identified with two F-14s

the previous day, the reception of such signals by the Vincennes furnished scant basis for identifying the plane as hostile."

This *New York Times* article on July 17, 1988 talks about a horrible mistake by the *US Navy*, when it shot down an *Iran Air* passenger flight from Tehran to Dubai killing 290 innocent people over Iran's own territorial waters. *Squawk* refers to the 4-digit code of the radar transponder in aircrafts that helps uniquely identify an aircraft for *Air Traffic Control* (ATC); This code is passed to pilots via radio communications from ATC and pilots then set the radio transponder to respond to this particular code, somewhat like giving a unique name to the aircraft within that airport coverage area.

RADAR (*RAdio Detecting And Ranging*) uses the reflection of radio waves from moving objects to detect them and calculate their speed, similar to how bats use *echolocation*, high frequency sound waves to detect objects and navigate in the middle of the night. In 1842, Austrian physicist Christian Doppler at Prague in Czech republic discovered the effect named for him, *doppler effect*: the change in frequency of a wave in relation to an observer who is moving relative to the wave source; Change in frequency is a function of observer and source relative speed and direction, with approaching observer seeing an increase in frequency and observer moving away noticing a decrease, also called *redshift* in astronomy, after the phenomenon of visible light moving towards red (lower wavelength) from galaxies hurtling away from us due to expanding universe. The doppler effect helps calculate the speed of an airplane or speeding car by the police radar. Heinrich Hertz at the *University of Karlsruhe* near Stuttgart in Germany discovered radio waves predicted by Maxwell's theory (see the chapter on *Radio*); He also discovered that metallic objects reflect radio waves. German inventor Christian Hülsmeyer at Düsseldorf built the first radar, a ship detection device using radio waves called "Telemobiloscope" in 1904. Robert Watson-Watt, a descendant of steam engine pioneer James Watt, worked at the *meteorological office* near London in the early 20th century and was an expert at using radio waves to observe weather patterns. In 1935, he led the first Radar effort in the UK, constructing a network of radar stations along the coast of England, called *Chain Home*. It was the first deployment of radar and used long wavelength (10 meter) radio waves that also required huge antennas of 110 meters, height of a 20 story building, tall. British engineers Harry Boot and John Randall invented the *cavity magnetron*, also the main component in the microwave oven, in the early 1940s. The magnetron generated microwaves at wavelengths about 10 centimeters, a factor of 100 times smaller than *Chain Home* wavelength. This vastly reduced the size of the antenna by the same factor, as it is a function of the wavelength and made radar more practical. First Air traffic control tower radar was introduced in 1946 at *Indianapolis* Airport near Chicago in the US.

Weather surveillance radars can detect most precipitation within around 150km of the radar, and intense rain or snow within around 250km. It is also able to detect and warn of hazards associated with severe storms that include hail, tornadoes, high winds, and intense precipitation. LIDAR (*LIght Detection And Ranging*) uses shorter wavelength visible light using the radar principle and is used heavily in autonomous vehicles, both on earth as self driving cars and on Mars as rovers, and in geographic mapping.

Radar ranks high for its role in air traffic control from the busiest airports in the world - *Atlanta* in the eastern US with *111 Million* passengers (⅓ of the US population), *Beijing* in China with *100 Million* and *Los Angeles* in the western US with *88 Million* passengers occupying top three in 2019 - to small airports around the world; In 2019, air traffic controllers at Atlanta Airport directed 909,431 flights through takeoff and landing that requires radar (and radio communications ranked much higher, see the chapter on *Radio*). Weather radars play an important role in weather prediction and planning for weather events such as hurricanes that would claim thousands of lives and billions in property damage without proper planning as happened in *Hurricane Katrina* in the Southeast US in 2005 that claimed around 2,000 lives.

RFID

"TORONTO, Oct.8 (Reuters) - The Toronto police said last Friday that they had cracked an international high-technology credit card fraud ring that uses electronically doctored credit cards. The Hong Kong-based scheme involves copying electronic information from a genuine card and attaching it to a stolen one.

The seized cards were stolen and doctored in Hong Kong and the four men caught using them were Chinese nationals who had immigrated to Canada, said Roy Teeft, detective inspector of the Toronto Police Department's intelligence services. The scheme affects all major credit cards. When the theft of a credit card is reported, the issuing company makes the card worthless by invalidating its two built-in safeguards - information recorded on a magnetic stripe and the multi-digit number.

In the case of the cards seized by the Toronto police, the last four digits of the number had been reimbossed and a special $1,200 coding machine, attached to a home computer, was used to alter the stripe. Both stripe and number then corresponded to data attached to valid cards."

The *New York Times* of Oct 9, 1990 reports on the interesting attack on *Magnetic stripe cards* commonly used in machine readable

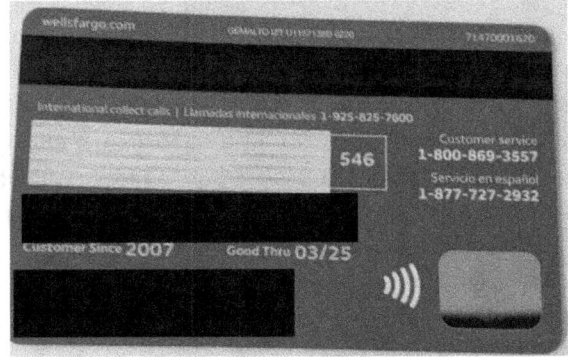

cards such as transit cards and credit and debit cards in the article, "High-Tech Credit Fraud". Magnetic stripe card was first used on transit tickets for the London underground in 1964 and California's newly opened *Bay Area Rapid Transit system* in 1972. *IBM, American Express* and *American Airlines* ran the first pilot of credit cards with magnetic stripes at *Chicago Airport* for issuing airline tickets in the US in 1970 based on a US standard for the magnetic stripe adopted in 1969. The first person to affix magnetic media to a plastic card for data storage was IBM engineer Forrest Parry in 1960. The magnetic stripe has three tracks, the first two tracks storing information such as account number, name, expiration date with the third track mostly unused. As the data is stored in unencrypted form it is easy to alter this information as our "Hong Kong thieves" have done. It is still used in cards only for compatibility with older readers.

Smart cards or *chip cards* have a microchip in the card. Mobile phone *SIM cards* are the most common chip cards with more than 7 Billion of them around the world; It was introduced in the *GSM specification* of 1987 with the first GSM service launched in 1991 (see the chapter on *Mobile phone*). In 1984, French banks started the first chip card trials for payments. In 1986, first chip cards were issued by the *Bank of Virginia* and the *Maryland National Bank* in the US. In 1987, Turkey introduced driving license cards with chips. In 1994, the first *Europay, MasterCard, and Visa* (EMV) specification, the popular standard now for chip and contactless payments, was adopted.

RFID (*Radio Frequency IDentification*) tags transmit data about an item through radio waves to the RFID reader such as the one in a payment terminal. RFID tags typically do not have a battery; instead, they receive energy from the radio waves generated by the reader. When the tag receives the transmission from the reader/antenna, the energy runs through the internal antenna to the tag's chip. The energy activates the chip, which modulates the energy with the desired information, and then transmits a signal back toward the reader. Modern cards contain a microchip that can be read using a chip reader and through an RFID based contactless terminal. *High Frequency* (HF) RFID uses radio waves at *13.56 MHz*, also called *Near Field Communication*

(NFC), for card based applications. HF tags, including an ATM card or a smart card used to open office doors, are passive tags and have a short read range, less than 3 feet. *Ultra High Frequency* (UHF) 860– 960 MHz RFID tags that can be read up to 25 meters, or 5 stories tall, are used in *supply chain tracking* such as the ones in the *US Department of Defense* or large retailers like Walmart, pharmaceuticals, electronic tolling and asset tracking such as the ones in library books. The RFID payment Indicator image from EMV consists of *4-circular chords of increasing radius from left to right*.

The British developed the first active *"identify friend or foe"* (IFF) system as part of *Chain Home* discussed in the Radar chapter that had a transmitter on each British plane. When it received signals from radar stations on the ground, it began broadcasting a signal back that identified the aircraft as friendly, the very first RFID tag (*transponder*). The USSR famously used a passive RFID powered microphone hidden under the US seal gifted to the US ambassador by them in 1945 to eavesdrop on the US embassy in Moscow. Mario Cardullo and William Parks near Washington DC in the US patented a passive radio transponder with memory in 1973. In 1983, the first patent for an RFID tag, "PORTABLE RADIO FREQUENCY EMITTING IDENTIFIER", was granted to Charles Walton at *Los Gatos* near San Francisco in the US. In the 1970s the *Los Alamos National Laboratories* developed an RFID system to track the transportation of nuclear materials. Both Google and Apple, two major smartphone OS providers, have introduced payment systems based on RFID in phones.

RFID ranks high as ubiquitous technology in all the financial system cards such as ATM cards and smartphones. Smartphone based payments will likely replace all physical card based payments in a decade and the magnetic stripe will become obsolete. RFID is used in access control in offices and hotels, automatic toll systems and asset tracking in stores. Almost everyone in the developed world and substantial numbers in the developing world have an RFID tag with them as payment cards or smartphones, billions of them around the world. More than 20 Billion RFID tags are sold every year (2019), ranking it high.

Laser

"Fiber optics involves the transmission of light and images through a flexible bundle of plastic-coated glass optical fibers, each no thicker than a human hair. The light is generated by a laser or by a light-emitting diode, which is similar to the light source used in electronic calculators and digital watches. The light may represent the conversion into light pulses of a voice, a television image or computer information. But it is potentially, and increasingly practically, better than telephone, television, computer and other communications transmitted over conventional metal wires.

A 144-strand glass cable the diameter of a thumb can carry at least five times as many telephone calls as an arm-thick 1,500-pair cable of copper wire. In television, a bundle of six fibers could make possible more than 1,000 channels. By contrast, a 3/4-inch coaxial cable can carry about 40 channels. Optical-quality glass, though tricky to manufacture, is not made from strategic metals. And copper cables need repeaters to amplify a fading signal every mile or so; optical-fiber systems may need them as infrequently as every six miles.

Most optical systems carry information in the form of a digital code by switching the laser or diode light on and off at stunning speeds. When it arrives at the other end of the cable, its quality is nearly flawless, for glass cables and

light signals are immune to distortion by distance, electromagnetic interference or the weather."

This article in The *New York Times* of Oct 17, 1982 describes fiber optical communications with laser light that had started only a few years back. Now optical fibers and laser light passing through it carry the majority of the traffic on the internet including inter-continental undersea cables, reminiscent of telegraph cables in the 19th century and the telephone cables of the 20th century. The biggest difference between telephone cables and optical fibers is that the latter carry information as digital data of 0s and 1s instead of analog data in telephone cable electric signals. Telegraph itself is a digital technology as the *Morse code* is clearly digital (see the chapter on *Telegraph*).

Laser (*Light Amplification by Stimulated Emission of Radiation*) light is coherent, that is all photons are in the same *phase* and *constructively interfere* to make a narrow and powerful light beam. Laser light is produced by putting electrons in a lasing material such as *Gallium Arsenide* in a laser diode into a higher energy level using an energy source such as an electrical current and then using emitted photons to force transition ("stimulate") to a lower energy level resulting in additional emitted photons; The new photons in turn force the transition of additional molecules of the lasing material continuing the *positive feedback* loop. All the photons emitted this way happen to be in the same *phase*, the important attribute that makes the laser possible. Charles Townes and Arthur Schawlow at *Bell Laboratories* in the US in 1958 proposed the basic structure for a device that would produce laser light. In 1960, Theodore Maiman at the *Hughes Aircraft Company* at Malibu near Los Angeles in the US made the first laser. He pumped a *ruby crystal* with energy in such a way that it emitted highly amplified light waves in *red* (694.3 nm wavelength). Earlier in 1953, Charles Townes and his students James Gordon and Herbert Zeiger at *Columbia University* at New York in the US produced the first *Maser*, a device operating on similar principles to the laser, but amplifying microwave radiation (higher wavelength) rather than infrared or visible radiation. In the USSR, Nikolay Basov and Aleksandr Prokhorov independently invented the Maser and Laser as well around the same time. In 1957, Japanese engineer Jun-ichi Nishizawa also independently invented a laser based on semiconductors, precursor to laser diodes. The first visible wavelength laser diode, and Light Emitting Diode (*LED*) that works on similar principles, were invented by Nick Holonyak at *General Electric* at Syracuse in the US near Toronto in 1962.

The first *transatlantic fiber-optic cable*, TAT-8 with a capacity of 280 Mbps, comparable to current home broadband capacity, was built in 1988. *South-East Asia - Middle East - Western Europe 3* (Sea-Me-We3 or SMW3) is a submarine cable linking 39 cable landing stations in 33 countries and 4 continents, including Asia, Astralia, Africa and Europe; Sea-Me-We 3 (SMW3) is the longest submarine cable system in the world with a total length of 39,000 km, close to the circumference of earth of 40,075 km. It spans all the way from England and Germany in Europe to South Korea and Japan in Asia via Turkey and India. Now there are an estimated *1.3 million kilometers* of submarine cables in service globally, enough to go around the earth more than 32 times. Some of the fiber optic cables such as *Google's transatlantic Dunant submarine cable* connecting France and the US can carry hundreds of terabytes of data.

The laser used at the checkout counter in a bar-code scanner emits a wavelength of around *650nm* and produces a power of at most 1 milliwatt. *Blu-ray* uses *gallium nitride* laser diodes that produce *405nm* near-ultraviolet light; *CD* uses *780nm* near-infrared lasers and DVD uses *650nm* red laser. The world's most powerful laser (as of 2021), developed by *Thales* of France and *Extreme Light Infrastructure for Nuclear Physics* (ELI-NP) in Romania, can achieve a power level of 10 PW (1 followed by 16 zeros worth of watts, this is a *mind bogglingly* big number for power as the total US electrical power capacity of around 1000 GW is 10,000 times smaller); It can literally vaporize matter.

Laser ranks high for its importance in optical fiber communications that carry internet and telephone traffic. Given the lag in satellite based communications almost all the data across the continents is carried by laser via submarine fiber optic cables. Laser is used in all barcode scanners in supermarkets. Laser is used in manufacturing, medicine and in the production of *microprocessors* as part of semiconductor *photolithography*. It is common in billions of household items such as DVD players, ranking it in the top 100.

Traffic light

"In a municipal traffic control system, intersection signals comprising four pairs of electric lights, each pair of different colors, constituting start and stop signals, two pairs positioned above so as to be visible along one roadway, in opposite directions, the other two pairs positioned above so as to be visible along the other or intersecting roadway in opposite directions, the first two pairs being on opposite sides of the intersection, and the opposite sides of the street to which they are appropriated, and the second two pairs being similarly on opposite sides of the intersection and on opposite sides of the street to which they are appropriated, whereby each pair of signals can be used to control traffic toward the intersection in one direction along one street only, a source of current and circuit connections therefore to all of the signals, separate circuit closers for each signal and for each pair, and a master circuit closer

supplying current alternately to similar signals of opposite pairs on the same street only, whereby switching together of 90° traffic is rendered impossible."

The patent of "MUNICIPAL TRAFFIC-CONTROL SYSTEM" filed in 1913 by James Hoge, at Cleveland across Detroit at *Lake Erie* in the US, describes the first traffic light in the world. By 1913, the US had 12.94 automobiles per 1000 people, more than India had 90 years later in 2002 (10.1 per 1000 people). In 1915, it grew to 24.77 per 1000 people, more than India had in 2012, almost 100 years later. When we see the traffic in many Indian towns now - bicycles, cars, cattles, people, animal drawn carriages - we can get a picture of the US in the 1910s and why it required the introduction of traffic signals.

First traffic signal powered by electricity in the world, based on James Hoge's patent, was installed in Cleveland in 1914; It had red and green lights and a buzzer to provide a warning for color changes. First three-colour signal was installed in New York city in 1918. Decades earlier, in 1868 the world's first attempt at the traffic light powered by gas was installed in the London borough of Westminster, close to *Westminster Bridge*; However it ended in failure within a month when it exploded on a policeman's face. In 1923, the first traffic light in Europe was installed in Paris; Europe's largest cities soon followed suit: Berlin in 1924, Rome in 1926, London in 1927, Barcelona in 1930 and Tokyo, first in Asia, in 1930. In 1951, the world's first crosswalk (*zebra crossing*) was installed in Slough near London in England. The New York Times of March 17, 1929, still early days for traffic signals, describes the traffic robot in glowing, reverential terms, what we nowadays would use for complex AI systems (particularly *unexplainable*, deep neural network ones that are serious problems for our society as no one *including their builders* understand why they make the decisions they do), in the article - AN UNASSUMING ROBOT COMMANDS "STOP AND GO":

"Machine rule of a new sort has taken possession of Manhattan. It is not the political kind this time, but the kind that anxious philosophers have been warning the world about since the steam engine was young. A soulless mechanism directs our street traffic. A robot of brass, copper, porcelain and rubber governs our going up and down in the city and our walking to and fro in it. Frankenstein Knickerbocker has created a monster, and now he bows to its will or he hears from the police."

The first *Convention on the Unification of Road Signals* was signed in Geneva on March 30, 1931 with three color (red, yellow, green) traffic lights becoming the standard. On the flip side, traffic lights come with their own problems:

- Congestion - Stoplights compress traffic together making them go in bursts, bunching them in a way that uses roads less efficiently resulting in wasted time.
- Safety - When vehicles crash at traffic signals, the results are violent and sometimes deadly as they require drivers to operate in binary mode, full stop or full speed unlike a roundabout, for instance. *T-bone accidents* kill more than 5,000 people (more than 20% of all fatalities, next only to *head on collisions*) every year in the US.
- Expense - Stoplights are expensive, costing *250K-500K* dollars per signal in the US.

Traffic lights rank in the list as one of the ubiquitous inventions around the world that affect people's daily lives. However it is not sure whether we should count it as an invention, anti-invention or simply a necessary evil when the intersection is neither big enough to build overpasses nor small enough to live with stop signs or roundabouts. We spend a lot of our time daily in traffic and a large percentage of it staring at these lights, praying to them to change; In Los Angeles, people spend an average of *119 hours per year* (or a full five days!) sitting in traffic (2017), although this includes freeway traffic moving at a snail's pace. People in the *San Francisco-Oakland* area spend *103 hr/year* followed by Washington DC at *102 hr/year*. *Bogotá*, Colombia, the worst in the world, with drivers on average losing about *191 hours* – nearly eight days – each year in congestion; Rome, Paris and London were the worst in Europe as expected. The US alone has 300,000 traffic signals with New York city alone having more than 12,000 of them;Japan has another 200,000 signals interestingly with a twist: *blue* for go. It has been with us for over a century, taking up our time whether we like it or not, earning a place in the list of inventions.

Barcode

"Then he grabbed a quart of milk, a light bulb and a bag of candy and ran them over an electronic scanner. The look of wonder flickered across his face again as he saw the item and price registered on the cash register screen.

"This is for checking out?" asked Mr. Bush. "I just took a tour through the exhibits here," he told the grocers later. "Amazed by some of the technology."

Marlin Fitzwater, the White House spokesman, assured reporters that he had seen the President in a grocery store. A year or so ago. In Kennebunkport.

Some grocery stores began using electronic scanners as early as 1976, and the devices have been in general use in American supermarkets for a decade."

This is the funny story about President George Bush being amazed by the barcode, an amazing technology, in the front page of The *New York Times* on Feb 5, 1992; Unfortunately for him he did it a decade or two late and he ended up losing the election later that year to Bill Clinton for being out of touch with the "common man". The barcode was invented by Norman Joseph Woodland of Ventnor near Philadelphia

and Bernard Silver of Philadelphia in the US in 1949. Their 1952 patent, "CLASSIFYING APPARATUS AND METHOD", explains the *standard barcode*, a *colored variation* and the *circular version* that they thought would be more accurate as follows:

"Fig. 1 *[a standard barcode]* shows a pattern of white lines, 2, 3 and 4 on a dark background 5. Line 1 is a datum line and the positions of lines 2, 3 and 4 are fixed with respect to line 1. There are, then, fixed places for three lines 2, 3 and 4 in the pattern, and these lines are termed information lines. While the lines have fixed places in the pattern, the lines do not necessarily fill the places. For example, line 4 is missing from its place in the pattern in Fig. 3. A Zero (0) is associated with a vacant line position and the numera one (1) With an occupied line position. The information pattern of Fig. 3 could therefore be replaced by the code number 110 and the pattern of Fig. 4 by 101. The information lines have particular Weights assigned to them; line 4 is 2^0, line 3 is 2^1, line 2 is 2^2. A digit (1 or 0) associated with a particular line position in the code number is used as the coefficient of the weight assigned to that line position. The information in Fig. 3 is therefore decoded as follows:

$$110 = 1.2^2 + 1.2^1 + 0.2^0 = 6$$

... The present embodiment of our invention utilizes a pattern of light and dark lines, but in certain of the modifications of our invention, a pattern utilizing several colors may be used with advantage. The straight line pattern of Fig. 1 is useful only where the pattern can be oriented with respect to the photosensitive apparatus. The straight line pattern is modified into the circular pattern of Fig. 10 *[a circular barcode]* in order that orientation of the pattern be made unnecessary. Lines 6, 7, 8 and 9 of Fig. 10 correspond respectively to lines 1, 2, 3 and 4 of Fig. 1."

In Troy near Chicago in the US, on June 26, 1974, that the first item marked with the *Universal Product Code* (UPC), a *Wrigley's Juicy Fruit chewing gum*, was scanned at the checkout of Troy's *Marsh Supermarket* using *National Cash Register* (NCR) barcode scanners. Earlier in 1973, *Uniform Product Code Council* (UPCC), an association of grocery industry companies, had selected the rectangular black and white barcode, developed by George Laurer at IBM, as the standard over RCA's circular barcode technology inventors themselves advocated. The original UPC carried an 11-digit code, six identifying the manufacturer and five identifying the product; a 12th digit was added later as the check digit. Now UPC has evolved into its superset EAN-13, an international version with 13 digits. The EAN-13 barcode number, also known as the *Global Trade Item Number* (GTIN), consists of the GS1 prefix, the manufacturer's code, the product code, and the checksum digit. Besides the national GS1 prefixes, typically used for standard retail items, there are prefixes for specialized purposes, such as coupons, refunds, serial publications (magazines and newspapers), books (ISBN), and sheet music (ISMN). For compatibility with UPC, EAN-13 codes beginning with 0 are 12-digit UPC codes with the prepended 0 digit. Now it is evolving into RFID (Radio Frequency ID) based technology (see the chapter on *RFID*), the technology used in smart cards and in libraries to embarrass us when we forget to check out a book, as it uses radio waves and requires no line of sight. (One of my early works as a computer engineer was to develop RFID technologies at Microsoft two decades back!)

Barcodes are one of the most ubiquitous items in the world and another offshoot of the computer revolution. In less than 50 years it has made marketplaces more productive and saved a lot of people's time around the world. It is also very critical to keep track of the inventory such as out of stock items in a grocery store. According to GS1, successor to UPCC, it is scanned 5 billion times every day around the world, ranking it high, in the top 100, as one of the most ubiquitous inventions in the world.

ATM

"On 27 June 1967, a summer on from England's footballers winning the World Cup and two years before man first set foot on the Moon, the first ATM was up and running at a branch of Barclays Bank in Enfield High Street, Middlesex, UK. It was one small step from pavement to cashpoint, but one giant leap for time-conscious members of the public after quick and easy access to their savings.

Actor Reg Varney (1916-2008), from the BBC1 sitcom Beggar My Neighbour and the soon-to-be star of On the Buses, became Barclays' first ATM user on that historic day, in the glare of the gathered media. After a celebrity-endorsed trial run, inquisitive customers were issued with paper cheques - in the days before plastic cards - that contained traces of the radioactive substance carbon-14 (aka radiocarbon), which the machine detected, enabling it to match a cheque to a four-digit pin number. Customers fed their cheques into the new device in return for a shiny, new £10 note -

"quite enough for a wild weekend" in those days, according to the Scottish-born inventor of the ATM, John Shepherd-Barron OBE."

Guinness World Records describes the story of the world's first *cash machine*, as the *Automatic Teller Machine* (ATM) is called in England, this way. However John Shepherd-Barron's system developed at *De La Rue* in England, now a major banknote and cheque printing company, was not the ATM card and PIN system we know today as it used *carbon-14 coated* paper cheques. James Goodfellow of the UK, working at the *Smiths Industries* in London, England, founded in 1851 and a major automotive and aerospace instruments manufacturer at the time, holds the first patent on the modern ATM; It was filed on May 2, 1966 in England, a full year before the inauguration described above. His corresponding US patent, "ACCESS-CONTROL EQUIPMENT" describes the modern ATM:

"A money-dispensing system dispenses a pack of money upon request by an authorized bank-customer, the request involving presentation to a card-reader of the customer's individually-allotted punched-card, and operation of a set of ten push-buttons in accordance with the customer's personal-identification number. The system dispenses a pack of money only if there is correspondence between this number entered by the push-buttons and a number that is read by the card reader from the card."

Goodfellow's card was similar to a computer punch card, used as input in early computers and not the current magnetic strip cards. Don Wetzel, Marion Karecki, Thomas Barnes and others in *Docutel* at Dallas, Texas in the US developed the ATM machine called *Docuteller* and ATM card with a now familiar magnetic strip and PIN. A *New York Times* article from May 14, 1976 describes the early history [emphasis added]:

"Although the first automatic teller was installed in 1969, at a Chemical Bank branch in New York City, **the machines have been in general use nationally only for two years or so.**

There are about 4,000 of the machines in use around the country, according to an industry source, about three quarters of them built and serviced by the Docutel Corporation of Irving, Texas. Docutel reports that it now has about 700 institutional customers, with the heaviest concentrations in Atlanta, Chicago, Detroit, Portland, Ore., Seattle and Columbus, Ohio. "

In 1970, there were fewer than 1,500 machines around the world, mainly in Europe, North America and Japan. IBM of the US was a critical player in the early evolution of the ATM. In developing countries ATMs started to appear in the 1980s. First ATM in Africa was installed in 1981 by *Standard bank* at Johannesburg, South Africa. First ATM in India was installed in 1987 by *HSBC bank* at Mumbai. The number of ATMs went from 40,000 machines in 1980 to a million by 2000; Now there are around 3.2 Million ATMs in the world. China has the largest number of ATMs at 1.1 Million followed by the US at 0.5 Million. *Diebold Nixdorf*, formed by the merger of *Diebold* founded in 1859 in *Cincinnati* near Chicago in the US and *Wincor Nixdorf* founded in 1952 in *Paderborn* near Cologne in Germany, has the largest ATM market with around 32% of ATMs in the world with NCR, founded in 1884 at *Dayton* near Cincinnati in the US, the second at 26%.

ATM is one among the many gifts to humanity that the computer and transistor revolution of the twentieth century provided. ATMs rank high as one of the most useful inventions in day to day life and one of the most important financial innovations after the money itself. In fact, former US central bank (*Federal Reserve*) chairman Paul Volcker said in 2009 that it was hard to think of a worthwhile financial innovation since the ATM. It for the first time allowed people to withdraw their own money during after hours and weekends and made banks more popular and useful. There were 5.1 Billion ATM withdrawals in 2018 in the US alone totalling 800 Billion dollars, almost 4% of US GDP, ranking it in the book.

Tank

British former Prime Minister Winston Churchill is known for momentous mistakes and callousness. When millions of Indians were dying in the *Bengal famine of 1943* in British India, he reportedly asked, adding insult to irresponsibility, "if the famine is so bad, why is Gandhi still alive?" Remember he was the prime minister of Britain at the time and hence the head of state of British India. As the *lord of Admiralty* he led a disastrous campaign in *Gallipoli,* at the mediteranian in Turkey overlooking Greece, that resulted in around 250,000 casualties including 46,000 dead and a humiliating defeat and retreat for the British. As *Chancellor of Exchequer*, In 1925, he returned Britain to the *gold standard* with the pre world war exchange rate resulting in massive unemployment and deflation and completely destroying the British economy; Britain eventually left the gold standard for good in 1931 finally stopping the bloodbath. In my view, british and allies won the second world war in spite of Churchill not because of him, a view corroborated by the british public's evaluation of him at the next available opportunity, when they kicked him out of office in the election of July 1945, immediately at the end of the world war. He left with a humiliating defeat but not before claiming in the election campaign that the opposition *Labour party* would bring in "some form of *Gestapo*, no doubt very humanely directed in the first instance." Instead the great Clement Atlee, Labour prime minister with *"all substance and no show"* in Margaret

Thatcher's words, ended up giving freedom to India and created the *National Health Service* (NHS) in 1948, one of the first successful universal healthcare systems in the world, which the US and other nations are still struggling to achieve after 70 years. (British democratic system should be thanked for not turning Churchill's hare-brained schemes into Mao's *"Great leap forward"*). One of the ideas from this "fertile" mind of Churchill is an anti-invention, Tank, the topic of this chapter.

As the Lord of Admiralty, Churchill formed the *Landship committee* to make an *Armoured Personnel Carrier*, the tank, in 1915. The reason the Lord of Admiralty had to form the committee in the first place was that no sane person in the British Army thought it was such a great idea. Fittingly for such an anti-invention, the first tank, *British Mark I*, was introduced in the *battle of the Somme*, one of the bloodiest episodes of World War I and the largest campaign on the *Western Front* that killed or injured *over a million* allied and German soldiers in 1916. Tanks ironically (or rather typical of Churchill's ideas), played their most important role in World War II at the hands of Nazis, when masses of German APCs overran much of Europe and parts of north Africa using the *Blitzkrieg* - lightning war - tactic. One of the enduring images of tanks in history is that of Chinese military mowing down its own [unarmed] citizens mercilessly using APCs at Tiananmen Square in 1989. BBC notes [emphasis added]:

"The Chinese army crackdown on the **1989 Tiananmen Square protests killed at least 10,000 people**, according to newly released UK documents.

The figure was given in a secret diplomatic cable from then British ambassador to China, Sir Alan Donald. Sir Alan said the source had been reliable in the past "and was careful to separate fact from speculation and rumour". The envoy wrote: "Students understood they were given one hour to leave square but after five minutes APCs attacked.

"Students linked arms but were mown down including soldiers. APCs then ran over bodies time and time again to make 'pie' and remains collected by bulldozer. Remains incinerated and then hosed down drains. "Four wounded girl students begged for their lives but were bayoneted."

Tanks typically are "impressive looking duds". The *National Interest magazine* literally ran an article in April 2021, "M103: Why Was the U.S. Army's Last Heavy Tank a Dud?". Tanks played a counter-productive role in Israel's surprising and humiliating defeat at the hands of *Hezbollah* in the *2006 Israel-Lebanon war*; Israel's defense minister and many top ranking generals resigned after the disastrous campaign; The world watched in shock as the 30,000 strong Israeli military became sitting ducks for few hundreds on Hezbellah's side as they made mincemeat out of their "advanced" tanks using anti-tank missiles such as shoulder-fired Rocket Propelled Grenade (*RPG-29*). The *Observer* notes in an article in Aug. 2006:

"Celebrated as one of the most heavily protected tanks in the world and the embodiment of the might of Israel's ground forces, **the Merkava tanks seemed to become practice targets over the last three weeks for Hezbollah anti-tank missile teams.** And as a ceasefire went into effect this week, footage of smoke billowing from paralyzed Merkava tanks are likely to remain burned in the collective memory as one of the dominant images of a war gone awry."

Tanks rank high as an unfortunate anti-invention that was used to kill millions and suppress freedom movements such as the ones in Tiananmen square. It likely prolonged the first world war, making it a "world war" in the first place. Ironically, it increased the power and destructiveness of Nazi Germany in world war II. There are around 100,000 tanks in the world, a whopping 1,000 Billion dollars "worth" of them (at a unit cost of $10 Million). As an anti-invention sucking up a lot of military resources and a useless one at that (except maybe for mowing down unarmed peaceful protesters as in China, Egypt and Myanmar) it ranks as one of humanity's top ranking anti-inventions, humanity at one of its stupidest moments.

(Image credit:PBS Newshour)

Zipper

"The invention was especially designed, for use as a shoe-fastener; but is capable of general application wherever clasps consisting of interlocking parts may be applied, as for example, to mail-bags, belts, and the closing of seams uniting flexible bodies. To these ends, the clasps are made with interlocking parts, which when in position, can only engage with each other when at an angle to the line of strain. The clasps have underreaching and overlapping projections or lips at their forward ends, which prevent the engagement or disengagement of the hook-portions of the clasps, except when thrown upward, so that the parts stand at an angle to each other of about ninety degrees.

... I therefore provide a hand device, consisting of a movable guide, having cam-ways for permitting the passage of the clasps, by the movement of the guide from one end to the other of the series; and the cam-ways are so shaped and related that by the passage of the guide in one direction, the clasps will be drawn together and engaged, while by the passage of the guide in the other direction, the clasps will be disengaged and separated. In other words, one end of the guide has two channels or grooves, for receiving the parts of the

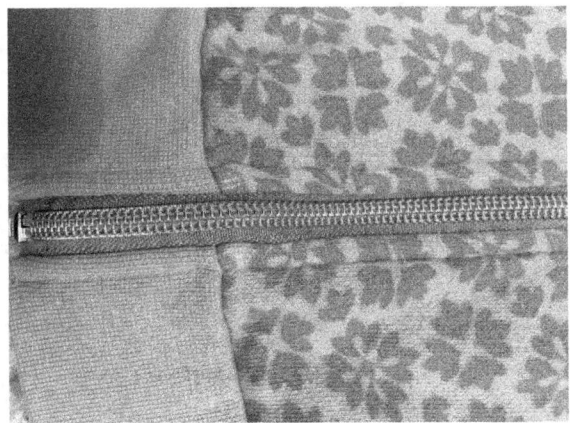

fasteners when open or disengaged, and this may be called the forward end of the guide. The other or back end of the guide has a single channel or cam-way, into which the two channels from the forward end converge over an angular center ridge or instep."

The first patent for zippers, "CLASP LOCKER OR UNLOCKER FOR SHOES" by Whitcomb Judson of Chicago in the US in 1893 describes the zipper this way. Gideon Sundback of Jersey city near New York working for the company founded by Judson himself, *Hookless fastener company* (later "Talon Zipper") of Meadville near Philadelphia in the US made crucial improvements for zippers by increasing fastening elements (per unit length) and made them into interlocking teeths instead of hooks clasping each other to produce the modern zipper found in clothes such as Jeans in 1917. This started the boom in zippers twenty five years after the invention of Judson.

Yoshida Kōgyō Kabushikigaisha (YKK, "Yoshida Manufacturing Shareholding Company") is the largest maker of zippers in the world with more than 10 Billion zippers sold in 2018; YKK was founded by Tadao Yoshida in 1934 at Tokyo in Japan in the early days of the zipper boom. *Levi Strauss*, founded by the namesake Levi Strauss in 1853 in San Francisco in the US, is the largest maker of jeans in the world with a revenue of 5.7 Billion dollars in 2019. Levi Strauss and Jacob Davis patented the blue jeans, "IMPROVEMENT IN FASTENING POCKET-OPENINGS" in 1873. Levi's brought out its first model with a zipper in 1947 as a way to expand the market to East Coast women. The *New York Times* of Feb. 21, 1947 writes in the article, "WHO INVENTED THE ZIPPER?":

"Dr. Frank B. Jewett once remarked in an address delivered at New York University that he has often thought "how infinitesimally small would have been the chance of any man or group of men, except the one who actually had the idea, planning to invent the common zipper." That made P.J. Frederico, an examiner in the Patent Office, wonder who this remarkable man was. After months of detective work in the files of the Patent Office he lays a wreath on the brow of Whitcomb L. Judson, who may have been mute all these years, so far as the public is concerned, but who is no longer inglorious, thanks to Mr. Frederico's revelations in the Journal of the Patent Office Society.

Judson's first patent goes back to 1891, his second to 1893, his last to 1905. Mr. Frederico would give much to know more about Judson and so would we, in view of Dr. Jewett's high opinion of the zipper. Like most good inventors, Judson was prolific. He invented apparatus for street railways, engines and variable speed transmissions and clutches for automobiles and machines to make his zippers. When he filed his application for the zipper the Patent Office was unable to find any anticipation, which is remarkable and which again bears out Dr. Jewett's conclusion."

Zipper ranks in the list as one of most useful inventions playing a big role in the 20th century. It is one of the most complex and original inventions in the list despite its apparent simplicity as the *New York Times* article points out, "how infinitesimally small would have been the chance of any man or group of men, except the one who actually had the idea, planning to invent the common zipper." In comparison, Watt's steam engine *condenser*, pasteurization or Hargreave's *spinning jenny* look more straightforward in the sense someone would have come up with it sooner or later given the obvious problem to be solved, very low efficiency of Newcomen's engine, for instance. However the zipper was so original that Judson had a hard time selling it to people for decades till Sundback came along. Put another way, if we were to replay human history in the universe, zippers may not even be invented again unlike the steam engine condenser. The worldwide *denim market* alone, with zip being an important part, is estimated to be around 100 Billion dollars per year. There are more than *15 Billion zippers* produced every year in the world making it one of the truly original, useful and ubiquitous inventions of humanity.

Ten principles of human innovation

1. Incremental inventions matter, a lot

There are two kinds of inventions. The first kind is the completely original, breakthrough invention, the *telegraph* for instance. There were no predecessors to telegraph and it was a completely original invention that changed the course of human communications and history. Another example is the *antibiotic* of Alexander Fleming that again changed the course of human health and human history; For instance, compared to the first world war, far fewer people died in the second world war due to infections thanks to the availability of the antibiotics. The second kind of invention is far more common and equally important. In this case, the precursor to the invention already existed in some form. The invention itself is typically an incremental, efficiency improvement that vastly improves the existing technology, making it more practical and affordable. Many important inventions fall into this second category; James Watt's improvement of Newcomen's steam engine by introducing a separate condenser that made steam engines practical is one; (Newcomen's invention itself could be considered to belong to the first kind of invention). Henry Bessemer's process for making steel from *pig iron* by removing carbon instead of making steel by adding carbon to *wrought iron* that started the revolution in steel is another. One of the greatest inventions ever, printing press of Johannes Gutenberg, itself is an incremental invention in the sense both the books and a way of producing them existed before the invention; Printing press automated the process using movable types and a press, enabling the production of books at scale and thereby making the books affordable to masses for the first time, bringing in the *renaissance* and knowledge revolution. Many other well-known inventions such as spinning Jenny over spinning wheel, transistor over vacuum tube belong to the second category.

The main difference is the first kind of inventions look like random events of history and we, as a society or nation, can only do so much about the rate of inventing them other than provide the solid foundations discussed in other principles below that apply to all the inventions in general. However the second kind looks more amenable to human agency and we can improve their rate with conscious effort; Starting with the awareness that many inventions we think of as original, breakthrough inventions (again, Watt's condenser in the steam engine is the prototypical example) are incremental inventions; Next realizing that they have been historically very important, as important as "purely" original inventions; How many inventions are more important than the printing press and the steam engine? Third, encouraging people to make the inventions public, even if they look hopelessly impractical now as it will very likely be improved by further advances going by the historical evidence. Fourth, making the problems needing incremental inventions widely known through books, articles and prizes; British were masters of the last method resulting in such important inventions as the *Chronometer* of John Harrison and *Detector lock* of Jeremiah Chubb. (That gives me an idea for the next book, *The 100 inventions waiting to happen*? that describes important problems awaiting incremental, efficiency inventions!)

2. Intellectual Property (IP) rights matter

Article I Section 8 | Clause 8 – Patent and Copyright Clause of the Constitution. [The Congress shall have power] "To promote the progress of science and useful arts, by securing for limited times to authors and inventors the exclusive right to their respective writings and discoveries."

The United States has been at the forefront of innovation producing geniuses such as Thomas Alva Edison with inventions as varied as light bulb, phonograph and cinema among thousands of others. The United States constitution itself protects the innovation that is a master stroke of innovation in itself by the founders of the US. There would have been geniuses like Edison in India and Nigeria who couldn't profit from their inventions as other people would simply have copied (or to use the technical term, *stolen*) it without any consequences, wasting away their talents as they went back to working in the mill or driving a taxi to make ends meet.

As mentioned in the chapter on Cotton gin, Eli Whitney's struggles in the 18th century United States illustrates the same point. After inventing the Cotton gin Whitney had to spend so much time defending his invention against thieves (many inventions look "simple" and "obvious" in hindsight) including state sponsored ones in the weak intellectual property rights era of the United States, similar to India or Egypt of today, that he had very little time to spend on further inventions. Edison, living in the late 19th century United States had an easier time and could spend his time far more productively, resulting in many more inventions such as the light bulb, cinema, electricity and battery benefiting humanity instead of spending his time defending the telephone transmitter patent for the rest of his life. The lives of these two master inventors of the United States clearly show the importance of strong intellectual property rights in the development of inventions and the nation (look at the number of inventions by country in Table: 2).

3. Immigrants make a big difference

America tops this book by a healthy margin. There is one big reason: *immigrants*. Immigrants tend to be more motivated and ambitious; For one, they crossed oceans and other hurdles to get to places as compared to people who live within 18 miles of their mothers. For all its faults, America has been the best country for immigrants for centuries, whether it is Mayflower pilgrims escaping established churches in the 17th century, Irish escaping the horrible potato famine in the 19th century or Jews escaping Nazis in the 20th century. While the US at times tried to turn away immigrants including Jews during the Nazi persecution and Asians in the 19th century, other countries have been consistently far worse and paid the price for it; This provided a huge competitive advantage on a platter to the US. Japan, a developed country, hardly admits any immigrants. India is probably one of the hardest countries for migrants to live. Only other country comparable to the US in terms of immigration is the UK and the results prove it; They get the second spot in this book.

Alexander Bell of Telephone came to the US from England via Canada. Henry Ford's father was Irish and mother was the child of Belgian immigrants. The Wright brothers' maternal grandfather emigrated from Germany. Leonard Kleinrock's parents were poor Jewish immigrants from Ukraine. Great Thomas Edison's father was from Canada. Isaac Singer's parents were German immigrants. Igor Sikorsky was from Russia. Martin Cooper's parents were Ukrainian Jewish Immigrants as well. Marconi went to England from Italy to make his invention, *radio*, a reality after Italy government officials *responsible for inventions* laughed at his idea. He didn't choose to go to Germany or Japan, a big competitive disadvantage for those countries. I think this is one of the reasons Germany is relatively far behind, far enough that I was surprised to see the final numbers in Table: 2, as compared to her reputation as an *invention power house* and her ability as a nation; After all it is the nation of Johannes Gutenberg and Albert Einstein. Major self-goals such as kicking all the Jews out, out of the nation or the world altogether, under the Nazis didn't do them any favor either in this matter.

4. Basic science matters

Fundamental scientific understanding is critical for inventions as tinkering and trial and error only goes so far. The 19th century produced an astonishing number of inventions related to electromagnetism, electricity itself, telegraph, telephone, radio to name a few; The main reason was electromagnetism was completely understood only in the 19th century thanks to the likes of Michael Faraday and James Maxwell. For the same reason, the 19th century produced refrigeration and automobiles instead of the 17th century mainly because thermodynamics was developed in that century thanks to the works of James Maxwell (again), Ludwig Boltzmann and Sadi Carnot. The 20th century produced transistors, the foundation of all modern electronic technology; Quantum mechanics, the foundation of transistors, was discovered in the 20th century. This also provides a simple prescription for educating the children to sustain productivity and innovation. School children should be taught fundamental science instead of the latest fads such as "coding", "typewriting" or "plastics".

This is also a good place to explain an important but somewhat tangential point. We, as a society, should be exceptionally careful about taking advice from corporations on what is good for society and what to teach our children, in particular. Let's take "coding" for an example I know very well and why teaching it in school is a bad idea. First point, technology companies can't even employ their own current workforce as after about 10 years they simply kick out the engineers as they are "overpaid" and recruit low wage entry level workers preferably with a visa with restrictions such as H1-B in the United States, a modern form of bonded labor where their stay in the country is tied to their jobs making them slaves in effect to the management. (This is the main reason technology companies want to increase these visas; In their current form these visas work against all workers in the long term, both domestic and foreign; I wouldn't be surprised if it turns out to be written by corporate lobbyists). The often cited soundbite of technology companies as well as other industries such as grocery chains that "we can't hire people" is merely a code for "we can't hire people for the low wages we are willing to pay without cutting into our exorbitant executive packages or shareholder profits". To quote a literal headline from an article, "CEO Compensation Has Grown 940% since 1978: Typical Worker Compensation Has Risen Only 12% during That Time". It is very unlikely that all the productivity increase in the last 40 years is due to the CEOs. Going back to the main point, their "coding" skill they spent years learning in the school will become obsolete in about 10 years after technology companies deem them "too overpaid" or "too old" (or both) by the time they are in their 30s, particularly when our life expectancy is increasing and we will work for 40-50 years.

Second point, as economist Milton Friedman nicely argues in the article, "The Social Responsibility Of Business Is to Increase Its Profits" (read it and come back if you have never read it before even if you disagree with him; More important: read it in full particularly if you disagree with him and come up with *logical reasons* why he is wrong; I tried and it is hard), it is silly of us to expect corporations to play a social role; They are here to make profits for the shareholders. For the same reason, despite the *lip service* they don't care about society, diversity, students or facts; If lies drive more eyeballs they (looking at you,

Facebook - "Delay, Deny and Deflect: How Facebook's Leaders Fought Through Crisis." *The New York Times*) will definitely promote lies even if unwittingly. They just care about money and profits. If it gives them profits and advantage they won't blink twice before giving all our data to governments after preaching to everyone else and their grandmothers about privacy (looking at you, *Apple* - "Censorship, Surveillance and Profits: A Hard Bargain for Apple in China." *The New York Times*). That also means governments and the rest of society should rein them in, including for tax avoidance, labor exploitation including through bonded labor visas and for exorbitant pay for executives at the cost of shareholders and workers. But more importantly for our purposes, to belabor the point, it is a bad idea to get advice for what is good for the country, people and humanity from corporations whose aim is to make money for its shareholders. This means we should concentrate on basic mathematics, science and critical thinking and leave out any job specific skills, in particular "coding", out of schools. Corporations like Google may want compliant *"code monkeys"* out of the box, but that doesn't make it the best strategy for the education of a society. By this logic, we should be teaching everyone how to bag a grocery just in case they end up being wal-mart grocers. It makes more sense, as there are around three million grocery store workers in the US, more than the number of computer programmers, at less than two million making the former more probable.

In summary, basic science and mathematics in schools and no profession specific skills as the former is important for sustaining productivity and inventions and would be still important 50 years from now.

5. Productivity is everything

Quoting economist Paul Krugman again:

> "Productivity isn't everything, but in the long run it is almost everything. A country's ability to improve its standard of living over time depends almost entirely on its ability to raise its output per worker."

Productivity and inventions are inextricably linked together. An invention worth its salt improves human productivity. Printing press made the production of books much faster compared to writing one book at a time by scribes, making the books affordable to people. Cotton gin enabled a worker to clean fifty pounds of cotton instead of one pound a day before the invention. Telegraph improved the speed of communication from 100 kilometers an hour to a billion kilometers an hour by a factor of 10 million.

The best way to raise human productivity is through inventions and automating away, the repetitive parts of the jobs, increasing the happiness of the workers in the process. Before the invention of Cotton gin, a cotton worker's job was incredibly repetitive, boring and unproductive. Before the invention of computers, the job of human "computers" would have been repetitive, error prone and relatively unproductive in hindsight given the speed of computers. These inventions increased human productivity and human happiness. It is not just *human* productivity. The invention of skyscrapers increased the productivity of scarce land in dense cities; The invention of fertilizers increased the productivity of scarce agricultural land.

As a day will always be twenty four hours, productivity has to improve, as Krugman says, for us to produce more in a day or in a year; Our inventions are the best way to raise our productivity.

6. Great inventions disappear

Quoting BBC in the article "The cheap pen that changed writing forever":

The ballpoint pen is testament, Sax says, to one of the tenets of "really great design – it almost disappears. If the ballpoint didn't exist and you launched one today on Kickstarter, it would be the biggest thing ever," he says.

Some of the greatest inventions are the ones we take for granted and we don't even know their inventors. Ballpoint pen is one of them, Bicycle is another; Road itself is one of the greatest inventions almost as important as the wheel. The greatest inventions are like water. We don't think about them much till we miss them; It disappears from our consciousness. Clothes, furniture, pot, numbers, writing, clock, thermometer, light and electricity in developed countries - hallmark of all these great inventions is they disappear completely from our minds and we can take them for granted.

7. Inventions create jobs in the long run

Original *"Luddites"* of England were afraid of the textile inventions of the industrial revolution. They broke the machines, protested and marched to stop the machines from taking over their jobs, which the machines did anyway. If they had succeeded in stopping the march of the industrial inventions, spinning Jenny, flying shuttle, power loom and so on, they might have been comfortable in the short run; But the country and indeed, the humanity would have lost a lot in the long run, including affordable clothes for millions who could afford them for the first time, including many of the weaver families themselves. To be fair to the "Luddites" themselves, they were very afraid of going to the poor workhouses where families would be separated, husbands from wives, parents from children, sisters from brothers; Bills have to be paid in the *short run*.

In 1900, about 40% of the US population lived on farms and worked in agriculture. Now, it is around 1%. It doesn't mean our unemployment rate is around 39%. It also doesn't mean we are struggling to feed people. The farms and agriculture have become vastly more productive due to agricultural inventions, fertilizer and automobiles topmost among them, requiring much less workers. The remaining 39% simply switched to more productive work for their time. In fact, the statistics are even better as the labor force participation rate for women was under 20% in 1900 and now it is 47%. Somehow, around 30% more women managed to find work even though we lost so many farm jobs.

Economists call this the *"lump of labor"* fallacy; The fallacy that the number of jobs in the economy is fixed and if one starts working then someone has to lose the job. When I started writing this book, I didn't make a writer or someone else lose their job. It may only be true of government jobs where we really don't want to increase the number of jobs drastically, as we the people have to pay for them with our taxes. (This is the reason retirement age in government jobs is a sensitive topic as two generations compete for the same, fixed number of jobs). It is not even remotely true of all the jobs in the economy. This is for the same reason *immigrants are not stealing your job*. (However as mentioned earlier, poorly written laws can give a lot of power to corporations over all the workers, whose only goal is to maximize profits and not social or worker welfare). Governments trying to slow down the pace of inventions to save jobs are doing the country, its people or its workers no favor in the long run.

8. People have minds

Pessimists from Thomas Malthus to Paul Ehlrich complained about a burgeoning population polluting pristine earth and bringing famine and catastrophe. They have always predicted that the next famine is around the corner and it never materialized, thankfully. Now the climate change warriors are doing the same. Some even go as far as saying that they don't have children because of climate change and to save the earth. What they don't realize is this burgeoning population consists of people with intelligent brains. In the 19th and 20th century the pace of inventions accelerated as there were more people (with trained minds) to do the inventions. If more people create climate change, more people also have more brains available to solve that problem. For some reason, some of these extreme "environmentalists" have the idea of people as zombies who simply consume resources on earth instead of members of a species of supreme problem solvers.

They have the view of people as cockroaches encroaching the earth and they would like to reduce them by a few billion. If there are no people and other animals and plants what would be so special about earth? It would be yet another barren rock just like Mars or Venus orbiting a middling star among billions in the universe. No one would notice if it completely disappeared. *We give earth meaning and special status.*

We will never solve any problem by eliminating people, directly or indirectly. For an example, China's horrific *one child policy* has produced many problems including the inverted social pyramid with very few workers to support its aging population and a very lopsided sex ratio. Best way to solve hard problems including climate change is to have as many minds as we can to work on the problems; Harder the problem, the more minds we need to increase the probability. This also means we need to provide the foundations for those minds to be productive and innovative which brings us to the last two principles, the most important of them all.

9. Basic literacy matters

Now we come to the two top most important factors for a country to be innovative more than anything else.

According to *Our World in Data*, in 1870, the US had 80% literacy rate; The UK and France had 76% and 69% literacy rate respectively. More than a hundred years later, in 1991, half the population in India was illiterate with a literacy rate of only 48%; Egypt in 1986 had a literacy rate of 44%. Both the countries at the turn of the 21st century were below that of the Netherlands in 1650 (53%), more than 300 years back. Now we can see why some of the former countries top this list of inventions. In 2001, less than 7% of Indians were graduates with around 5% for women according to the 2001 census. In the US, one in every four Americans was a graduate according to the 2000 census. (In the recent 2020 census, it is even better at one in every three adult Americans).

In India, a staggering 350 Million people are still illiterate, more than the population of the US and around 4.5 times the population of Germany; 350 Million minds are a terrible thing to waste. (Unfortunately, my own father in India and his four brothers never went to school, not even to first grade). One of them would have found a better alternative to Cement, one of the biggest emitters of greenhouse gases; Another could have written an enjoyable book on *Thirukkural*; Yet another may have found a universal coronavirus vaccine a decade back.

If any developing countries are looking for a magic potion of innovation, it is staring them right in the face; It is not launching a spaceship to the moon or mars; It is not building a statue at half a billion dollars for bragging rights; It is not building nuclear weapons; It is unglamorous; It is consistent hard work. The single biggest thing a country can do to make the country innovative is to address the foundations of education particularly at the school level.

10. Educating women

I was struck and saddened by one thing in this book as I have two girls of my own. Most of the inventors in the book were men. One simple way humanity as a whole could have been more innovative was to educate the girls and double the number of inventions. If 350 Million minds are a terrible thing to waste, it is even more so in the case of billions of minds. It reminds me of a story about Bill Gates in a *New York Times* article:

Bill Gates recalls once being invited to speak in Saudi Arabia and finding himself facing a segregated audience. Four-fifths of the listeners were men, on the left. The remaining one-fifth were women, all covered in black cloaks and veils, on the right. A partition separated the two groups. Toward the end, in the question-and-answer session, a member of the audience noted that Saudi Arabia aimed to be one of the Top 10 countries in the world in technology by 2010 and asked if that was realistic. **"Well, if you're not fully utilizing half the talent in the country,"** Gates said, **"you're not going to get too close to the Top 10."** The small group on the right erupted in wild cheering.

In the end analysis, inventions are about improving human productivity. Educating all women may be the single biggest productivity booster in human history and foster a new era in the history of human inventions. This is the single biggest and most important principle I learned from writing this book. I hope a 22nd century version of this book, with much better representation of women in it, will note this trend of women's education in the 20th and 21st century that proved a turning point in our inventions and productivity.

Table 1: When were they invented?

Century	Inventions
21st century	0
20th century	22 [computers, internet, mobile phone, GPS, antibiotic, airplane, fertilizer, transistor, television, rocket, satellite, barcode, ATM, bleach, nuclear weapon, washing machine, vacuum tube, radar, RFID, laser, traffic light, tank]
19th century	30 [telegraph, telephone, phonograph, cinema, photography, railways, bicycle, light bulb, sewing machine, radio, internal combustion engine, refrigerator, automobile, pain killer, skyscraper, elevator, anesthetic, petroleum, plastic, rubber, electric motor, pasteurization, x-ray, soap, cement, dynamite, kerosene lamp, safety match, zipper, pesticide]
18th century	8 [electricity, battery, thermometer, vaccine, steam engine, flying shuttle, spinning jenny, cotton gin]
17th century	4 [Newspaper, clock, microscope, telescope]
1-16 century	7 [firearm, compass, paper, toilet, eye glasses, printing press, tobacco]
history-1 BCE	16 [writing, numbers, map, wheel, tea, iron metallurgy, letter, pen, road, calendar, money, glass, lock, ruler, canal, dam]
prehistory	13 [cloth, fire, ship, agriculture, religion, alcohol, bridge, pot, furniture, stove, shoe, paint, knife]

Table 2: They were invented in which countries?

Country	Inventions
Argentina	1 [pen]
Australia	1 [stove]
Canada	3 [petroleum, kerosene lamp, clock]
China	4 [paper, compass, firearm, money]
Egypt	4 [writing, lock, map, calendar]
England	25 [steam engine, railways, spinning jenny, plastics, telegraph, iron metallurgy, telephone, antibiotic, ATM, vacuum tube, flying shuttle, tank, electric motor, television, safety match, cement, clock, toilet, stove, bleach, computer, internet, petroleum, stove, vaccine]
France	9 [internal combustion engine, vaccine, photography, sewing machine, ruler, pasteurization, bleach, computer, flying shuttle]
Germany	14 [book, internal combustion engine, automobile, newspaper, fertilizer, bicycle, pain killer, refrigerator, rocket, x-ray, clock, pesticide, map, microscope]
Hungary	1 [pen]
Netherlands	4 [thermometer, microscope, clock, telescope]
India	1 [numbers]
Italy	7 [eye glasses, radio, battery, dynamite, electricity, calendar, letter]
Poland	2 [petroleum, kerosene lamp]
Russia	2 [satellite, rocket]
Sweden	2 [dynamite, stove]
Switzerland	2 [pesticide, map]
Turkey	1 [money]
USA	38 [computer, transistor, antibiotic, telephone, phonograph, light bulb, cinema, GPS, mobile phone, internet, airplane, nuclear weapon, sewing machine, plastics, skyscrapor, elevator, rocket, telegraph, photography, automobile, firearm, barcode, vacuum tube, electricity, cotton gin, anesthetic, rubber, petroleum, electric motor, zipper, traffic light, laser, RFID, washing machine, soap, clock, kerosene lamp, stove]
World	27 [cloth, fire, wheel, religion, glass, alcohol, agriculture, ship, bridge, calendar, money, iron metallurgy, pen, map, knife, pot, furniture, letter, shoe, road, tobacco, dam, soap, ruler, toilet, paint, canal]

Bibliography

Cloth

1. Lobell, Jarrett A. "World's Oldest Dress." *Archaeology Magazine*, 2017, www.archaeology.org/issues/241-features/top10/5113-egypt-tarkhan-dress.
2. Lobell, Jarrett A. "Dressing for the Ages." *Archaeology Magazine*, 2016, www.archaeology.org/issues/215-1605/trenches/4349-trenches-egypt-predynastic-period-tarkhan-dress.
3. Hogenboom, Melissa. "Earth - We Did Not Invent Clothes Simply to Stay Warm." *BBC*, BBC, 19 Sept. 2016, www.bbc.com/earth/story/20160919-the-real-origin-of-clothes.
4. Saraceni, Jessica Esther. "Denisova Cave Yields a 50,000-Year-Old Needle." *Archaeology Magazine*, 23 Aug. 2016, www.archaeology.org/news/4784-160823-denisova-cave-needle.
5. Lavoie, Amy. "Oldest-Known Fibers to Be Used by Humans Discovered." *Harvard Gazette*, Harvard Gazette, 10 Sept. 2009, news.harvard.edu/gazette/story/2009/09/oldest-known-fibers-discovered/.
6. "The Woman Who Cut off Her Breasts to Protest a Tax." *BBC News*, BBC, 27 July 2016, www.bbc.com/news/world-asia-india-36891356.
7. Hajar, Rachel. "History of medicine timeline." Heart views : the official journal of the Gulf Heart Association vol. 16,1 (2015): 43-5. doi:10.4103/1995-705x.153008
8. Backwell, L; d'Errico, F; Wadley, L (2008). "Middle Stone Age bone tools from the Howiesons Poort layers, Sibudu Cave, South Africa". Journal of Archaeological Science. 35 (6): 1566–1580. doi:10.1016/j.jas.2007.11.006
9. Balter M. Archaeology. Clothes make the (hu) man. *Science*. 11 Sept. 2009;325(5946):1329. doi:10.1126/science.325_1329a.
10. "GDP per Capita, PPP (Current International $) - India." *World Bank*, 2019, data.worldbank.org/indicator/NY.GDP.PCAP.PP.CD?locations=IN.
11. Melissa A. Toups, Andrew Kitchen, Jessica E. Light, David L. Reed, Origin of Clothing Lice Indicates Early Clothing Use by Anatomically Modern Humans in Africa, Molecular Biology and Evolution, Volume 28, Issue 1, January 2011, Pages 29–32, doi:/10.1093/molbev/msq234.
12. Gilligan, Ian. Climate, Clothing, and Agriculture in Prehistory: Linking Evidence, Causes, and Effects. United Kingdom, Cambridge University Press, 2018, https://www.google.com/books/edition/Climate_Clothing_and_Agriculture_in_Preh/Ux50DwAAQBAJ?hl=en.
13. "Genesis Chapter 3 1611 KJV (King James Version)." *OFFICIAL KING JAMES BIBLE ONLINE: AUTHORIZED KING JAMES VERSION (KJV)*, www.kingjamesbibleonline.org/1611_Genesis-Chapter-3/.

Fire

1. Adler, Jerry. "Why Fire Makes Us Human." Smithsonian.com, Smithsonian Institution, 1 June 2013, www.smithsonianmag.com/science-nature/why-fire-makes-us-human-72989884/.

2. Darwin, Charles. The Descent of Man, and Selection in Relation to Sex. United Kingdom, D. Appleton, 1872, https://www.google.com/books/edition/The_Descent_of_Man_and_Selection_in_Rela/LYEQAAAAYAAJ?hl=en.
3. Pinker, Steven. The Language Instinct. United Kingdom, William Morrow, 2008, https://www.google.com/books/edition/_/cEQIPQAACAAJ?sa=X&ved=2ahUKEwi5IpmOo_fwAhUiIDQIHbZVAgsQre8FMBI6BAgmEFE.
4. Koebnick C, Strassner C, Hoffmann I, Leitzmann C. Consequences of a long-term raw food diet on body weight and menstruation: results of a questionnaire survey. Ann Nutr Metab. 1999;43(2):69-79. doi:10.1159/000012770.
5. Wrangham, Richard W.. Catching Fire: How Cooking Made Us Human. United Kingdom, Profile, 2010, https://www.google.com/books/edition/Catching_Fire/BVzV9reo0bYC?hl=en.
6. Berna, Francesco, et al. "Microstratigraphic Evidence of in Situ Fire in the Acheulean Strata of Wonderwerk Cave, Northern Cape Province, South Africa." *PNAS*, National Academy of Sciences, 15 May 2012, www.pnas.org/content/109/20/E1215.
7. Gugliotta, Guy. "The Great Human Migration." *Smithsonian.com*, Smithsonian Institution, 1 July 2008, www.smithsonianmag.com/history/the-great-human-migration-13561/.
8. Gowlett, J A J. "The discovery of fire by humans: a long and convoluted process." Philosophical transactions of the Royal Society of London. Series B, Biological sciences vol. 371,1696 (2016): 20150164. doi:10.1098/rstb.2015.0164.
9. Gowlett, John AJ, and Richard W. Wrangham. "Earliest fire in Africa: towards the convergence of archaeological evidence and the cooking hypothesis." Azania: Archaeological Research in Africa 48.1 (2013): 5-30, doi:10.1080/0067270X.2012.756754.
10. Clark, J. Desmond. "The Influence of Environment in Inducing Culture Change at the Kalambo Falls Prehistoric Site." The South African Archaeological Bulletin, vol. 19, no. 76, 1964, pp. 93–101. JSTOR, www.jstor.org/stable/3888549. Accessed 1 June 2021.
11. Bradish, Sarah Powers. Old Norse Stories. United States, American Book Company, 1900, https://www.google.com/books/edition/Old_Norse_Stories/4m0XAAAAIAAJ?hl=en.

Wheel

1. Diehl, Richard A., and Margaret D. Mandeville. "Tula, and Wheeled Animal Effigies in Mesoamerica." Antiquity, vol. 61, no. 232, 1987, pp. 239–246., doi:10.1017/s0003598x00052054.
2. Adams, Cecil. "Why Did the Peoples of the New World Fail to Invent the Wheel?" *The Straight Dope*, The Straight Dope, 3 Sept. 1983, www.straightdope.com/21341399/why-did-the-peoples-of-the-new-world-fail-to-invent-the-wheel.
3. Anthony, David W.. The Horse, the Wheel, and Language: How Bronze-Age Riders from the Eurasian Steppes Shaped the Modern World. United Kingdom, Princeton University Press, 2010, https://www.google.com/books/edition/The_Horse_the_Wheel_and_Language/nLIufwC4szwC?hl=en.

4. "Pottery Amphora Decorated with a Chariot Procession." *The British Museum*, -700, www.britishmuseum.org/collection/object/G_1936-1017-1.
5. "Amphoroid Krater." *The British Museum*, www.britishmuseum.org/collection/object/G_1911-0428-1.
6. Gambino, Megan. "A Salute to the Wheel." *Smithsonian.com*, Smithsonian Institution, 17 June 2009, www.smithsonianmag.com/science-nature/a-salute-to-the-wheel-31805121/.
7. Kennedy, Randy. "King Tut's Chariot Arrives in Times Square." *The New York Times*, The New York Times, 2 Aug. 2010, www.nytimes.com/2010/08/03/arts/design/03chariot.html.
8. "When Was the Wheel Invented?" *New Scientist*, www.newscientist.com/definition/the-wheel/.
9. "PHARAOH'S CHARIOT TAKEN FROM TOMB; Spectators Thrilled by Beauty and Brilliance of Its Color- Ing and Ornaments. WOOD EMBOSSED WITH GOLD Small Parts Show Decay After 3,000 Years, but All Can Be Pre- Served -- Noted Visitors There." *The New York Times*, The New York Times, 5 Feb. 1923, timesmachine.nytimes.com/timesmachine/1923/02/05/105902096.html?pageNumber=3.
10. Alex, Bridget. "Archaeologists Have Long Sought - But Never Found - The Very First Wheel." *Discover Magazine*, Discover Magazine, 14 Feb. 2020, www.discovermagazine.com/planet-earth/archaeologists-have-long-sought-but-never-found-the-very-first-wheel.
11. Balter, Michael. "Mysterious Indo-European Homeland May Have Been in the Steppes of Ukraine and Russia." *Science*, American Association for the Advancement of Science, 10 Dec. 2017, www.sciencemag.org/news/2015/02/mysterious-indo-european-homeland-may-have-been-steppes-ukraine-and-russia.

Agriculture

1. Stevens, William K. "Dry Climate May Have Forced Invention of Agriculture." *The New York Times*, The New York Times, 2 Apr. 1991, www.nytimes.com/1991/04/02/science/dry-climate-may-have-forced-invention-of-agriculture.html.
2. Smith, Adam. The Wealth of Nations. United States, Dover Publications, 2019, https://www.google.com/books/edition/The_Wealth_of_Nations/gZeWDwAAQBAJ?hl=en.
3. Diamond, Jared. Guns, Germs, and Steel: The Fates of Human Societies (20th Anniversary Edition). United States, W. W. Norton, 2017, https://www.google.com/books/edition/Guns_Germs_and_Steel_The_Fates_of_Human/XLo9DgAAQBAJ?hl=en.
4. Mishel, Lawrence, and Julia Wolfe. "CEO Compensation Has Grown 940% since 1978: Typical Worker Compensation Has Risen Only 12% during That Time." *Economic Policy Institute*, 14 Aug. 2019, www.epi.org/publication/ceo-compensation-2018/.
5. Diamond, Jared. "The Worst Mistake in the History of the Human Race." *Discover Magazine*, Discover Magazine, May 1987, pp. 64-66,

www.discovermagazine.com/planet-earth/the-worst-mistake-in-the-history-of-the-human-race.

6. Fuller, Dorian Q, et al. "The Domestication Process and Domestication Rate in Rice: Spikelet Bases from the Lower Yangtze." Science, American Association for the Advancement of Science, 20 Mar. 2009, science.sciencemag.org/content/323/5921/1607.full.
7. Garris, Amanda J et al. "Genetic structure and diversity in Oryza sativa L." Genetics vol. 169,3 (2005): 1631-8. doi:10.1534/genetics.104.035642.
8. Hopf, Maria, et al. Domestication of Plants in the Old World: The Origin and Spread of Domesticated Plants in Southwest Asia, Europe, and the Mediterranean Basin. United Kingdom, OUP Oxford, 2012, https://www.google.com/books/edition/Domestication_of_Plants_in_the_Old_World/1hHSYoqY-AwC?hl=en.
9. DAVIS, S., VALLA, F. Evidence for domestication of the dog 12,000 years ago in the Natufian of Israel. Nature 276, 608–610 (1978). doi.org/10.1038/276608a0.
10. Driscoll, Carlos A et al. "The Taming of the cat. Genetic and archaeological findings hint that wildcats became housecats earlier--and in a different place--than previously thought." Scientific American vol. 300,6 (2009): 68-75, www.ncbi.nlm.nih.gov/pmc/articles/PMC5790555/.
11. Zeder, Melinda A. "THE DOMESTICATION OF ANIMALS." Journal of Anthropological Research, vol. 68, no. 2, 2012, pp. 161–190. JSTOR, www.jstor.org/stable/23264664. Accessed 31 May 2021.
12. Xiang, Hai, et al. "Early Holocene Chicken Domestication in Northern China." *PNAS*, National Academy of Sciences, 9 Dec. 2014, www.pnas.org/content/111/49/17564.full.
13. "Horses Tamed Earlier than Thought." *BBC News*, BBC, 5 Mar. 2009, news.bbc.co.uk/2/hi/science/nature/7926235.stm.

Writing

1. Clayton, Ewan. "The Evolution of Writing – 5000 BC to Today." *The British Library*, The British Library, 19 Feb. 2019, www.bl.uk/history-of-writing/articles/the-evolution-of-writing.
2. Clayton, Ewan. *Where Did Writing Begin?* The British Library, 9 Apr. 2019, www.bl.uk/history-of-writing/articles/where-did-writing-begin.
3. Clayton, Ewan. *The Evolution of the Alphabet*. The British Library, 10 Apr. 2019, www.bl.uk/history-of-writing/articles/the-evolution-of-the-alphabet.
4. Wheelan, Charles J. *Naked Economics: Undressing the Dismal Science*. W.W. Norton & Company, 2019, https://www.google.com/books/edition/Naked_Economics_Undressing_the_Dismal_Sc/sT1kDwAAQBAJ?hl=en.
5. Quinn, Ben. "Isis Destruction of Palmyra's Temple of Bel Revealed in Satellite Images." *The Guardian*, Guardian News and Media, 1 Sept. 2015, www.theguardian.com/world/2015/sep/01/satellite-images-reveal-isis-destruction-of-palmyras-temple-of-bel.
6. The Code of Hammurabi, King of Babylon, about 2250 B.C.: Autographed Text, Transliteration, Translation, Glossary, Index of Subjects, Lists of Proper Names, Signs, Numerals, Corrections and Erasures, with Map, Frontispiece and Photograph of Text. United Kingdom, University of Chicago Press, Callaghan, 1904,

https://www.google.com/books/edition/The_Code_of_Hammurabi_King_of_Babylon_ab/aSANAAAAIAAJ?hl=en.
7. "Stèle: Code of Hammurabi." *Musée Du Louvre*, 27 Mar. 2021, collections.louvre.fr/en/ark:/53355/cl010174436.
8. Mason, William Albert. A History of the Art of Writing. United States, Macmillan, 1920, https://www.google.com/books/edition/A_History_of_the_Art_of_Writing/UxAuVMMccs4C?hl=en.
9. Taylor, Isaac. The Alphabet: An Account of the Origin and Development of Letters. United Kingdom, Kegan Paul, Trench, 1883, https://www.google.com/books/edition/The_Alphabet/0AsYAAAAYAAJ?hl=en.
10. Taylor, Isaac. Aryan alphabets. United Kingdom, Kegan Paul, Trench, 1883, https://www.google.com/books/edition/Aryan_alphabets/ETIbAAAAQAAJ?hl=en.
11. "The Phoenicians (1500–300 B.C.)." *The Metropolitan Museum of Art*, Oct. 2004, www.metmuseum.org/toah/hd/phoe/hd_phoe.htm.
12. "A To Z: The First Alphabet." *PBS*, Public Broadcasting Service, 24 Sept. 2020, www.pbs.org/wgbh/nova/video/a-to-z-the-first-alphabet/.

Numbers

1. O'Connor, J J, and E F Robertson. "Babylonian Numerals." *Maths History*, Dec. 2000, mathshistory.st-andrews.ac.uk/HistTopics/Babylonian_numerals/.
2. Charles, Mike. "Ancient Numeration Systems." *Ancient Numeration Systems*, Feb. 2007, fg.ed.pacificu.edu/charlesm/courses/mathsci/numsys/numsys.html.
3. "Stèle Pancarte ; Stèle De Néfertiabet." *Musée Du Louvre*, 4 Mar. 2021, collections.louvre.fr/en/ark:/53355/cl010005261#.
4. "The Dresden Codex." *World Digital Library*, Saxon State and University Library, Dresden, www.wdl.org/en/item/11621/.
5. "Royal Belgian Institute of Natural Sciences." *250: Ishango | Royal Belgian Institute of Natural Sciences*, www.naturalsciences.be/en/content/250-ishango.
6. Bhāskarācārya, and Brahmagupta. Algebra: With Arithmetic and Mensuration. United Kingdom, J. Murray, 1817, https://www.google.com/books/edition/Algebra/wYI_AAAAcAAJ?hl=en.
7. Casselman, Bill. "Reading the Bakhshali Manuscript." *American Mathematical Society*, June 2018, www.ams.org/publicoutreach/feature-column/fc-2018-06.
8. Encyclopaedia of the History of Science, Technology, and Medicine in Non-Westen Cultures. Germany, Springer Netherlands, 2013, pp. 72-73, https://www.google.com/books/edition/Encyclopaedia_of_the_History_of_Science/GzjpCAAAQBAJ?hl=en.
9. Durham, John W. "THE INTRODUCTION OF 'ARABIC' NUMERALS IN EUROPEAN ACCOUNTING." The Accounting Historians Journal, vol. 19, no. 2, 1992, pp. 25–55. JSTOR, www.jstor.org/stable/40698081. Accessed 30 May 2021.
10. Beery, Janet L., and Frank J. Swetz. "The Best Known Old Babylonian Tablet?" *Mathematical Association of America*, www.maa.org/press/periodicals/convergence/the-best-known-old-babylonian-tablet.
11. Swetz, Frank J. "Mathematical Treasure: Maya Numeration in Dresden Codex." *Mathematical Association of America*, www.maa.org/press/periodicals/convergence/mathematical-treasure-maya-numeration-in-dresden-codex.
12. Singleton, Esther. Ancient, 4004 B.C. to 70 A.D. United States, P.F. Collier & Son, 1903, https://www.google.com/books/edition/Ancient_4004_B_C_to_70_A_D/A3uuws3ylh4C?hl=en.

13. "A 3D-Print of Ancient History: One of the Most Famous Mathematical Texts from Mesopotamia." *Institute for the Preservation of Cultural Heritage*, Yale University, 16 Jan. 2016, ipch.yale.edu/news-events/3d-print-ancient-history-one-most-famous-mathematical-texts-mesopotamia.

Paper

1. "Clay Cuneiform Tablet." *The British Museum*, -330, www.britishmuseum.org/collection/object/W_1880-0617-496.
2. "Parchment - Full Hides." *Pergamena*, www.pergamena.net/parchment.
3. "Parchment". Encyclopedia Britannica, 4 Apr. 2014, www.britannica.com/topic/parchment.
4. "Papyrus". Encyclopedia Britannica, 5 Mar. 2020, www.britannica.com/topic/papyrus-writing-material.
5. Munsell, Joel. A Chronology of Paper and Paper-making. United States, J. Munsell, 1870, https://www.google.com/books/edition/A_Chronology_of_Paper_and_Paper_making/6MkS16clglAC?hl=en.
6. Du Halde, Jean Baptiste. The General History of China: Containing A Geographical, Historical, Chronological, Political and Physical Description of the Empire of China, Chinese-Tartary, Corea and Thibet, Including an Extract and Particular Accaount of Their Customs, Manners, Ceremonies, Religion, Arts and Sciences. United Kingdom, Watts, 1736, pp. 415-420, https://www.google.com/books/edition/The_General_History_of_China/6ehAAAAAcAAJ?hl=en.
7. D. B. H. "HISTORY OF PAPER AND ITS MANUFACTURE." The Massachusetts Teacher (1873-1874), vol. 26, no. 2, 1873, pp. 47–52. JSTOR, www.jstor.org/stable/45020228. Accessed 29 May 2021.
8. Ancient Libraries. United States, Cambridge University Press, 2013, https://www.google.com/books/edition/Ancient_Libraries/IawMdyHJxNIC?hl=en.
9. Johnston, Harold Whetstone. Latin Manuscripts: An Elementary Introduction to the Use of Critical Editions for High School and College Clases. United States, Scott, Foresman, 1897, https://www.google.com/books/edition/Latin_Manuscripts/PEJCAQAAIAAJ?hl=en.
10. Taylor, William Cooke. The Natural History of Society in the Barbarous and Civilized State: An Essay Towards Discovering the Origin and Course of Human Improvement. United Kingdom, Longman, 1840, https://www.google.com/books/edition/The_Natural_History_of_Society_in_the_Ba/A-hAAAAAcAAJ?hl=en.
11. Encyclopedia of Chinese History. United Kingdom, Taylor & Francis, 2016, p. 516, https://www.google.com/books/edition/Encyclopedia_of_Chinese_History/2UAIDwAAQBAJ?hl=en.

Book

1. "Codex Sinaiticus." *The British Library*, The British Library, 15 Jan. 2015, www.bl.uk/collection-items/codex-sinaiticus.
2. "Gutenberg Bible." *The British Library*, The British Library, 16 Jan. 2015, www.bl.uk/collection-items/gutenberg-bible.
3. "Bible De Gutenberg [Texte Imprimé] : [Biblia Latina]." *The Bibliothèque Nationale De France Catalogue Général*, [Mainz, Johannes Gutenberg Et Johannes Fust, Ca 1455], catalogue.bnf.fr/ark:/12148/cb35951126v.

4. "Gutenberg Bible (Biblia Latina)." *The Library of Congress*, 1454, www.loc.gov/item/52002339/.
5. "Germany: Memories of a Nation, Gutenberg: In the Beginning Was the Printer." *BBC Radio 4*, BBC, 21 Oct. 2014, www.bbc.co.uk/programmes/b04k6sjj.
6. "A BEAM OF PROTONS ILLUMINATES GUTENBERG'S GENIUS." *The New York Times*, The New York Times, 12 May 1987, timesmachine.nytimes.com/timesmachine/1987/05/12/973987.html?pageNumber=53.
7. "Coréen 109. 백운화상초록불조직지심체요절. 白雲和尚抄錄佛祖直指心體要節. Päk Un Hoa Sañ Č'orok Pulčo Č'ikč'i Simč'e Yočŏl (1377)." *Archives Et Manuscrits*, Bibliothèque Nationale De France. Département Des Manuscrits, 1377, archivesetmanuscrits.bnf.fr/ark:/12148/cc78021m.
8. Rolevinck, Werner. Fasciculus temporum. Belgium, Johannes Veldener, 1475, https://www.google.com/books/edition/_/F7GFDwI0OCcC?hl=en.
9. Townsend, Howard. The Sinai Bible Or Bibliorum Codex Sinaiticus Petropolitanus: Read Before the Albany Institute, December 15, 1863. United States, J. Munsell, 1866, https://www.google.com/books/edition/The_Sinai_Bible_Or_Bibliorum_Codex_Sinai/e09OAQAAMAAJ?hl=en.
10. Smith, Adam. The Wealth of Nations. United States, Dover Publications, 2019, https://www.google.com/books/edition/The_Wealth_of_Nations/gZeWDwAAQBAJ?hl=en.
11. *The World Book Encyclopedia. Volume 2, B*, World Book, Inc., 2015, pp. 465-468.
12. "History." *Koenig & Bauer*, www.koenig-bauer.com/en/holding/history/.
13. "Parchment". Encyclopedia Britannica, 4 Apr. 2014, www.britannica.com/topic/parchment.
14. "Papyrus". Encyclopedia Britannica, 5 Mar. 2020, www.britannica.com/topic/papyrus-writing-material.
15. Johnston, Harold Whetstone. Latin Manuscripts: An Elementary Introduction to the Use of Critical Editions for High School and College Clases. United States, Scott, Foresman, 1897, https://www.google.com/books/edition/Latin_Manuscripts/PEJCAQAAIAAJ?hl=en.

Clock

1. Huygens, Christiaan. "The Hague Clock, Pendulum Clock According to Christiaan Huygens, Salomon Coster The Hague, 1657." *Rijksmuseum Boerhaave*, 1657, boerhaave.adlibhosting.com/Details/collect/18139.
2. "SI Base Unit: Second (s)." *Bureau International Des Poids Et Mesures (BIPM)*, www.bipm.org/en/si-base-units/second.
3. "SI Base Unit: Metre (m)." *Bureau International Des Poids Et Mesures (BIPM)*, www.bipm.org/en/si-base-units/metre.
4. Huygens, Christiaan. Horologium oscillatorium sive de motu pendulorum ad horologia aptato demonstrationes geometricae. France, apud F. Muguet, 1673, https://www.google.com/books/edition/Horologium_oscillatorium_sive_de_motu_pe/SvkuAAAAQAAJ?hl=en.
5. Bickel, Susanne and Gautschy, Rita. "Eine ramessidische Sonnenuhr im Tal der Könige" ("A Ramesside Sundial in the Valley of the Kings") *Zeitschrift für Ägyptische Sprache und Altertumskunde*, vol. 141, no. 1, 2014, pp. 3-14. https://doi.org/10.1515/zaes-2014-0001.
6. "CLOCK AND WATCHES.; A Storehouse of Information About Old and Curious Ones." *The New York Times*, The New York Times, 5 Aug. 1899,

timesmachine.nytimes.com/timesmachine/1899/08/05/102412881.html?pageNumber=26.
7. Britten, Frederick James. Old Clocks and Watches and Their Makers. United Kingdom, B.T. Batsford, 1899, https://www.google.com/books/edition/Old_Clocks_and_Watches_and_Their_Makers/vJ22Tp_zK4IC?hl=en.
8. "ELECTRIFIED QUARTZ CRYSTAL DISPLACES CLOCK PENDULUM." *The New York Times*, The New York Times, 13 Oct. 1929, timesmachine.nytimes.com/timesmachine/1929/10/13/91971816.html?pageNumber=176.
9. "FIND MOON AFFECTS SPEED OF CLOCKS; Yale Scientists at Meeting of Astronomers Tell of Study Made With Loomis Chronograph. STAR TEMPERATURE GAUGE Wisconsin Professor Tells of Use of Photo-Electric Cell to Learn Star's Color Also." *The New York Times*, The New York Times, 2 Jan. 1931, timesmachine.nytimes.com/timesmachine/1931/01/02/96378597.html?pageNumber=14.
10. "Accuracy Stressed For Crystal Device In Japanese Watch; JAPANESE STRESS WATCH ACCURACY." *The New York Times*, The New York Times, 5 Jan. 1970, timesmachine.nytimes.com/timesmachine/1970/01/05/80013325.html?pageNumber=55.
11. JANSON, DONALD. "NATION TESTS USE OF ATOMIC CLOCK; New Standard, Increasing Accuracy, Expected to Be Official Time Measure." *The New York Times*, The New York Times, 11 Aug. 1957, timesmachine.nytimes.com/timesmachine/1957/08/11/90833924.html?pageNumber=37.
12. "Science/Nature | Atomic Ticker Clocks up 50 Years." *BBC News*, BBC, 2 June 2005, news.bbc.co.uk/2/hi/science/nature/4587919.stm.
13. ESSEN, L., PARRY, J. An Atomic Standard of Frequency and Time Interval: A Cæsium Resonator. Nature 176, 280–282 (1955), doi.org/10.1038/176280a0.
14. Essen, L., and J. V. L. Parry. "The Caesium Resonator as a Standard of Frequency and Time." *Philosophical Transactions of the Royal Society of London. Series A, Mathematical and Physical Sciences*, vol. 250, no. 973, 1957, pp. 45–69. JSTOR, www.jstor.org/stable/91638. Accessed 28 May 2021.
15. *The World Book Encyclopedia*. Volume 4, C, World Book, Inc., 2015, pp. 682i-682k.

Calendar

1. Plait, Phil. "Another 31,556,941 Seconds Around the Sun? Happy Recycled Earth Orbital Time Period!" Slate Magazine, 1 Jan. 2014, slate.com/technology/2014/01/new-year-2014-how-astronomers-define-the-year.html.
2. USEFUL CONSTANTS. International Earth Rotation and Reference Systems Service (IERS), 13 Feb. 2014, hpiers.obspm.fr/eop-pc/models/constants.html.
3. Manning, Evan M. "How Many Days Are in a Year?" NASA PUMAS (Practical Uses of Math And Science), 5 Sept. 1997, pumas.nasa.gov/sites/default/files/examples/04_21_97_1.pdf.

4. "CONFORMS ON CALENDAR.; Orthodox Church Will Adopt Gregorian System Oct. 1." *The New York Times*, The New York Times, 13 June 1923, timesmachine.nytimes.com/timesmachine/1923/06/13/104962935.html?pageNumber=3.
5. Lewis, Danny. "Happy Leap Day! Brought to You by Julius Caesar." *Smithsonian.com*, Smithsonian Institution, 29 Feb. 2016, www.smithsonianmag.com/smart-news/happy-leap-day-brought-to-you-by-Julius-Caesar-180958242/.
6. VAIL, W.H. "CHANGES IN THE CALENDAR.; Old Style and New Style in Relation to the Greek System." *The New York Times*, The New York Times, 20 June 1923, timesmachine.nytimes.com/timesmachine/1923/06/20/105995648.html?pageNumber=18.
7. "POPE GREGORY'S TRIUMPH NOW COMPLETE." *The New York Times*, The New York Times, 14 June 1923, timesmachine.nytimes.com/timesmachine/1923/06/14/104963866.html?pageNumber=18.
8. Morland, Samuel. "Perpetual Calendar Medal." *The British Museum*, 1650, www.britishmuseum.org/collection/object/H_1891-0217-8.
9. "Papyrus Written with Black Ink with Demotic Legal Document, Dated Month of Mesore, Year 4 of Ptolemy III Euergetes I: 6 Ll." *The British Museum*, -243, www.britishmuseum.org/collection/object/Y_EA10389.
10. "Babylonian Calendar Tablet." *The British Museum*, www.britishmuseum.org/collection/object/W_1880-0617-739.
11. Vizza, Francesco. "ALOYSIUS LILIUS AUTHOR OF THE GREGORIAN REFORM OF THE CALENDAR." PhilSci Archive, University of Pittsburgh, 13 Oct. 2018, philsci-archive.pitt.edu/id/eprint/15151.
12. *The World Book Encyclopedia*. Volume 3, C, World Book, Inc., 2015, pp. 28-32.

Money

1. "Functions of Money, Economic Lowdown Podcasts: Education: St. Louis Fed." *Federal Reserve Bank of St. Louis*, 3 Nov. 2020, www.stlouisfed.org/education/economic-lowdown-podcast-series/episode-9-functions-of-money.
2. "The Origin of Money." *The Money Question*, 17 June 2019, themoneyquestion.org/the-origin-of-money/.
3. Schaps, David M. "The Invention of Coinage in Lydia, in India, and in China". XIV International Economic History Congress, Helsinki 2006. Session 30; http://www.helsinki.fi/iehc2006/papers1/Schaps.pdf.
4. Lowther, Ed. "A Short History of the Pound." *BBC News*, BBC, 14 Feb. 2014, www.bbc.com/news/uk-politics-26169070.
5. "The Euro." *European Central Bank*, 28 Jan. 2021, www.ecb.europa.eu/euro/html/index.en.html.
6. "Lydia Coin." *The British Museum*, 1866, www.britishmuseum.org/collection/object/C_1866-1201-3671.
7. Von Glahn, Richard. "Monies of Account and Monetary Transition in China, Twelfth to Fourteenth Centuries." Journal of the Economic and Social History of the Orient, vol.

53, no. 3, 2010, pp. 463–505. JSTOR, www.jstor.org/stable/20789802. Accessed 27 May 2021.
8. Wolla, Scott A. "Would a Gold Standard Brighten Economic Outcomes?" *Economic Research - Federal Reserve Bank of St. Louis*, 1 Jan. 2015, research.stlouisfed.org/publications/page1-econ/2015/01/01/would-a-gold-standard-brighten-economic-outcomes/.
9. Polo, Marco. "Chapter XXIV. *How the Great Kaan Causeth the Bark of Trees, made into something like Paper, to pass for Money over all his Country.*", The Book of Ser Marco Polo: The Venetian. United Kingdom, J. Murray, 1903, pp. 423-430, https://www.google.com/books/edition/The_Book_of_Ser_Marco_Polo/VxMEAAAAMAAJ?hl=en.
10. Wheelan, Charles. Naked Money: A Revealing Look at Our Financial System. United States, W. W. Norton, 2016, https://www.google.com/books/edition/Naked_Money_A_Revealing_Look_at_Our_Fina/12suCgAAQBAJ?hl=en.
11. Smith, Adam. The Wealth of Nations. United States, Dover Publications, 2019, https://www.google.com/books/edition/The_Wealth_of_Nations/gZeWDwAAQBAJ?hl=en.
12. Becker, Gary S.. Human Capital: A Theoretical and Empirical Analysis, with Special Reference to Education. United Kingdom, University of Chicago Press, 1993, https://www.google.com/books/edition/Human_Capital/REERnwEACAAJ?hl=en.

Ruler

1. materese, Robin. "Busting Myths about the Metric System." *NIST*, 6 Oct. 2020, www.nist.gov/blogs/taking-measure/busting-myths-about-metric-system.
2. "The International System of Units (SI): Base Units." *BIPM*, www.bipm.org/en/measurement-units/si-base-units.
3. *The World Book Encyclopedia*. Volume 13, M, World Book, Inc., 2015, p. 542.
4. *The World Book Encyclopedia*. Volume 7, F, World Book, Inc., 2015, p. 352.
5. *The World Book Encyclopedia*. Volume 21, W, World Book, Inc., 2015, pp. 184-188.
6. "History of the 42-Gallon Oil Barrel." *American Oil & Gas Historical Society*, 10 Mar. 2021, www.aoghs.org/transportation/history-of-the-42-gallon-oil-barrel/.
7. Pollack, Andrew. "Missing What Didn't Add Up, NASA Subtracted an Orbiter." *The New York Times*, The New York Times, 1 Oct. 1999, www.nytimes.com/1999/10/01/us/missing-what-didn-t-add-up-nasa-subtracted-an-orbiter.html.
8. "THE WEIGHTS AND MEASURES TREATY." *The New York Times*, The New York Times, 31 Oct. 1878, timesmachine.nytimes.com/timesmachine/1878/10/31/80733643.html?pageNumber=4.
9. Lowther, Ed. "A Short History of the Pound." *BBC News*, BBC, 14 Feb. 2014, www.bbc.com/news/uk-politics-26169070.
10. Ghosh, Pallab. "Kilogram Gets a New Definition." *BBC News*, BBC, 16 Nov. 2018, www.bbc.com/news/science-environment-46143399#.

Map

1. Freak, Jeff, and Shannon Holloway. "Taking the Most Direct Route to Straddie." *Traveller*, 15 Mar. 2012, www.traveller.com.au/taking-the-most-direct-route-to-straddie-1v85m.
2. "Imago Mundi: British Museum." *The British Museum*, www.britishmuseum.org/collection/object/W_1882-0714-509.
3. Thompson, Clive. "From Ptolemy to GPS, the Brief History of Maps." *Smithsonian.com*, Smithsonian Institution, 1 July 2017, www.smithsonianmag.com/innovation/brief-history-maps-180963685/?page=2.
4. *A Land beyond the Stars*. Museo Galileo - Institute and Museum of the History of Science, exhibits.museogalileo.it/waldseemuller/index.html.
5. Desjardins, Jeff. "Mapped: Visualizing the True Size of Africa." *Visual Capitalist*, 19 Feb. 2020, www.visualcapitalist.com/map-true-size-of-africa/.
6. Taylor, Adam. "This Interactive Map Shows How 'Wrong' Other Maps Are." *The Washington Post*, WP Company, 2 May 2019, www.washingtonpost.com/news/worldviews/wp/2015/08/18/this-interactive-map-shows-how-wrong-other-maps-are/.
7. Sobel, Dava. Longitude: The True Story of a Lone Genius Who Solved the Greatest Scientific Problem of His Time. United States, Bloomsbury Publishing, 2010, https://www.google.com/books/edition/Longitude/lW8DYSPa6fEC?hl=en.
8. "GERMAN PROFESSOR FIRST SUGGESTED AMERICA'S NAME; Martin Waldseemuller, Teacher of Geography at St. Die. in Lorraine, Germany, Was Its Originator." *The New York Times*, The New York Times, 10 Apr. 1910, timesmachine.nytimes.com/timesmachine/1910/04/10/102037454.html?pageNumber=84.
9. Mason, Betsy. "Why Your Mental Map of the World Is Wrong." *Culture*, National Geographic, 3 May 2021, www.nationalgeographic.com/culture/article/all-over-the-map-mental-mapping-misconceptions.
10. Pentecost, Kate. "Where on Earth Are You? A Beginners Guide to Longitude." *Australian National Maritime Museum*, 8 June 2016, www.sea.museum/2016/06/08/where-on-earth-are-you-a-beginners-guide-to-longitude/.
11. Walters, Joanna. "Boston Public Schools Map Switch Aims to Amend 500 Years of Distortion." *The Guardian*, Guardian News and Media, 23 Mar. 2017, www.theguardian.com/education/2017/mar/19/boston-public-schools-world-map-mercator-peters-projection.
12. Müller, Karl, and Fischer, Curt Theodor. Geōgraphikē hyphēuēsis: pars 1. Lib. I-III. France, A. Firmin Didot, 1883, https://www.google.com/books/edition/Ge%C5%8Dgraphik%C4%93_hyph%C4%93u%C4%93sis_pars_1_Lib_I/qmw0AQAAMAAJ?hl=en.

Ship

1. "Historical Estimates of World Population." *The United States Census Bureau*, 5 July 2018, www.census.gov/data/tables/time-series/demo/international-programs/historical-est-worldpop.html.

2. "Jar with Boat Designs." *Brooklyn Museum*, www.brooklynmuseum.org/opencollection/objects/3276.
3. *How Much Water Is There on Earth?*, www.usgs.gov/special-topic/water-science-school/science/how-much-water-there-earth.
4. *Who Made America? | Innovators | Robert Fulton*. Public Broadcasting Service, www.pbs.org/wgbh/theymadeamerica/whomade/fulton_hi.html.
5. *Crossing the Atlantic by Steamship*. Smithsonian National Postal Museum, postalmuseum.si.edu/exhibition/long-may-it-wave-conflict-and-exploration/crossing-the-atlantic-by-steamship.
6. Tharoor, Shashi. "'But What about the Railways ...?' The Myth of Britain's Gifts to India." *The Guardian*, Guardian News and Media, 8 Mar. 2017, www.theguardian.com/world/2017/mar/08/india-britain-empire-railways-myths-gifts.
7. Santora, Marc. "Syria Says It Will Ration Fuel as the Economic Toll of the Blockage Grows." *The New York Times*, The New York Times, 28 Mar. 2021, www.nytimes.com/2021/03/28/world/middleeast/syria-fuel-suez-canal.html.
8. Diamond, Jared. Guns, Germs, and Steel: The Fates of Human Societies (20th Anniversary Edition). United States, W. W. Norton, 2017, pp. 283-307, 320-338 https://www.google.com/books/edition/Guns_Germs_and_Steel_The_Fates_of_Human/XLo9DgAAQBAJ.
9. Kaul, Chandrika. "History - British History in Depth: From Empire to Independence: The British Raj in India 1858-1947." *BBC*, BBC, 3 Mar. 2011, www.bbc.co.uk/history/british/modern/independence1947_01.shtml.
10. Weisman, Steven R. "India Widow's Death at Pyre Creates a Shrine." *The New York Times*, The New York Times, 19 Sept. 1987, www.nytimes.com/1987/09/19/world/india-widow-s-death-at-pyre-creates-a-shrine.html.
11. Bruce, William Napier. Life of general sir Charles Napier. United Kingdom, n.p, 1885, https://www.google.com/books/edition/Life_of_general_sir_Charles_Napier/JHgIAAAAQAAJ.
12. "Who Made America? | Innovators | Robert Fulton." *PBS*, Public Broadcasting Service, www.pbs.org/wgbh/theymadeamerica/whomade/fulton_hi.html.
13. "Steaming Across the Atlantic." *Connecticut History | a CTHumanities Project*, 10 May 2021, connecticuthistory.org/steaming-across-the-atlantic/.
14. "Is the Airplane to Supplant the Battleship?; New Method of Naval Warfare Seems Indicated by Abandonment of All Big-Gun Building Programs and Substitution of Wholesale Flying-Machine Carriers." *The New York Times*, The New York Times, 11 July 1920, timesmachine.nytimes.com/timesmachine/1920/07/11/112662982.html?pageNumber=36.

Iron metallurgy

1. Schultz, Colin. *The Ancient Egyptians Had Iron Because They Harvested Fallen Meteors*. Smithsonian Institution, 30 May 2013, www.smithsonianmag.com/smart-news/the-ancient-egyptians-had-iron-because-they-harvested-fallen-meteors-86153874/.

2. Britannica, The Editors of Encyclopaedia. "Iron Age". Encyclopedia Britannica, 10 Sep. 2020, https://www.britannica.com/event/Iron-Age.
3. "Global Crude Steel Output Increases by 3.4% in 2019." *Worldsteel Association*, 27 Jan. 2020, www.worldsteel.org/media-centre/press-releases/2020/Global-crude-steel-output-increases-by-3.4--in-2019.html.
4. "FAO Cereal Supply and Demand Brief | World Food Situation | Food and Agriculture Organization of the United Nations." *FAO.org*, 4 Feb. 2021, www.fao.org/worldfoodsituation/csdb/en/.
5. "2020 World Steel in Figures." *Worldsteel Association*, 2020, www.worldsteel.org/en/dam/jcr:f7982217-cfde-4fdc-8ba0-795ed807f513/.
6. Morgan, J W, and E Anders. "Chemical composition of Earth, Venus, and Mercury." Proceedings of the National Academy of Sciences of the United States of America vol. 77,12 (1980): 6973-7. doi:10.1073/pnas.77.12.6973
7. Hamilton, Elizabeth G. *Adventures in Experimental Smelting*. Penn Museum, 15 Nov. 2007, www.penn.museum/sites/expedition/adventures-in-experimental-smelting/.
8. "Golden Gate Bridge Fast Facts." *CNN*, Cable News Network, 26 July 2020, www.cnn.com/2013/11/01/us/golden-gate-bridge-fast-facts/index.html.
9. "BUILDING BIG: Databank: Empire State Building." *PBS*, Public Broadcasting Service, www.pbs.org/wgbh/buildingbig/wonder/structure/empire_state.html.
10. "Bucharest International Conference Centre." *Palace of Parliament - Bucharest International Conference Centre*, cic.cdep.ro/en/general-presentation/palace-of-parliament-the-building.
11. "AS1085 50kg AS Rail." *Railway Rail Supplier - Supply International Standard Steel Rails*, www.railwayrail.com/products/as1085-australia-standard-50kg-as-railway-steel-rail/.
12. "Indian Railways Civil Engineering Portal." *About Us*, 31 Mar. 2019, ircep.gov.in/AboutUs.html.
13. "Appliances." *American Iron and Steel Institute*, 13 July 2020, www.steel.org/steel-markets/appliances/.
14. *IRON ORE*. U.S. Geological Survey, Mineral Commodity Summaries, January 2021, Jan. 2021, pubs.usgs.gov/periodicals/mcs2021/mcs2021-iron-ore.pdf.
15. BESSEMER, HENRY. "US49055A - IMPROVEMENT IN MACHINERY FOR THE MANUFACTURE OF IRON AND STEEL." United States Patent, 25 July 1865, patents.google.com/patent/US49055A/en.
16. *The World Book Encyclopedia*. Volume 10, I, World Book, Inc., 2015, pp. 436-452.
17. Thilo Rehren et.al, "5,000 years old Egyptian iron beads made from hammered meteoritic iron", Journal of Archaeological Science, Volume 40, Issue 12, 2013, pp 4785-4792, https://doi.org/10.1016/j.jas.2013.06.002.

Knife

1. "Company: Victorinox Swiss Army (USA)." *Victorinox Swiss Army*, www.swissarmy.com/us/en/Victorinox/Welcome/Company/cms/company.
2. Foulkes, Imogen. "From Humble Tool to Global Icon." *BBC News*, BBC, 30 July 2009, news.bbc.co.uk/2/hi/europe/8172917.stm.

3. "Tutankhamun's Knife Was 'Made from Meteorite Iron'." *BBC News*, BBC, 2 June 2016, www.bbc.com/news/world-middle-east-36432635.
4. "Knife Blade, Dark Grey Flint, Mousterian, from Le Moustier, Peyzae, France, 10000-5001 BC." *Science Museum Group*, collection.sciencemuseumgroup.org.uk/objects/co106386/knife-blade-dark-grey-flint-mousterian-from-le-knife.
5. "Bifacial Knife with Handle Ca. 2949 B.C." *Metropolitan Museum of Art*, www.metmuseum.org/art/collection/search/547438.
6. So, Jenny F., et al. The Great Bronze Age of China: An Exhibition from the People's Republic of China. United Kingdom, *Metropolitan Museum of Art*, 1980.
7. "Mycenaean Dagger." *The British Museum*, 1887, www.britishmuseum.org/collection/object/G_1887-0501-7.
8. "Mycenaean Dagger." *The British Museum*, 1897, www.britishmuseum.org/collection/object/G_1897-0401-1484.
9. Wu, Katherine J. "Graduate Student Discovers One of World's Oldest Swords in Mislabeled Monastery Display." *Smithsonian.com*, Smithsonian Institution, 16 Mar. 2020, www.smithsonianmag.com/smart-news/graduate-student-discovers-one-worlds-oldest-swords-mislabeled-monastery-display-180974420/.
10. "Late Cypriot III Knife." *The British Museum*, 1897, www.britishmuseum.org/collection/object/G_1897-0401-994.
11. "Iron Age Iran Sword." *The British Museum*, 1968, www.britishmuseum.org/collection/object/W_1968-1012-20.
12. Wu, Katherine J. "Archaeology Intern Unearths Spectacular, 2,000-Year-Old Roman Dagger." *Smithsonian.com*, Smithsonian Institution, 2 Mar. 2020, www.smithsonianmag.com/smart-news/archaeology-intern-unearths-spectacular-2000-year-old-roman-dagger-180974310.
13. "Honjo Masamune – The Best Katana Ever Made." *BladesPro US*, 17 Oct. 2019, www.americanbladespro.com/blogs/news/honjo-masamune-best-katana-ever-made.
14. 文化財・史跡 - 小村神社と国宝・金銅荘環頭大刀ほか, www.kochinet.ed.jp/hidaka-v/sub4_1.html.
15. "ZWILLING History: Modern Company since 1731." *Zwilling*, www1.zwilling.com/uk/en/pages/about-zwilling.html.
16. "Our History." *WÜSTHOF*, www.wuesthof.com/en-fi/discover/our-history/.
17. "History: Victorinox Swiss Army (USA)." *Victorinox Swiss Army*, www.swissarmy.com/us/en/history/cms/history.
18. "History: International." *Wenger*, www.wenger.ch/global/en/About-us/History/cms/history.
19. "History: Manufactory." *Boker USA*, www.bokerusa.com/history.
20. "Company History." *Puma Knife Company USA: History*, pumaknifecompanyusa.com/companyhistory.aspx.
21. "The Canterbury Tales by Geoffrey Chaucer." *The British Library*, The British Library, 10 Dec. 2014, www.bl.uk/collection-items/the-canterbury-tales-by-geoffrey-chaucer.
22. Chaucer, Geoffrey. *1.4 The Reeve's Prologue and Tale*, 1400, chaucer.fas.harvard.edu/pages/reeves-prologue-and-tale.
23. "History." *Kikuichi Cutlery*, kikuichi.net/pages/about-kikuichi-cutlery.
24. "OUR ARMY SWORDS 'MADE IN GERMANY'; Port Appraiser Kracke Reports 1,000 Imported From Solingen in the Past Year. CLUB MEMBERS INDIGNANT Name

Committee to Investigate -- Dealers Say No Officers' Swords Are Made Here." *The New York Times*, The New York Times, 13 June 1923, timesmachine.nytimes.com/timesmachine/1923/06/13/104962977.html?pageNumber=10.

Pot

1. Graber, Cynthia, and Nicola Twilley. "How Pots and Pans Turned Us Into Creative Cooks." *The Atlantic*, Atlantic Media Company, 21 June 2018, www.theatlantic.com/science/archive/2018/06/pots-and-pans/563330/.
2. Huysecom, E., et al. "The Emergence of Pottery in Africa during the Tenth Millennium Cal BC: New Evidence from Ounjougou (Mali)." Antiquity, vol. 83, no. 322, 2009, pp. 905–917, doi.org/10.1017/S0003598X00099245.
3. Craig, O., Saul, H., Lucquin, A. et al. Earliest evidence for the use of pottery. Nature 496, 351–354 (2013). doi.org/10.1038/nature12109.
4. "A PURITAN'S WORLDLY GOODS.; INVENTORY OF THE POSSESSION OF A BOSTON MERCHANT IN 1712." *The New York Times*, The New York Times, 17 July 1882, timesmachine.nytimes.com/timesmachine/1882/07/17/103418013.html?pageNumber=2.
5. Wu, Xiaohong, et al. "Early Pottery at 20,000 Years Ago in Xianrendong Cave, China." *Science*, American Association for the Advancement of Science, 29 June 2012, DOI: 10.1126/science.1218643.
6. Nakamura, Toshio, et al. "Radiocarbon Dating of Charred Residues on the Earliest Pottery in Japan." Radiocarbon, vol. 43, no. 2B, 2001, pp. 1129–1138., doi:10.1017/S0033822200041783.
7. "History." *Lodge Cast Iron*, www.lodgecastiron.com/about-lodge/history.
8. "ZWILLING History: Modern Company since 1731." *Zwilling*, www1.zwilling.com/uk/en/pages/about-zwilling.html.
9. Virginia. "Griswold Cast Iron – History, Value, Identify Guide In 2021." *The Pan Handler*, 2021, thepan-handler.com/griswold-cast-iron/.
10. Hicks, Boonie. "Wagner Cast Iron: Wagner Ware History, Dates and Logos." *Boonie Hicks*, 18 June 2020, www.booniehicks.com/wagner-cast-iron/.
11. "Cooking Utensils and Nutrition: MedlinePlus Medical Encyclopedia." *MedlinePlus*, U.S. National Library of Medicine, 30 Apr. 2019, medlineplus.gov/ency/article/002461.htm.
12. "The Manufacture." *Mauviel 1830*, www.mauviel.com/en/the-manufacture/.
13. "Company Profile." *Gaggenau*, www.gaggenau.com/press/company-profile.

Furniture

1. Friedman, Uri. "The 5,000-Year History of the Chair." *The Atlantic*, Atlantic Media Company, 30 Aug. 2016, www.theatlantic.com/international/archive/2016/08/chairs-history-witold-rybczynski/497657/.
2. Wayman, Erin. "The World's Oldest Mattress." *Smithsonian.com*, Smithsonian Institution, 14 Dec. 2011, www.smithsonianmag.com/science-nature/the-worlds-oldest-mattress-7513279/.

3. Rawsthorn, Alice. "Celebrating the Everychair of Chairs, in Cheap Plastic." *The New York Times*, The New York Times, 4 Feb. 2007, www.nytimes.com/2007/02/04/style/04iht-design5.html.
4. Rashid, Karim. "A Brief History of the Humble Plastic Chair (and Its Oil-Free Future)." *CNN*, Cable News Network, 14 Sept. 2015, edition.cnn.com/style/article/history-of-plastic-chairs-karim-rashid/index.html.
5. "The Global Object - The Monobloc." *Deutsche Welle*, 20 Jan. 2014, www.dw.com/en/the-global-object-the-monobloc/av-17373066.
6. King, Rodney S. "Phrygian Furniture From Gordion." *Expedition Magazine - Volume 16 Issue 3*, Penn Museum, 8 May 1974, www.penn.museum/sites/expedition/phrygian-furniture-from-gordion/.
7. Pinsker, Joe. "Something Is Changing in the Way People Eat at Home." *The Atlantic*, Atlantic Media Company, 22 May 2019, www.theatlantic.com/family/archive/2019/05/meals-couches-bedrooms-kitchen-table/590026/.
8. "Hatnefer's Chair." *The Metropolitan Museum of Art* , www.metmuseum.org/art/collection/search/543868.
9. "Chair of Reniseneb." *The Metropolitan Museum of Art* , www.metmuseum.org/art/collection/search/547687.
10. "Queen Hetepheres' Throne." *Harvard Museums of Science & Culture*, hmsc.harvard.edu/queen-hetepheres-throne.
11. "Trône : [Quatre Protomés De Panthères Forment Les Pieds; Les Accoudoirs Sont Constitués De Deux Bandeaux Ajourés, Décorés De Rosettes (Partie Inférieure) Et De Motifs Végétaux (Registre Supérieur); Le Dossier, De Forme Triangulaire, Est Orné De Trois Cercles Et De Rinceaux. Ce Siège, à L'origine Pliant, Serait Une Oeuvre Du VIIe Siècle, Ou Une 'Réplique Carolingienne']." *Bibliothèque Nationale De France*, gallica.bnf.fr/ark:/12148/btv1b77001621.
12. "New Kingdom Rulers Tutankhamun." *New Kingdom Rulers*, egyptianmuseum.org/explore/new-kingdom-ruler-tutankhamun.
13. Burton, Harry, et al. Tutankhamun, His Tomb and Its Treasures. United Kingdom, Metropolitan Museum of Art, 1976.

Toilet

1. Ritchie, Hannah, and Max Roser. "Sanitation." *Our World in Data*, 25 Sept. 2019, ourworldindata.org/sanitation#what-share-of-people-practice-open-defecation.
2. Alex, Bridget. "What the Earliest Toilets Say About How Human Civilization Has Evolved." *Discover Magazine*, Discover Magazine, 9 Apr. 2020, www.discovermagazine.com/planet-earth/what-the-earliest-toilets-say-about-how-human-civilization-has-evolved.
3. "Evolution of Sanitation and Wastewater Technologies Through the Centuries." United Kingdom, IWA Publishing, 2014, https://www.google.com/books/edition/Evolution_of_Sanitation_and_Wastewater_T/mbgrBQAAQBAJ
4. *ROCA 2019*, June 2020, www.roca.com/memoria_roca_2019/#/home.
5. "Company History." *Hansgrohe Group*, www.hansgrohe-group.com/en/about-us/history.

6. Frerichs, Ralph R. "John Snow and the Broad Street Pump: On the Trail of an Epidemic." *Department of Epidemiology*, UCLA, Nov. 2003, www.ph.ucla.edu/epi/snow/snowcricketarticle.html.
7. "United Nations Millennium Development Goals." *United Nations*, United Nations, www.un.org/millenniumgoals/.
8. "THE 17 GOALS | Sustainable Development." *United Nations*, United Nations, sdgs.un.org/goals.
9. Stamp, Jimmy. "From Turrets to Toilets: A Partial History of the Throne Room." *Smithsonian.com*, Smithsonian Institution, 20 June 2014, www.smithsonianmag.com/history/turrets-toilets-partial-history-throne-room-180951788/.
10. "A Brief History of The Flush Toilet." *The British Association of Urological Surgeons*, www.baus.org.uk/museum/164/a_brief_history_of_the_flush_toilet.
11. "Burj Khalifa Features Sanitary Fittings from Hansgrohe." *DesignCurial*, 12 Jan. 2010, www.designcurial.com/news/burj-khalifa-features-sanitary-fittings-from-hansgrohe.
12. "Roca's History." *ROCA Bathrooms*, www.uk.roca.com/about-roca/roca-history.
13. "Kohler Co. History." *Kohler*, www.kohlercompany.com/who-we-are/our-history/.
14. "Kohler." *Forbes*, Forbes Magazine, 2020, www.forbes.com/companies/kohler/.
15. "200 Years of 'Living Bathrooms.'" *Duravit*, www.duravit.us/service/company/history.us-en.html.
16. "Company Information." *American Standard*, www.americanstandard-us.com/about/company-info.
17. Murakami, Sakura. "Japan's next Restroom Revolution? Phasing out Squat Toilets for Tokyo 2020." *The Japan Times*, 24 Feb. 2020, www.japantimes.co.jp/news/2020/02/24/national/squat-toilets-tokyo-2020/.

Religion

1. Ambrosino, Brandon. "How and Why Did Religion Evolve?" *BBC Future*, 18 Apr. 2019, www.bbc.com/future/article/20190418-how-and-why-did-religion-evolve.
2. "Qafzeh: Oldest Intentional Burial ." *The Smithsonian Institution's Human Origins Program*, 27 Apr. 2021, humanorigins.si.edu/evidence/behavior/burial/qafzeh-oldest-intentional-burial.
3. Curry, Andrew. *Gobekli Tepe: The World's First Temple?* Smithsonian Institution, 1 Nov. 2008, www.smithsonianmag.com/history/gobekli-tepe-the-worlds-first-temple-83613665/.
4. Bergman, Ronen. "A High Holy Whodunit." *The New York Times*, The New York Times, 25 July 2012, www.nytimes.com/2012/07/29/magazine/the-aleppo-codex-mystery.html.
5. Sadeghi, Behnam. "The Origins of the Koran: From Revelation to Holy Book." *BBC News*, BBC, 23 July 2015, www.bbc.com/news/world-middle-east-33631745.
6. "Rig Veda." *The British Library*, The British Library, 17 Oct. 2018, www.bl.uk/collection-items/rig-veda.
7. Sadeghi, Behnam, and Mohsen Goudarzi. "Ṣanʻāʼ 1 and the Origins of the Qurʼān." *Der Islam*, vol. 87, no. 1-2, 2012, pp. 1–129., doi:10.1515/islam-2011-0025.
8. "The Dead Sea Scrolls." *The Israel Museum, Jerusalem*, 23 Feb. 2021, www.imj.org.il/en/wings/shrine-book/dead-sea-scrolls.

9. "Leningrad Codex." *West Semitic Research Project*, USC Dana and David Dornsife College of Letters, Arts and Sciences, dornsife.usc.edu/wsrp/leningrad-codex/.
10. Gugliotta, Guy. "The Great Human Migration." *Smithsonian.com*, Smithsonian Institution, 1 July 2008, www.smithsonianmag.com/history/the-great-human-migration-13561/.
11. Scham, Sandra. "The World's First Temple." *Archaeology Magazine Archive*, 2008, archive.archaeology.org/0811/abstracts/turkey.html.
12. "Religions - Buddhism: The Buddha." *BBC*, BBC, 2 Oct. 2002, www.bbc.co.uk/religion/religions/buddhism/history/history.shtml.
13. "Religions - Jainism: Mahavira." *BBC*, BBC, 10 Sept. 2009, www.bbc.co.uk/religion/religions/jainism/history/mahavira.shtml.
14. Jacobs, Frank. "These Are All the World's Major Religions in One Map." *World Economic Forum*, 26 Mar. 2019, www.weforum.org/agenda/2019/03/this-is-the-best-and-simplest-world-map-of-religions/.

Pen

1. Smith, David. "It's 70 Today, but Our Favourite Pen Just Keeps Rolling Along." *The Guardian*, Guardian News and Media, 14 June 2008, www.theguardian.com/technology/2008/jun/15/news.
2. Dowling, Stephen. "The Cheap Pen That Changed Writing Forever." *BBC Future*, BBC, 29 Oct. 2020, www.bbc.com/future/article/20201028-history-of-the-ballpoint-pen.
3. LOUD, JOHN J. "US392046A - PEN." *United States Patent*, 30 Oct. 1888, patents.google.com/patent/US392046A/en.
4. "The History of Ink." *The Week UK*, The Week, 5 Dec. 2014, www.theweek.co.uk/innovation-at-work/63310/the-history-of-ink.
5. "Ink Chemistry." *Chemistry World*, Chemistry World, 20 May 2020, www.chemistryworld.com/news/ink-chemistry/3002158.article.
6. Erhan, Sevin Z., and Marvin O. Bagby. "US5122188A - Vegetable Oil-Based Printing Ink." *United States Patent*, 16 June 1992, patents.google.com/patent/US5122188A/en.
7. "OLD EGYPT SOUGHT A FOUNTAIN PEN." *The New York Times*, The New York Times, 6 Feb. 1927, timesmachine.nytimes.com/timesmachine/1927/02/06/96634058.html?pageNumber=140.
8. Bayley, Caroline. "Staedtler and Faber-Castell's Productive Pencil Rivalry." *BBC News*, BBC, 14 Apr. 2011, www.bbc.com/news/business-13019777.
9. Bennett, Howard J. "Ever Wondered about the Lead in Pencils?" *The Washington Post*, WP Company, 30 Nov. 2014, www.washingtonpost.com/lifestyle/kidspost/ever-wondered-about-the-lead-in-pencils/2014/11/26/f8b5869c-548a-11e4-809b-8cc0a295c773_story.html.
10. "Pencil Leads - What Does HB, 2B Etc Mean?" *Pencils Direct*, 3 Mar. 2017, www.pencilsdirect.co.uk/blog/pencil-leads-what-does-hb-2b-etc-mean-362/.

Letter

1. "Valentine's Day Love Letter." *The British Library*, Feb. 1477, www.bl.uk/learning/timeline/item126579.html.
2. Woolley, Leonard. "Tablet: British Museum." *The British Museum*, 1953, www.britishmuseum.org/collection/object/W_1953-0411-71.
3. "POSTAL FACTS." *United States Postal Service*, 2020, facts.usps.com/.
4. "Significant Dates - Who We Are." *USPS*, about.usps.com/who-we-are/postal-history/significant-dates.htm.
5. *Postal History, India Post*, www.indiapost.gov.in/VAS/Pages/PostalHistory.aspx.
6. Dutta, Sandipan. "Travel - The World's Highest Post Office." *BBC*, BBC, 7 June 2018, www.bbc.com/travel/story/20180606-the-worlds-highest-post-office.
7. "ANNUAL REPORT 2019-20." *DEPARTMENT OF POSTS, INDIA*, Ministry of Communications, Government of India, 2020, www.indiapost.gov.in/VAS/DOP_PDFFiles/AnnualReportEng2019_20.pdf.
8. "Clock Ticking for World's Oldest Post Office in Sanquhar." *BBC News*, BBC, 4 Feb. 2020, www.bbc.com/news/uk-scotland-south-scotland-51371100.
9. Alexander, Ruth, and Polly Hope. "Which Country Has the Most Expensive Postal Charges?" *BBC News*, BBC, 6 Apr. 2012, www.bbc.com/news/magazine-17614367.
10. "Domestic Postal Rates." *An Official Website of India Post - Maharashtra and Goa*, www.maharashtrapost.gov.in/htmldocs/postalchart.htm.
11. "Price List 2021." *Posten Norge*, www.posten.no/en/price-list.
12. "The 1840 Penny Black." *National Postal Museum*, postalmuseum.si.edu/exhibition/the-queen%E2%80%99s-own-postal-reforms-that-transformed-the-mail-the-worlds-first-postage-stamps.
13. "History." *Deutsche Post*, www.dpdhl.com/en/about-us/history.html.

Shoe

1. "Oregon Caves Yield Artifacts Of Habitation 10,000 Years Ago; Particles of Bison Hide and Sagebrush Foot Coverings Are Dug Out of Pumice Attributed to Near-By Craters." *The New York Times*, The New York Times, 5 Sept. 1938, timesmachine.nytimes.com/timesmachine/1938/09/05/98187124.html?pageNumber=17.
2. Roetzel, Bernhard. "'Wir Brauchen Mehr Hausverstand' - Ein Interview Mit Peter Eduard Meier." *Der Feine Herr*, 21 Sept. 2020, feineherr.de/eduard-meier-interview/.
3. Guzy, Tanner. "Understanding Shoe Construction: Goodyear Welt, Blake Stitch, & Cementing." *Primer Magazine*, 17 Apr. 2014, www.primermagazine.com/2014/spend/understanding-shoe-construction-goodyear-welt-blake-stich-cementing.
4. BALACHANDRAN, MANU. "VKC Group: From Regional To Global." *Forbes India*, ForbesIndia, 17 Mar. 2021, www.forbesindia.com/article/regional-goliaths/vkc-group-from-regional-to-global/66649/1.
5. "'Oldest Leather Shoe' Discovered." *BBC News*, BBC, 10 June 2010, www.bbc.com/news/10281908.
6. "These Boots Were Made for Romans." *Museum of London*, Museum of London, 17 Mar. 2021, www.museumoflondon.org.uk/discover/these-boots-were-made-romans.
7. "Shoe -20004." *Museum of London*, collections.museumoflondon.org.uk/online/object/9017.html.

8. Murawski, Carl. "What Is A Goodyear Welt?" *YouTube*, YouTube, 14 Dec. 2017, www.youtube.com/watch?v=G9xSnu69Qtg.
9. Madelaine, Eve. "How Shoes Are Made: Step by Step." *Italian Shoe Factory*, 15 May 2019, italianshoefactory.com/blog/how-shoes-are-made/.
10. Blakemore, Erin. "Only 30 Dutch Wooden Shoe Makers Remain." *Smithsonian.com*, Smithsonian Institution, 6 Apr. 2017, www.smithsonianmag.com/smart-news/only-30-dutch-wooden-shoe-makers-remain-180962804/.
11. *Marble Gravestone of Xanthippos, the Shoemaker*. The British Museum, artsandculture.google.com/asset/marble-gravestone-of-xanthippos-the-shoemaker/ngHbckvVPUCsPQ.
12. Cooke, Todd. "Everything You Ever Wanted to Know About Footwear Lasts." *Heddels*, 9 May 2018, www.heddels.com/2014/12/everything-ever-wanted-know-footwear-lasts/.

Stove

1. Lindqvist, F. W. "Brännare Fotogenkök, Som Arbeta Med i Ett Rörsystem Förgasad Fotogen." *Swedish Patent*, 25 Feb. 1892, tc.prv.se/spd/p/pdf/vHZVE6t-M0DP0tTkAJNoTg/SE3944.C1.pdf.
2. Chen, Ingfei. "Open-Fire Stoves Kill Millions. How Do We Fix It?" *Smithsonian.com*, Smithsonian Institution, 1 Dec. 2012, www.smithsonianmag.com/science-nature/open-fire-stoves-kill-millions-how-do-we-fix-it-132348165/.
3. "Household Air Pollution and Health." *World Health Organization*, World Health Organization, 8 May 2018, www.who.int/news-room/fact-sheets/detail/household-air-pollution-and-health.
4. Aldrich, Mark. "The Rise and Decline of the Kerosene Kitchen: A Neglected Energy Transition in Rural America, 1870-1950." Agricultural History, vol. 94, no. 1, 2020, pp. 24–60. JSTOR, www.jstor.org/stable/10.3098/ah.2020.094.1.024. Accessed 23 Apr. 2021.
5. "GAS PLANT IN STEEL BOTTLE.; Dr. Snelling's Process Gives Month's Supply in Liquid Form." *The New York Times*, The New York Times, 1 Apr. 1912, timesmachine.nytimes.com/timesmachine/1912/04/01/104894410.html?pageNumber=9.
6. Woodward, Maggie, and Bill McNary. *Thanksgiving Week: EIA Data Highlight How Energy Is Used in the Kitchen - Today in Energy*. U.S. Energy Information Administration - EIA - Independent Statistics and Analysis, 19 Nov. 2018, www.eia.gov/todayinenergy/detail.php?id=37552.
7. "Timeline." *Primus*, primus.us/pages/timeline.
8. "ECONOMICAL COOKING." *The New York Times*, The New York Times, 24 Apr. 1887, timesmachine.nytimes.com/timesmachine/1887/04/24/103141739.html?pageNumber=3.
9. "Kitchen Stoves: 1900-1919 Steel, Gas & Electricity." *The Evolution of Home Appliances in the U.S.*, evolutionhomeappliances.weebly.com/kitchen-stoves-1900-1919-steel-gas--electricity.html.

10. "How Much Carbon Dioxide Is Produced When Different Fuels Are Burned?" *American Geosciences Institute*, 4 Jan. 2019, www.americangeosciences.org/critical-issues/faq/how-much-carbon-dioxide-produced-when-different-fuels-are-burned.
11. "Propane vs. Natural Gas: A Comparison for Homeowners." *Santa Energy Corporation*, 2 Sept. 2019, www.santaenergy.com/blog/propane-vs-natural-gas/.
12. "How Potent Is Methane?" *FactCheck.org*, 7 Dec. 2018, www.factcheck.org/2018/09/how-potent-is-methane/.
13. "A PIONEERING APPROACH." *Bharat Petroleum Corporation Limited*, www.bharatpetroleum.com/About-BPCL/Our-Journey/A-pioneering-approach.aspx.
14. Kramer, Stephanie. "With Billions Confined to Their Homes Worldwide, Which Living Arrangements Are Most Common?" *Pew Research Center*, Pew Research Center, 30 July 2020, www.pewresearch.org/fact-tank/2020/03/31/with-billions-confined-to-their-homes-worldwide-which-living-arrangements-are-most-common.
15. Samanta, Koustav. "India's LPG Use to Surge from Record as Govt Promotes Cleaner Fuel." *Reuters*, Thomson Reuters, 3 May 2019, www.reuters.com/article/india-lpg/indias-lpg-use-to-surge-from-record-as-govt-promotes-cleaner-fuel-idUSL3N22B0SM.
16. SPENCER, PERCY L. "US2495429A - Method of Treating Foodstuffs." *United States Patent*, 8 Oct. 1945, patents.google.com/patent/US2495429A/en.

Road

1. Stromberg, Joseph. *"Roads Were Not Built for Cars": How Cyclists, Not Drivers, First Fought to Pave US Roads.* Vox, 19 Mar. 2015, www.vox.com/2015/3/19/8253035/roads-cyclists-cars-history.
2. "The Asphalt Industry from the 1800s to World War II." *Asphalt Magazine*, asphaltmagazine.com/asphalthistory_two/.
3. Wilford, John Noble. "World's Oldest Paved Road Found in Egypt." *The New York Times*, The New York Times, 8 May 1994, www.nytimes.com/1994/05/08/world/world-s-oldest-paved-road-found-in-egypt.html.
4. "BRO Builds World's Highest Motorable Road in Ladakh at 19,300 Feet." *Mint*, 2 Nov. 2017, www.livemint.com/Politics/YpojBHOzLBVypMDOylGPXJ/BRO-builds-worlds-highest-motorable-road-in-Ladakh-at-1930.html.
5. Harmon, Maureen. "World's Longest Road: Story Behind the Pan-American Highway." *Pegasus Magazine*, University of Central Florida, 2019, www.ucf.edu/pegasus/pan-american-highway/.
6. Wilkinson, Clifton. "Is the Pan-American Highway the Ultimate Road Trip?" *Lonely Planet*, Lonely Planet, 22 July 2020, www.lonelyplanet.com/articles/the-pan-american-highway-the-ultimate-road-trip.
7. "Majority Commuted Less than 30 Minutes in 2019." *Products Eurostat News - Eurostat*, Eurostat, 21 Oct. 2020, ec.europa.eu/eurostat/web/products-eurostat-news/-/DDN-20201021-2.
8. "Indians Spend 7% of Their Day Getting to Their Office." *The Economic Times*, 3 Sept. 2019,

economictimes.indiatimes.com/jobs/indians-spend-7-of-their-day-getting-to-their-office/articleshow/70954228.cms.
9. Morris, Roderick Conway. "Strolling Back Into Rome's Past Along the Appian Way." *The New York Times*, The New York Times, 20 Feb. 1998, www.nytimes.com/1998/02/20/style/IHT-strolling-back-into-romes-past-along-the-appian-way.html.
10. "Traffic Congestion Ranking: TomTom Traffic Index." *TomTom*, 2020, www.tomtom.com/en_gb/traffic-index/ranking/.
11. Berechman, Joseph. "Transportation—Economic Aspects of Roman Highway Development: the Case of Via Appia." *Transportation Research Part A: Policy and Practice*, vol. 37, no. 5, 2003, pp. 453–478., doi:10.1016/s0965-8564(02)00056-3.

Thermometer

1. Becker, Adam. "Earth - Why Does Time Always Run Forwards and Never Backwards?" *BBC*, BBC, 9 Mar. 2015, www.bbc.com/earth/story/20150309-why-does-time-only-run-forwards.
2. Bragg, Melvyn. "In Our Time - The Second Law of Thermodynamics - BBC Sounds." *BBC News*, BBC, 16 Dec. 2004, www.bbc.co.uk/sounds/play/p004y2bm.
3. Chavez, Isabel. "SI Units – Temperature." *National Institute of Standards and Technology (NIST)*, 26 Jan. 2021, www.nist.gov/pml/weights-and-measures/si-units-temperature.
4. Planck, M. "Zur Theorie das Gesetzes der Energieverteilung im Normalspektrum / On the Theory of Energy Distribution Law of the Normal Spectrum Radiation." *Verhandl. Dtsch. phys. Ges. (German Physical Society),* vol.2, no. 17,1900, pp. 237-245, hermes.ffn.ub.es/luisnavarro/nuevo_maletin/Planck%20(1900),%20Distribution%20Law.pdf.
5. Eddington, Arthur Stanley. The Nature of the Physical World. United Kingdom, Macmillan, 1928, pp.74, https://www.google.com/books/edition/The_Nature_of_the_Physical_World/8svaAAAAMAAJ.
6. "Temperature & Energy." *Energy Foundations for High School Chemistry*, 2014, highschoolenergy.acs.org/content/hsef/en/how-do-we-use-energy/temperature-energy.html.
7. Daniel Gabriel Fahrenheit (1686–1736). Nature 138, 428–429 (1936), doi.org/10.1038/138428a0.
8. Celsius, Linnæus and the Centigrade Thermometer. Nature 136, 365–366 (1935), doi.org/10.1038/136365c0.
9. Materese, Robin. "Kelvin: Boltzmann Constant." *National Institute of Standards and Technology (NIST)*, 5 June 2019, www.nist.gov/si-redefinition/kelvin/kelvin-boltzmann-constant.
10. Ghosh, Pallab. "Kilogram Gets a New Definition." *BBC News*, BBC, 16 Nov. 2018, www.bbc.com/news/science-environment-46143399.

Steam Engine

1. WATT, JAMES. "GB176900913A - Steam Engines, &c." *British Patent*, 29 Apr. 1769, patents.google.com/patent/GB176900913A/en.
2. Gudde, Nick. "James Watt, Office Equipment and High-Street Fashion." *Science Museum Blog*, 29 Jan. 2021, blog.sciencemuseum.org.uk/james-watt-high-street-fashion/.
3. "Devon - Discover Devon - Newcomen's Steam Revolution." *BBC*, BBC, 30 Jan. 2008, www.bbc.co.uk/devon/discovering/famous/thomas_newcomen.shtml.
4. Museum, Deutsches. "Savery's Pistonless Steam Pump." *Deutsches Museum: Savery's Pistonless Steam Pump*, www.deutsches-museum.de/en/collections/machines/power-engines/steam-engines/precursors/savery-pump/.
5. "James Watt and the Separate Condenser." *Science Museum Blog*, 28 Dec. 2018, blog.sciencemuseum.org.uk/james-watt-and-the-separate-condenser/.
6. "Titanic vs Today's Engines." *British Gas Business Blog*, British Gas Business, 21 Nov. 2014, www.britishgas.co.uk/business/blog/titanic-vs-todays-engines/.
7. "Global Warming of 1.5 °C." *Intergovernmental Panel on Climate Change*, 2018, www.ipcc.ch/sr15/chapter/chapter-1/.
8. "Nuclear Power Plants." *U.S. Energy Information Administration (EIA)*, 16 Apr. 2020, www.eia.gov/energyexplained/nuclear/nuclear-power-plants.php.
9. "Newcomen Atmospheric Engine." *National Museums Scotland*, www.nms.ac.uk/explore-our-collections/stories/science-and-technology/newcomen-engine/.
10. "Steam Turbines (DOE CHP Technology Fact Sheet Series) – Fact Sheet, 2016." *Energy.gov*, July 2016, www.energy.gov/sites/default/files/2016/09/f33/CHP-Steam%20Turbine.pdf.
11. "U.S. Energy Information Administration - EIA - Independent Statistics and Analysis." *How Electricity Is Generated - U.S. Energy Information Administration (EIA)*, 9 Nov. 2020, www.eia.gov/energyexplained/electricity/how-electricity-is-generated.php.
12. "This Month in Physics History | Thomas Savery Patents an Early Steam Engine." *American Physical Society*, July 2018, www.aps.org/publications/apsnews/201807/history.cfm.

Railways

1. "Locomotion No 1 by Colin Bainbridge." *PBase*, 28 Oct. 2012, www.pbase.com/csdesign/image/23664260.
2. Soffel, Jenny. "5 Things You Need to Know about the Future of India's Railways." *World Economic Forum*, 16 June 2015, www.weforum.org/agenda/2015/06/5-things-you-need-to-know-about-the-future-of-indias-railways/.
3. "Population, Total." *The World Bank*, data.worldbank.org/indicator/SP.POP.TOTL?most_recent_value_desc=true.
4. Eden, Caroline. "On Board the Trans-Siberian Railway for a Centenary Ride." *The Guardian*, Guardian News and Media, 12 Nov. 2016, www.theguardian.com/travel/2016/nov/12/trans-siberian-railway-russia-100-centenary.
5. Kundu, Rhik. "Growth in Air Passenger Traffic Plummets in 2019." *Mint*, 20 Jan. 2020,

www.livemint.com/news/india/domestic-air-passenger-traffic-growth-falls-to-3-74-in-2019-11579510228941.html
6. "GDP per Capita, PPP (Current International $)." *The World Bank*, data.worldbank.org/indicator/NY.GDP.PCAP.PP.CD.
7. "GDP and Spending - Gross Domestic Product (GDP) - OECD Data." *OECD*, data.oecd.org/gdp/gross-domestic-product-gdp.htm.
8. Kalia, Arnav. "Interesting Facts About Indian Railways." *Facts About Indian Railways – Longest Railway Platform| Invest India*, 5 July 2019, www.investindia.gov.in/team-india-blogs/interesting-facts-about-indian-railways.
9. "Freight Rail & Preserving the Environment." *Association of American Railroads*, Feb. 2021, www.aar.org/wp-content/uploads/2020/06/AAR-Sustainability-Fact-Sheet.pdf.
10. "Highlights of the Automotive Trends Report." *EPA*, Environmental Protection Agency, 6 Jan. 2021, www.epa.gov/automotive-trends/highlights-automotive-trends-report.
11. Marks, Paul. "NASA Criticised for Sticking to Imperial Units." *New Scientist*, 22 June 2009, www.newscientist.com/article/dn17350-nasa-criticised-for-sticking-to-imperial-units/.
12. "2016 Presidential Election Results." *The New York Times*, The New York Times, 9 Aug. 2017, www.nytimes.com/elections/2016/results/president.
13. "Maps and Data - Average Annual Fuel Use by Vehicle Type." *Alternative Fuels Data Center: Maps and Data - Average Annual Fuel Use by Vehicle Type*, US Department of Energy, Feb. 2020, afdc.energy.gov/data/10308.
14. Fuller, Thomas, et al. "Can America Still Build Big? A California Rail Project Raises Doubts." *The New York Times*, The New York Times, 25 Feb. 2019, www.nytimes.com/2019/02/25/us/california-high-speed-rail.html.

Internal combustion engine

1. LENOIR, ETIENNE JEAN-JOSEPH. "FR43624A - Un Moteur à Air Dilaté Par La Combustion Des Gaz." *French Patent*, 24 Jan. 1860, patents.google.com/patent/FR43624A.
2. DIESEL, RUDOLF. "DE67207C - Arbeitsverfahren Und Ausführungsart Für Verbrennungskraftmaschinen." *German Patent*, 23 Feb. 1893, patents.google.com/patent/DE67207C/.
3. "Stationary Gas Engine 'Lenoir Engine', 1861." *Deutsches Museum*, www.deutsches-museum.de/en/collections/machines/power-engines/combustion-engines/precursors-of-the-combustion-engine/lenoir-engine-1861/.
4. Bryant, Lynwood. "The Origin of the Automobile Engine." Scientific American, vol. 216, no. 3, 1967, pp. 102–113., www.jstor.org/stable/24931437.
5. "Gasoline Prices around the World, 08-Feb-2021." *GlobalPetrolPrices.com*, 8 Feb. 2021, www.globalpetrolprices.com/gasoline_prices/.
6. "Diesel Prices around the World, 08 Feb 2021." *GlobalPetrolPrices.com*, 8 Feb. 2021, www.globalpetrolprices.com/diesel_prices/.
7. *The World Book Encyclopedia. Volume 5, D*, World Book, Inc., 2015, pp. 198–199.
8. *The World Book Encyclopedia. Volume 8, G*, World Book, Inc., 2015, pp. 61–66.
9. *World of Change: Global Temperatures*. NASA, 29 Jan. 2020, earthobservatory.nasa.gov/world-of-change/global-temperatures.

10. Eschner, Kat. "Why Did People Think Steam-Powered Cars Were a Good Idea?" *Smithsonian.com*, Smithsonian Institution, 26 Jan. 2017, www.smithsonianmag.com/smart-news/why-did-people-think-steam-powered-cars-were-good-idea-180961872/.
11. "Diesel vs. Gasoline: Everything You Need to Know." *Car and Driver*, Car and Driver, 13 Nov. 2020, www.caranddriver.com/research/a31515330/diesel-vs-gasoline/.
12. Afework, Bethel, and Jason Donev. "Four Stroke Engine." *Four Stroke Engine - Energy Education*, 4 Jan. 2019, energyeducation.ca/encyclopedia/Four_stroke_engine.

Automobile

1. Johnson, Ben. "The Great Horse Manure Crisis of 1894." *Historic UK*, www.historic-uk.com/HistoryUK/HistoryofBritain/Great-Horse-Manure-Crisis-of-1894/.
2. Davies, Stephen. "The Great Horse-Manure Crisis of 1894: Stephen Davies." *FEE Freeman Article*, Foundation for Economic Education, 1 Sept. 2004, fee.org/articles/the-great-horse-manure-crisis-of-1894/.
3. Staples, Sarah. "Travel - How Germany Became the Country of Cars." *BBC*, BBC, 22 Aug. 2019, www.bbc.com/travel/story/20190821-how-germany-became-the-country-of-cars.
4. *The World Book Encyclopedia*. Volume 1, A, World Book, Inc., 2015, pp. 946-973.
5. Roots, Roger. "The Dangers of Automobile Travel: A Reconsideration." The American Journal of Economics and Sociology 66, no. 5 (2007): 959-76, www.jstor.org/stable/27739679.
6. Benz, Carl. "DE37435C - FAHRZEUG MIT GASMOTORENBETRIEB." *German Patent*, 2 Nov. 1886, patents.google.com/patent/DE37435C/en.
7. "Who Invented the Automobile?" *The Library of Congress*, 9 Feb. 2020, www.loc.gov/everyday-mysteries/item/who-invented-the-automobile/.
8. "FORD CARS PASS 10,000,000.; Last Million Were Turned Out In 132 Working Days." *The New York Times*, The New York Times, 5 June 1924, timesmachine.nytimes.com/timesmachine/1924/06/05/104039878.html?pageNumber=36.
9. "Ford Car No. 10,000,000 Starts on Its Career; More Than Eight Million of Its Predecessors Still in Working Order, Including the Inventor's First Model -- Every Land Knows Their Cry." *The New York Times*, The New York Times, 13 July 1924, timesmachine.nytimes.com/timesmachine/1924/07/13/98801530.html?pageNumber=134.
10. FORD, HENRY. "SYSTEM THE SECRET OF FORD'S SUCCESS; Building in Large Quantities Reduces Original Cost, Which Is Prohibitive. BIG SAVING ON MATERIALS Simplicity of Design Makes Further Reduction Possible, While Economies in Marketing Aid as Well." *The New York Times*, The New York Times, 3 Jan. 1909, timesmachine.nytimes.com/timesmachine/1909/01/03/101032619.html?pageNumber=39.
11. "New Cars Which Are Already the Market for Next Year.; Keen Interest in 1909 Types That Arrived in This City During the Past Week. Motorists Study Low Priced Cars Showing Thorough Knowledge of Automobile Detail." *The New York Times*, The New York Times, 8 Nov. 1908,

timesmachine.nytimes.com/timesmachine/1908/11/08/104812475.html?pageNumber=36.

Electricity

1. "Electricity Access." *Our World in Data*, https://ourworldindata.org/grapher/share-of-the-population-with-access-to-electricity?time=1990&region=World
2. "History." *America's Electric Cooperatives*, 17 Nov. 2016, www.electric.coop/our-organization/history.
3. "The History of Energy in Germany." *Planète Énergies*, 29 Apr. 2015, www.planete-energies.com/en/medias/saga-energies/history-energy-germany.
4. "A MAGIC GLOW: THE ELECTRIFICATION OF GERMANY." *Deutsches Historisches Museum Blog*, 25 Apr. 2016, www.dhm.de/blog/2016/08/25/the-electrification-of-germany.
5. Muthiah, S. "The Sporting Architect." *The Hindu*, The Hindu, 26 July 2016, www.thehindu.com/news/cities/chennai/chen-columns/the-sporting-architect/article2686899.ece.
6. "The IRFCA Photo Gallery." *Madras Electric Tramway - 2*, 12 Sept. 2014, www.irfca.org/gallery/Heritage/internet-archive/madras-electric-tramway-2.jpg.html.
7. Iea. "India Heading for the Centre of the Global Energy Stage, IEA Says - News." *IEA*, 27 Nov. 2015, www.iea.org/news/india-heading-for-the-centre-of-the-global-energy-stage-iea-says.
8. "History of Electricity Transmission in Britain." *History of Electricity Transmission in Britain | National Grid Group*, www.nationalgrid.com/group/about-us/our-history/history-electricity-transmission-britain.
9. EDISON, THOMAS A. "US287516A - SYSTEM OF ELECTRICAL DISTRIBUTION." *United States Patent*, 30 Oct. 1883, patents.google.com/patent/US287516A/en.
10. EDISON, THOMAS A. "US369441A - SYSTEM OF ELECTRICAL DISTRIBUTION." *United States Patent*, 6 Sept. 1887, patents.google.com/patent/US369441A/en.
11. TESLA, NIKOLA. "US390721A - DYNAMO-ELECTRIC MACHINE." *United States Patent*, 9 Oct. 1888, patents.google.com/patent/US390721/en.
12. TESLA, NIKOLA. "US382280A - ELECTRICAL TRANSMISSION OF POWER." *United States Patent*, 1 May 1888, patents.google.com/patent/US382280A/en.
13. "Direct Current Line Still Hot after 40 Years." *BPA.gov - Bonneville Power Administration*, 26 May 2010, www.bpa.gov/news/newsroom/Pages/Direct-current-line-still-hot-after-40-years.aspx.
14. King, Gilbert. "Edison vs. Westinghouse: A Shocking Rivalry." *Smithsonian.com*, Smithsonian Institution, 11 Oct. 2011, www.smithsonianmag.com/history/edison-vs-westinghouse-a-shocking-rivalry-102146036/.

Light bulb

1. EDISON, THOMAS A. "US223898A - ELECTRIC LAMP." *United States Patent*, 27 Jan. 1880, patents.google.com/patent/US223898A/en.

2. "Thomas Edison: How the Light Bulb Changed Everything." MPR News, MPR News, 28 Oct. 2019, www.mprnews.org/story/2019/10/28/how-the-light-bulb-changed-nearly-everything.
3. Roser, Max. "Light at Night." *Our World in Data*, 10 Dec. 2013, ourworldindata.org/light#satellite-images-of-the-earth-at-night.
4. Pinker, Steven. Enlightenment Now: The Case for Reason, Science, Humanism, and Progress. United States, Penguin Publishing Group, 2018, pp. 253–254, https://books.google.com/books?id=J6grDwAAQBAJ.
5. "A VIENNA THEATRE BURNED; THREE HUNDRED LIVES REPORTED LOST." *The New York Times*, The New York Times, 9 Dec. 1881, timesmachine.nytimes.com/timesmachine/1881/12/09/102916969.html?pageNumber=1.
6. "OVER 500 DIE IN CHICAGO THEATRE; Fire Panic in the Iroquois Causes Frightful Loss. DISASTER AT A MATINEE. Women and Children Trampled in the Wild Rush. FIRE STARTED ON THE STAGE " Mr. Blue Beard' Was Playing and an Electrical Appliance Is Supposed to Have Caused the Fire." *The New York Times*, The New York Times, 31 Dec. 1903, timesmachine.nytimes.com/timesmachine/1903/12/31/105071992.html?pageNumber=1.
7. "THE BROOKLYN CALAMITY; The Extent of the Disaster Underesimated. Two Hundred and Eighty-Three Bodies Recovered. Over Three Hundred and Fifty Lives Probably Lost. The Accounts Given by Those Who Escaped. TW0 ACTORS AMONG THE VICTIMS, Widespread Grief in the City of Churches. Scenes and Incidents at the Ruins of the Theatre." *The New York Times*, The New York Times, 7 Dec. 1876, timesmachine.nytimes.com/timesmachine/1876/12/07/80351848.html?pageNumber=1.
8. "THE VIENNA DEATH-ROLL." *The New York Times*, The New York Times, 19 Dec. 1881, www.nytimes.com/1881/12/19/archives/the-vienna-deathroll.html.
9. Nordhaus, William D. "Do Real-Output and Real-Wage Measures Capture Reality? The History of Lighting Suggests Not." *The Economics of New Goods*, by Timothy F. Bresnahan and Robert James Gordon, University of Chicago Press, 1997, pp. 29–70, www.google.com/books/edition/The_Economics_of_New_Goods/-P1X-XPpmCEC?hl=en.
10. "PRESIDENT SETS UP RURAL POWER UNIT; REA, Designed to End Farm 'Drudgery,' Gives Wide Field for Federal Activity." *The New York Times*, The New York Times, 12 May 1935, timesmachine.nytimes.com/timesmachine/1935/05/12/94606232.html?pageNumber=8.

Fertilizer

1. Ritchie, Hannah. "How Many People Does Synthetic Fertilizer Feed?" *Our World in Data*, 7 Nov. 2017, ourworldindata.org/how-many-people-does-synthetic-fertilizer-feed.
2. Ritter, Steven K. "The Haber-Bosch Reaction: An Early Chemical Impact On Sustainability." *American Chemical Society*, 18 Aug. 2008, cen.acs.org/articles/86/i33/Haber-Bosch-Reaction-Early-Chemical.html.

3. Tullo, Alexander H. "C&EN's Global Top 50 for 2020." *Chemical & Engineering News*, American Chemical Society, 27 July 2020, cen.acs.org/business/finance/CENs-Global-Top-50-2020/98/i29.
4. "Grain Yields Starting to Plateau." *Full Planet, Empty Plates: the New Geopolitics of Food Scarcity*, by Lester R. Brown, W.W. Norton, 2012.
5. "World Population with and without Synthetic Nitrogen Fertilizers." *Our World in Data*, 2015, ourworldindata.org/grapher/world-population-with-and-without-fertilizer.
6. Swaminathan, M. S. "From Bengal Famine to Right to Food." *The Hindu*, The Hindu, 4 Nov. 2016, www.thehindu.com/opinion/lead/From-Bengal-Famine-to-Right-to-Food/article12342992.ece.
7. Safi, Michael. "Churchill's Policies Contributed to 1943 Bengal Famine – Study." *The Guardian*, Guardian News and Media, 29 Mar. 2019, www.theguardian.com/world/2019/mar/29/winston-churchill-policies-contributed-to-1943-bengal-famine-study.
8. "Fritz Haber | The Nobel Prize in Chemistry 1918." *NobelPrize.org*, Nobel Media AB, www.nobelprize.org/prizes/chemistry/1918/haber/biographical/.
9. "Carl Bosch | The Nobel Prize in Chemistry 1931." *NobelPrize.org*, Nobel Media AB, www.nobelprize.org/prizes/chemistry/1931/bosch/biographical/.
10. Boerner, Leigh Krietsch. "Industrial Ammonia Production Emits More CO2 than Any Other Chemical-Making Reaction. Chemists Want to Change That." *Chemical & Engineering News*, American Chemical Society, 15 June 2019, cen.acs.org/environment/green-chemistry/Industrial-ammonia-production-emits-CO2/97/i24.
11. "About Sugar." *International Sugar Organization*, www.isosugar.org/sugarsector/sugar.
12. BOSCH, CARL. "US990191A - PROCESS OF PRODUCING AMINIONA." *United States Patent*, 18 Apr. 1911, patents.google.com/patent/US990191/en.

Sewing machine

1. Chan, Emily. "Why Do We Still Know So Little About How Our Clothes Are Made?" *British Vogue*, 8 Dec. 2019, www.vogue.co.uk/fashion/article/how-are-our-clothes-actually-made.
2. Manik, Julfikar Ali, and Jim Yardley. "Building Collapse in Bangladesh Leaves Scores Dead." *The New York Times*, The New York Times, 24 Apr. 2013, www.nytimes.com/2013/04/25/world/asia/bangladesh-building-collapse.html.
3. Manik, Julfikar Ali, and Jim Yardley. "17 Days in Darkness, a Cry of 'Save Me,' and Joy." *The New York Times*, The New York Times, 10 May 2013, www.nytimes.com/2013/05/11/world/asia/bangladesh-collapse-death-toll.html.
4. Hoskins, Tansy. "Reliving the Rana Plaza Factory Collapse: a History of Cities in 50 Buildings, Day 22." *The Guardian*, 23 Apr. 2015, www.theguardian.com/cities/2015/apr/23/rana-plaza-factory-collapse-history-cities-50-buildings.
5. Burke, Jason. "Rana Plaza: One Year on from the Bangladesh Factory Disaster." *The Guardian*, 19 Apr. 2014, www.theguardian.com/world/2014/apr/19/rana-plaza-bangladesh-one-year-on.

6. Wheeler, Jane. "Clothing Of The 1830s." *Conner Prairie*, 6 May 2019, www.connerprairie.org/educate/indiana-history/clothing-in-the-1800s/.
7. Cunningham, John. "Sewing Machine." *Encyclopædia Britannica*, 2 Apr. 2020, www.britannica.com/technology/sewing-machine.
8. Bogomolny, Bert. "Victorian Era Women's Fashion. Dresses, Clothing And Cost." *WardrobeShop*, 5 July 2020, www.wardrobeshop.com/blogs/vintage-style-fashion/victorian-era-womens-outfit-cost.
9. "History." *The Singer Brand History - 160+ Years of Sewing | Singer.com*, www.singer.com/history.
10. "Sewing Machine by Elias Howe: Science Museum Group Collection." *Sewing Machine by Elias Howe | Science Museum Group*, collection.sciencemuseumgroup.org.uk/objects/co44744/sewing-machine-by-elias-howe-sewing-machine.
11. "Sewing Machine Anatomy: How a Stitch Is Made." *Threads Sewing*, YouTube, 11 Mar. 2015, www.youtube.com/watch?v=2681yeSrsM0.

Vaccine

1. The New York Times. "Coronavirus World Map: Tracking the Global Outbreak." *The New York Times*, The New York Times, 28 Jan. 2020, www.nytimes.com/interactive/2020/world/coronavirus-maps.html.
2. Enlightenment Now the Case for Reason, Science, Humanism and Progress, by Steven Pinker, Penguin Books, 2019, pp. 63-64 and pp. 386; https://books.google.com/books?id=J6grDwAAQBAJ
3. "Deaths Caused by Smallpox as a Share of All Deaths in London." *Our World in Data*, 12 Feb. 2021, ourworldindata.org/grapher/deaths-from-smallpox-in-london?time=1701..1900.
4. *14 Diseases You Almost Forgot About (Thanks to Vaccines)*. Centers for Disease Control and Prevention, 8 May 2020, www.cdc.gov/vaccines/parents/diseases/forgot-14-diseases.html.
5. "Vaccine Timeline." *Historic Dates and Events Related to Vaccines and Immunization*, 4 Feb. 2021, www.immunize.org/timeline/.
6. Newman, Laura. "Maurice Hilleman." BMJ : British Medical Journal vol. 330,7498 (2005): 1028.
7. Bakalar, Nicholas. "A Diphtheria Cure, 1894." *The New York Times*, The New York Times, 10 May 2010, www.nytimes.com/2010/05/11/health/11first.html.
8. "Changing the Face of Medicine | Leila Alice Daughtry Denmark." *U.S. National Library of Medicine*, National Institutes of Health, 3 June 2015, cfmedicine.nlm.nih.gov/physicians/biography_78.html.
9. "Understanding MRNA COVID-19 Vaccines." *Centers for Disease Control and Prevention*, Centers for Disease Control and Prevention, 4 Mar. 2021, www.cdc.gov/coronavirus/2019-ncov/vaccines/different-vaccines/mrna.html.
10. Rhp. "Iron Lungs for Polio Victims, 1930s-1950s." *Rare Historical Photos*, 14 Oct. 2017, rarehistoricalphotos.com/iron-lungs-polio-1930s-1950s/.
11. "WHO Model List of Essential Medicines." *World Health Organization*, World Health Organization, 23 July 2019, www.who.int/publications/i/item/WHOMVPEMPIAU2019.06.

12. BEHRING, EMIL. "US606042A - DIPHTHERIA ANTITOXIN AND PROCESS OF MAKING SAME." *United States Patent*, 21 June 1898, patents.google.com/patent/US606042A/en.
13. DAUGHTRY-DENMARK L. WHOOPING COUGH VACCINE. Am J Dis Child. 1942;63(3):453–466. doi:10.1001/archpedi.1942.02010030023002.

Antibiotic

1. Bernard, Diane. "How a Miracle Drug Changed the Fight against Infection during World War II." *The Washington Post*, 29 July 2020, www.washingtonpost.com/history/2020/07/11/penicillin-coronavirus-florey-wwii-infection/.
2. "International Leprosy Association - History of Leprosy." *Dr Robert Greenhill Cochrane | International Leprosy Association - History of Leprosy*, leprosyhistory.org/database/person4.
3. Markel, Howard. "The Real Story behind Penicillin." *PBS*, Public Broadcasting Service, 27 Sept. 2013, www.pbs.org/newshour/health/the-real-story-behind-the-worlds-first-antibiotic.
4. "Amoxicillin - an Antibiotic." *Animalresearch.info*, 4 Aug. 2016, www.animalresearch.info/en/drug-development/drug-prescriptions/amoxicillin.
5. "Outpatient Antibiotic Prescriptions - United States, 2016." *Centers for Disease Control and Prevention*, Centers for Disease Control and Prevention, 29 Oct. 2018, www.cdc.gov/antibiotic-use/community/programs-measurement/state-local-activities/outpatient-antibiotic-prescriptions-US-2016.html.
6. Miller, Kelli. "Fluoroquinolone Antibiotics Linked to Serious Nerve Damage." *WebMD*, WebMD, 27 Aug. 2013, www.webmd.com/brain/news/20130826/fda-strengthens-fluoroquinolone-warning.
7. Pham, Thu D M et al. "Quinolone antibiotics." MedChemComm vol. 10,10 1719-1739. 28 Jun. 2019, doi.org/10.1039/c9md00120d.
8. Singh, Pushpendra, and Stewart T Cole. "Mycobacterium leprae: genes, pseudogenes and genetic diversity." Future microbiology vol. 6,1 (2011): 57-71. doi:10.2217/fmb.10.153.
9. Bui T, Preuss CV. Cephalosporins. [Updated 2021 Feb 17]. In: StatPearls [Internet]. Treasure Island (FL): StatPearls Publishing; 2021 Jan-, www.ncbi.nlm.nih.gov/books/NBK551517/.
10. "FDA Updates Warnings for Fluoroquinolone Antibiotics on Risks of Mental Health and Low Blood Sugar Adverse Reactions." *U.S. Food and Drug Administration*, FDA, 10 July 2018, www.fda.gov/news-events/press-announcements/fda-updates-warnings-fluoroquinolone-antibiotics-risks-mental-health-and-low-blood-sugar-adverse.
11. "All Antibiotic Classes." *Centers for Disease Control and Prevention*, Centers for Disease Control and Prevention, arpsp.cdc.gov/profile/antibiotic-use/all-classes.

Bleach

1. "GD510/1/90 - Papers Relating to Charles Tennant's Works at St Rollox. - 1798-1876." *Catalogue - The National Archives of Scotland*, 1798,

catalogue.nrscotland.gov.uk/nrsonlinecatalogue/details.aspx?reference=GD510%2F1%2F90.
2. "CHOLERA DEATH TOLL IN RUSSIA 100,000; Spreading Now Through Asiatic Russia and Threatening Regions Near Manchurian Border." *The New York Times*, The New York Times, 17 Sept. 1910, timesmachine.nytimes.com/timesmachine/1910/09/17/104951624.html?pageNumber=6.
3. "Cholera: The Forgotten Pandemic." *World Health Organization*, World Health Organization, 22 Oct. 2018, www.who.int/cholera/the-forgotten-pandemic/en/.
4. "History - Historic Figures: John Snow (1813 - 1858)." *BBC*, BBC, www.bbc.co.uk/history/historic_figures/snow_john.shtml.
5. IARC Working Group on the Evaluation of Carcinogenic Risks to Humans. Chlorinated Drinking-Water; Chlorination by-Products; Some Other Halogenated Compounds; Cobalt and Cobalt Compounds. Lyon (FR): International Agency for Research on Cancer; 1991. (IARC Monographs on the Evaluation of Carcinogenic Risks to Humans, No. 52.) Chlorinated drinking-water, https://www.ncbi.nlm.nih.gov/books/NBK506911/
6. "230 Years Young and Still Going Strong: Chlorinated Bleach Disinfectant." *Eurochlor*, 18 Feb. 2018, www.eurochlor.org/news/230-years-young-and-still-going-strong-chlorinated-bleach-disinfectant/.
7. "History of Drinking Water Treatment." *Centers for Disease Control and Prevention*, Centers for Disease Control and Prevention, 26 Nov. 2012, www.cdc.gov/healthywater/drinking/history.html.
8. "Cholera." *History.com*, A&E Television Networks, 12 Sept. 2017, www.history.com/topics/inventions/history-of-cholera.
9. McGREW, R. E. "THE FIRST CHOLERA EPIDEMIC AND SOCIAL HISTORY." Bulletin of the History of Medicine, vol. 34, no. 1, 1960, pp. 61–73. JSTOR, www.jstor.org/stable/44446659.
10. Ojeda Rodriguez JA, Kahwaji CI. Vibrio Cholerae. [Updated 2020 Jun 6]. In: StatPearls [Internet]. Treasure Island (FL): StatPearls Publishing; 2021 Jan-. www.ncbi.nlm.nih.gov/books/NBK526099/.
11. Katz, Jonathan M. "U.N. Admits Role in Cholera Epidemic in Haiti." *The New York Times*, The New York Times, 18 Aug. 2016, www.nytimes.com/2016/08/18/world/americas/united-nations-haiti-cholera.html.

Radio

1. MARCONI, GUGLIELMO. "GB189612039A - Improvements in Transmitting Electrical Impulses and Signals, and in Apparatus Therefor." *Great Britain Patent*, 2 July 1897, patents.google.com/patent/GB189612039A/en.
2. "A Short History of Radio." *Federal Communication Commission*, 2004, transition.fcc.gov/omd/history/radio/documents/short_history.pdf.
3. "Titanic, Marconi and the Wireless Telegraph." *Science Museum, London*, 24 Oct. 2018, www.sciencemuseum.org.uk/objects-and-stories/titanic-marconi-and-wireless-telegraph.

4. Manvell, Roger and Camacho, Jorge A.. "Broadcasting". Encyclopedia Britannica, 10 Aug. 2018, https://www.britannica.com/technology/broadcasting.
5. McDonough, John. "First Radio Commercial Hit Airwaves 90 Years Ago." *National Public Radio*, 29 Aug. 2012, www.npr.org/2012/08/29/160265990/first-radio-commercial-hit-airwaves-90-years-ago.
6. "Statistics on Radio: United Nations Educational, Scientific and Cultural Organization." *United Nations Educational, Scientific and Cultural Organization*, 2013, www.unesco.org/new/en/unesco/events/prizes-and-celebrations/celebrations/international-days/world-radio-day-2013/statistics-on-radio/.
7. "Why Is Energy in a Wave Proportional to Amplitude Squared?" *Physics Stack Exchange*, 9 Nov. 2011, physics.stackexchange.com/questions/16755/why-is-energy-in-a-wave-proportional-to-amplitude-squared.
8. "How Radio Works." *Georgia State University Library*, exhibits.library.gsu.edu/current/exhibits/show/georgiaradio/radio1920s/howradioworks.
9. "FM Station, Technical Standards Modulation Rules, Amendment." *FCC*, 9 Apr. 1984, docs.fcc.gov/public/attachments/FCC-84-113A1.pdf.
10. "Guglielmo Marconi." *Lemelson*, Massachusetts Institute of Technology, lemelson.mit.edu/resources/guglielmo-marconi.
11. "The Birth of Pioneering Electrical Engineer Guglielmo Marconi." *The Telegraph*, Telegraph Media Group, 28 Apr. 2017, www.telegraph.co.uk/technology/connecting-britain/guglielmo-marconi-birth/.

Telegraph

1. MORSE, SAMUEL F. B. "US1647A - IMPROVEMENT IN THE MODE OF COMMUNICATING INFORMATION BY SIGNALS BY THE APPLICATION OF ELECTRO-MAGNETISM." *United States Patent*, 20 June 1840, patents.google.com/patent/US1647A/en.
2. Mitchell, Gareth. "How Fast Does Electricity Flow?" *BBC Science Focus Magazine*, www.sciencefocus.com/science/how-fast-does-electricity-flow/.
3. *The World Book Encyclopedia. Volume 19, T*, World Book, Inc., 2015, pp. 90–92.
4. Morse, Samuel Finley Breese. First telegraphic message in the Library of Congress. 24 May, 1844. Image. https://www.loc.gov/item/mmorse000107/.
5. Eveleth, Rose. "After 163 Years, India Sends Its Last Telegram." *Smithsonian.com*, Smithsonian Institution, 15 July 2013, www.smithsonianmag.com/smart-news/after-163-years-india-sends-its-last-telegram-11115704/.
6. V., Sriram. "The Telegraph Comes to Town." *The Hindu*, The Hindu, 18 June 2013, www.thehindu.com/news/cities/chennai/chen-columns/the-telegraph-comes-to-town/article4824123.ece.
7. "How Perseverance Laid the First Transatlantic Telegraph Cable." *Science Museum*, 26 Sept. 2018, www.sciencemuseum.org.uk/objects-and-stories/how-perseverance-laid-first-transatlantic-telegraph-cable.

8. "Western Union Completes the First Transcontinental Telegraph Line." *History.com*, A&E Television Networks, 16 Nov. 2009, www.history.com/this-day-in-history/western-union-completes-the-first-transcontinental-telegraph-line.
9. Klein, Christopher. "How Abraham Lincoln Used the Telegraph to Help Win the Civil War." *History.com*, A&E Television Networks, 9 July 2020, www.history.com/news/abraham-lincoln-telegraph-civil-war.
10. "Pony Express." *National Parks Service*, U.S. Department of the Interior, 10 July 2020, www.nps.gov/poex/learn/historyculture/index.htm.
11. LIFFEN, JOHN. "Revealing The Real Cooke And Wheatstone Telegraph Dial." *Science Museum Blog*, 21 Oct. 2014, blog.sciencemuseum.org.uk/revealing-the-real-cooke-and-wheatstone-telegraph-dial/.
12. "Morse Code's Vanquished Competitor: The Dial Telegraph." *IEEE Spectrum: Technology, Engineering, and Science News*, 31 Aug. 2018, spectrum.ieee.org/telecom/standards/morse-codes-vanquished-competitor-the-dial-telegraph.

Telephone

1. Bell, Alexander Graham. "US174465A - Improvement in Telegraphy." *Google Patents*, Google, 7 Mar. 1876, patents.google.com/patent/US174465A/en.
2. "1870s – 1940s: Telephone." *Imagining the Internet*, Elon University, www.elon.edu/u/imagining/time-capsule/150-years/back-1870-1940.
3. "Telephones: The Connections Museum Seattle - Telecom History Exhibits." The Connections Museum Seattle, www.telcomhistory.org/connections-museum-seattle-exhibits/telephones/.
4. "India Number of Subscriber Fixed Line, 1960 – 2021: CEIC Data." , *1960 – 2021 Data*, www.ceicdata.com/en/indicator/india/number-of-subscriber-fixed-line.
5. "UK Number of Subscriber Fixed Line, 1960 – 2021: CEIC Data." *UK Number of Subscriber Fixed Line, 1960 – 2021 Data*, www.ceicdata.com/en/indicator/united-kingdom/number-of-subscriber-fixed-line.
6. "Germany Number of Subscriber Fixed Line, 1960 – 2021: CEIC Data." *Germany Number of Subscriber Fixed Line, 1960 – 2021 Data*, www.ceicdata.com/en/indicator/germany/number-of-subscriber-fixed-line.
7. *The World Book Encyclopedia. Volume 19, T*, World Book, Inc., 2015, pp. 94-101.
8. Eschner, Kat. "Long Before Siri, Emma Nutt's Voice Was on the Other End of the Line." *Smithsonian.com*, Smithsonian Institution, 1 Sept. 2017, www.smithsonianmag.com/smart-news/long-siri-emma-nutts-voice-was-other-end-line-180964651/.
9. "Site of the First Telephone Exchange." *National Parks Service*, U.S. Department of the Interior, 29 Aug. 2018, www.nps.gov/subjects/nationalhistoriclandmarks/site-of-the-first-telephone-exchange.htm.
10. "Almon Strowger." *Kansas Historical Society*, Feb. 2019, www.kshs.org/kansapedia/almon-strowger/16911.
11. "Alexander Graham Bell." *PBS*, Public Broadcasting Service, 1999, www.pbs.org/transistor/album1/addlbios/bellag.html.

Flying shuttle

1. Humphries, Richard. "The Flying Shuttle." *HUMPHRIES WEAVING*, 2 Feb. 2018, www.humphriesweaving.co.uk/flying-shuttle-richard-humphries/.
2. Crawford, Jason. "Learning the Loom." *The Roots of Progress*, 27 May 2018, rootsofprogress.org/learning-the-loom.
3. "Shuttle and Bobbin." *National Museum of American History*, americanhistory.si.edu/collections/search/object/nmah_640517.
4. Baldwin, James. "1840 Baldwin's Patent Model of a Loom Shuttle." *National Museum of American History*, americanhistory.si.edu/collections/search/object/nmah_1069662.
5. Jacquard, Joseph Marie. "Jacquard Loom." *National Museum of American History*, americanhistory.si.edu/collections/search/object/nmah_645517.
6. *Old Spitalfields Hand Loom with Jacquard Mechanism*. Science Museum Group, 1810, collection.sciencemuseumgroup.org.uk/objects/co45152/old-spitalfields-hand-loom-with-jacquard-mechanism-hand-loom-jacquard-machine.
7. *Ribbon Loom with Jacquard Head*. Science Museum Group, 1900, collection.sciencemuseumgroup.org.uk/objects/co8404795/ribbon-loom-with-jacquard-head-ribbon-loom-with-jacquard-head.
8. "John Kay Inventor of the Flying Shuttle." *Kay Family Association UK*, kayfamilyassociationuk.com/projects/john-kay-flying-shuttle.
9. "Joseph Marie Jacquard." *Centre For Computing History*, www.computinghistory.org.uk/det/19901/Joseph-Marie-Jacquard/.
10. "Types of Looms." *Toyota Industries Corporation*, www.toyota-industries.com/products/textile/variety/.
11. Kate. "What Does Warp Mean? Basic Weaving Terminology: The Weaving Loom." *The Weaving Loom | The Modern Lap Loom Weaver's Resource*, 4 May 2015, www.theweavingloom.com/what-does-warp-mean-basic-weaving-terminology/.

Spinning Jenny

1. Duignan, Brian. *Inventors and Inventions of the Industrial Revolution*. Encyclopædia Britannica, Inc., www.britannica.com/list/inventors-and-inventions-of-the-industrial-revolution.
2. "Spinning Jenny." *The National Archives*, The National Archives, 1 Feb. 2021, www.nationalarchives.gov.uk/education/resources/georgian-britain-age-modernity/spinning-jenny/.
3. BASSETT, FRANCIS JOSEPH. "MACHINES VS. MEN; New Mechanical Cotton Picker Renews Old Controversy." *The New York Times*, The New York Times, 27 Sept. 1936, timesmachine.nytimes.com/timesmachine/1936/09/27/85427290.html?pageNumber=79.
4. "The Spinning Jenny." *The British Library - The British Library*, www.bl.uk/learning/timeline/item107855.html.
5. Bourke-White, Margaret. "Gandhi and the Spinning Wheel | 100 Photographs | The Most Influential Images of All Time." *Time*, Time, 1946, 100photos.time.com/photos/margaret-bourke-white-gandhi-spinning-wheel.

6. Hills, Richard L. "Sir Richard Arkwright and His Patent Granted in 1769." *Notes and Records of the Royal Society of London*, vol. 24, no. 2, 1970, pp. 254–260. JSTOR, www.jstor.org/stable/531292. Accessed 22 May 2021.
7. "Spinning Mule." *Science Museum Group*, 1927, collection.sciencemuseumgroup.org.uk/objects/co8405253/spinning-mule-spinning-mule.
8. Arkwright, Richard. "Arkwright's Water Frame." *Science Museum Group*, 1775, collection.sciencemuseumgroup.org.uk/objects/co8411373/arkwrights-water-frame-water-frame.
9. "17th Century Spindle Wheel." *Science Museum Group*, collection.sciencemuseumgroup.org.uk/objects/co44844/17th-century-spindle-wheel-spinning-wheel.
10. "Spinning Mule Demonstration at Leeds Industrial Museum." *Leeds Industrial Museum*, 21 May 2019, www.youtube.com/watch?v=81PydkdYQws.

Cotton gin

1. Yafa, Stephen. Cotton: The Biography of a Revolutionary Fiber. United States, Penguin Books, 2006. pp. 82-83, https://www.google.com/books/edition/Cotton/DZoxe0aO9AEC?hl=en.
2. "Cotton Gin and Eli Whitney." *History.com*, A&E Television Networks, 4 Feb. 2010, www.history.com/topics/inventions/cotton-gin-and-eli-whitney.
3. Schur, Joan Brodsky. *Eli Whitney's Patent for the Cotton Gin*. National Archives and Records Administration, 23 Sept. 2016, www.archives.gov/education/lessons/cotton-gin-patent.
4. "An Idea in the Mind of a Genius; WHITTLING BOY. The Story of Eli Whitney, 1765-1825. By Roger Burlingame. 370 Pp. New York: Harcourt, Brace & Co. $3." *The New York Times*, The New York Times, 9 Feb. 1941, timesmachine.nytimes.com/timesmachine/1941/02/09/85373516.html?pageNumber=57.
5. "Ginning Machine." *Indiamart.com*, www.indiamart.com/proddetail/ginning-machine-1704483188.html.
6. "Reproduction of Eli Whitney's Cotton Gin Model." *National Museum of American History*, americanhistory.si.edu/collections/search/object/nmah_625483.
7. "Inventing Change: the Whitney Legacy." *The Eli Whitney Museum and Workshop*, www.eliwhitney.org/7/museum/about-eli-whitney/inventing-change.
8. "The Moving Assembly Line and the Five-Dollar Workday." *Ford*, corporate.ford.com/articles/history/moving-assembly-line.html.
9. "Eli Whitney Patents the Cotton Gin." *Mass Moments*, www.massmoments.org/moment-details/eli-whitney-patents-the-cotton-gin.html.

Newspaper

1. "Relation: Aller Fuernemmen Und Gedenckwuerdigen Historien: so Sich Hin Und Wider in Hoch- Und Nieder-Teutschland, Auch in Verlauffen Und Zutragen Möchte." *UB Heidelberg: Heidelberger Historische Bestände – Digital*, 27 Feb. 2021, digi.ub.uni-heidelberg.de/diglit/relation1609.

2. "Courante uyt Italien, Duytsland & at National library of Netherlands." *Nationale Bibliotheek, Netherlands*, www.kb.nl/organisatie/onderzoek-expertise/digitaliseringsprojecten-in-de-kb/project-databank-digitale-dagbladen/geselecteerde-titels-en-selectieprocedure/selectie-van-titels/1618-1800.
3. Tharoor, Shashi. "There's One Country in the World Where the Newspaper Industry Is Still Thriving." *World Economic Forum*, 24 May 2017, www.weforum.org/agenda/2017/05/despite-the-decline-of-printed-papers-theres-one-place-that-is-bucking-the-trend/.
4. "The Daily Courant." *The Library of Congress*, www.loc.gov/item/sn85020234/.
5. "The Bengal Gazette, an Anglo-Indian Newspaper." *The British Library*, The British Library, 23 Nov. 2017, www.bl.uk/collection-items/the-bengal-gazette-an-anglo-indian-newspaper.
6. "PRESS IN INDIA HIGHLIGHTS." *RNI*, 31 Mar. 2018, rni.nic.in/all_page/press_india.aspx.
7. "China Press, Media, TV, Radio, Newspapers - television, circulation, stations, papers, number, print, freedom", *Press Reference*, www.pressreference.com/Be-Co/China.html.
8. "MHS Collections Online: The New-England Courant." *Massachusetts Historical Society.* , www.masshist.org/database/viewer.php?item_id=55&pid=15.
9. "AFP in Dates." *AFP*, 23 Apr. 2012, www.afp.com/en/afp-dates.
10. "Reuters: a Brief History." *The Guardian*, Guardian News and Media, 4 May 2007, www.theguardian.com/media/2007/may/04/reuters.pressandpublishing.
11. "Timeline : Associated Press News Dispatches, 1915-1930." *The Library of Congress*, www.loc.gov/collections/associated-press-news-dispatches-1915-to-1930/articles-and-essays/timeline/.

Glass

1. "TYPES OF GLASS." *Corning Museum of Glass*, 8 Dec. 2011, www.cmog.org/article/types-glass.
2. "THE ORIGINS OF GLASSMAKING." *Corning Museum of Glass*, 1 Dec. 2011, www.cmog.org/article/origins-glassmaking.
3. *The World Book Encyclopedia. Volume 8, G*, World Book, Inc., 2015, pp. 211-222.
4. Moriarty, Nora I., and Debby Sneed. "How Glass Was Made in the Ancient Roman World." *Department of Classics*, 26 July 2019, www.colorado.edu/classics/2018/10/11/how-glass-was-made-ancient-roman-world.
5. "Core-Forming Technique." *Museum of Cycladic Art, Athens*, cycladic.gr/en/page/techniki-tou-pirina.
6. "This 1,600-Year-Old Goblet Shows That the Romans Were Nanotechnology Pioneers." *Smithsonian.com*, Smithsonian Institution, 1 Sept. 2013, www.smithsonianmag.com/history/this-1600-year-old-goblet-shows-that-the-romans-were-nanotechnology-pioneers-787224/.
7. "Amphora; Vessel (Closed); Cameo: British Museum." *The British Museum*, www.britishmuseum.org/collection/object/G_1945-0927-1.
8. "Drinking-Cup: British Museum." *The British Museum*, www.britishmuseum.org/collection/object/H_1958-1202-1.

9. Cartwright, Mark. "The Stained Glass Windows of Chartres Cathedral." *Ancient History Encyclopedia*, Ancient History Encyclopedia, 21 Mar. 2021, www.ancient.eu/article/1277/the-stained-glass-windows-of-chartres-cathedral/.
10. Bragg, William Henry, and William Lawrence Bragg. "The Structure of the Diamond." *Proceedings of the Royal Society of London. Series A, Containing Papers of a Mathematical and Physical Character*, 22 Sept. 1913, royalsocietypublishing.org/doi/10.1098/rspa.1913.0084.
11. Lipson, Henry Solomon, and A. R. Stokes. "The Structure of Graphite." *Proceedings of the Royal Society of London. Series A. Mathematical and Physical Sciences*, 24 Sept. 1942, royalsocietypublishing.org/doi/10.1098/rspa.1942.0063.
12. Anonyme. "Gobelet Sur Pied." *Musée Du Louvre*, 14 Dec. 2017, collections.louvre.fr/en/ark:/53355/cl010114697.
13. "Lump of Translucent Blue Glass." *The British Museum*, www.britishmuseum.org/collection/object/W_1919-1011-4039.

Anaesthetic

1. Robinson, Daniel H, and Alexander H Toledo. "Historical development of modern anesthesia." Journal of investigative surgery : the official journal of the Academy of Surgical Research vol. 25,3 (2012): 141-9. doi:10.3109/08941939.2012.690328, https://doi.org/10.3109/08941939.2012.690328
2. "James Young Simpson." *Royal College of Physicians of Edinburgh*, www.rcpe.ac.uk/heritage/college-history/james-young-simpson.
3. "Discovery and Development of Propofol, a Widely Used Anesthetic." *The Lasker Foundation*, www.laskerfoundation.org/awards/show/discovery-and-development-propofol-widely-used-anesthetic/.
4. *The Medical News*. Vol. 69, Lea Brothers & Company, 1896.
5. "Can a Drug Make You Tell the Truth?" *BBC News*, BBC, 3 Oct. 2013, www.bbc.com/news/magazine-24371140.
6. Markel, Howard. "The Painful Story behind Modern Anesthesia." *PBS*, Public Broadcasting Service, 16 Oct. 2013, www.pbs.org/newshour/health/the-painful-story-behind-modern-anesthesia.
7. "History of Anesthesia." *Wood Library-Museum of Anesthesiology*, 2 Dec. 2020, www.woodlibrarymuseum.org/history-of-anesthesia/.
8. Baraniuk, Chris. "The People Who Can't Go Numb at the Dentist's." *BBC Future*, BBC, 9 Jan. 2017, www.bbc.com/future/article/20170106-the-people-who-cant-go-numb-at-the-dentists.
9. Belluck, Pam. "With High-Profile Death, Focus on High-Risk Drug." *The New York Times*, The New York Times, 7 Aug. 2009, www.nytimes.com/2009/08/07/us/07propofol.html.
10. Dreifus, Claudia. "Call It a Reversible Coma, Not Sleep." *The New York Times*, The New York Times, 28 Feb. 2011, www.nytimes.com/2011/03/01/science/01conv.html.
11. Schembri, Frankie. "If You've Had Anesthesia, You Can Likely Thank This Veterinarian Who Just Won a Top Science Prize." *Science*, 11 Sept. 2018, www.sciencemag.org/news/2018/09/if-you-ve-had-anesthesia-you-can-likely-thank-veterinarian-who-just-won-top-science.

12. Marsh, Henry. "What Happens When You Go Under." *The New York Times*, The New York Times, 3 Jan. 2018, www.nytimes.com/2018/01/03/books/review/anesthesia-kate-cole-adams-counting-backwards-henry-jay-przybylo.html.
13. Levy, A. Goodman. "Sudden Death under Light Chloroform Anæsthesia." *Proceedings of the Royal Society of Medicine*, vol. 7, no. Sect_Anaesth, 1914, pp. 57–84., doi:10.1177/003591571400701421.
14. "Market Growth: General Anesthesia Drugs." *OBS Anesthesia Management Groups*, 14 Oct. 2019, radiusanesthesia.com/market-growth-general-anesthesia-drugs/.
15. Brown, Burnell Jr., Sevoflurane, Anesthesia & Analgesia: December 1995 - Volume 81 - Issue 6S - p 1S-3S, https://journals.lww.com/anesthesia-analgesia/fulltext/1995/12001/sevoflurane__introduction_and_overview.1.aspx.
16. WALLIN, RICHARD., REGAN, BERNARD M. et.al, Sevoflurane, Anesthesia & Analgesia: November 1975 - Volume 54 - Issue 6 - p 758-766, https://journals.lww.com/anesthesia-analgesia/Abstract/1975/11000/Sevoflurane__A_New_Inhalational_Anesthetic_Agent.21.aspx.
17. Geddes, Linda. "Banishing Consciousness: the Mystery of Anaesthesia." New Scientist, 23 Nov. 2011, www.newscientist.com/article/mg21228402-300-banishing-consciousness-the-mystery-of-anaesthesia/.
18. Torsten Gordh, Torsten E. Gordh, Kjell Lindqvist, David S. Warner; Lidocaine: The Origin of a Modern Local Anesthetic. Anesthesiology 2010; 113:1433–1437, https://doi.org/10.1097/ALN.0b013e3181fcef48.

Computer

1. "Algorithms and Turing Machines." *The Emperor's New Mind Concerning Computers, Minds and the Laws of Physics*, by Roger Penrose, Oxford University Press, 1999, pp. 40–60.
2. Kennedy, T.R. "Electronic Computer Flashes Answers, May Speed Engineering; NEW ALL-ELECTRONIC COMPUTER AND ITS INVENTORS." *The New York Times*, The New York Times, 15 Feb. 1946, timesmachine.nytimes.com/timesmachine/1946/02/15/93052340.html?pageNumber=1.
3. "Computer History Museum." *The Stored Program*, www.computerhistory.org/revolution/birth-of-the-computer/4/87.
4. "The Engines." *The Engines | Babbage Engine | Computer History Museum*, www.computerhistory.org/babbage/engines/.
5. De Mol, Liesbeth. "Turing Machines." *Stanford Encyclopedia of Philosophy*, Stanford University, 24 Sept. 2018, plato.stanford.edu/entries/turing-machine/.
6. Copeland, B. Jack. "The Church-Turing Thesis." *Stanford Encyclopedia of Philosophy*, Stanford University, 10 Nov. 2017, plato.stanford.edu/entries/church-turing/.
7. Turing, A.M. (1937), On Computable Numbers, with an Application to the Entscheidungsproblem. Proceedings of the London Mathematical Society, s2-42: 230-265. https://doi.org/10.1112/plms/s2-42.1.230

8. Von Neumann, John. First draft of a report on the EDVAC. Moore School of Electrical Engineering, University of Pennsylvania, 1945, https://dx.doi.org/10.5479/sil.538961.39088011475779.
9. Cruz, Frank da. The Jacquard Loom, 6 Apr. 2021, www.columbia.edu/cu/computinghistory/jacquard.html.
10. Geselowitz, Michael N. "The Jacquard Loom: A Driver of the Industrial Revolution." *IEEE Spectrum: Technology, Engineering, and Science News*, 1 Jan. 2019, spectrum.ieee.org/the-institute/ieee-history/the-jacquard-loom-a-driver-of-the-industrial-revolution.

Transistor

1. Kawahara, Jamie. *How Does a Computer Physically Store Binary Code?*, Androidgrl's Blog, 1 Jan. 2019, androidgrl.github.io/2019/01/01/binary/.
2. "ENIAC." *National Museum of American History*, americanhistory.si.edu/collections/search/object/nmah_1297478.
3. "How to Squeeze Billions of Transistors onto a Computer Chip." *IBM*, www.ibm.com/thought-leadership/innovation-explanations/mukesh-khare-on-smaller-transistors-analytics.
4. "Special Purpose Tubes." *Electrical Note*, 20 Feb. 2021, electricalnote.com/special-purpose-tubes/.
5. Fox, Chris. "Intel's next-Generation 7nm Chips Delayed until 2022." *BBC News*, BBC, 24 July 2020, www.bbc.com/news/technology-53525710.
6. Knight, Will. "World's Fastest Transistor Operates at Blinding Speed." *New Scientist*, 11 Apr. 2005, www.newscientist.com/article/dn7253-worlds-fastest-transistor-operates-at-blinding-speed/.
7. "1955: Photolithography Techniques Are Used to Make Silicon Devices." *1955: Photolithography Techniques Are Used to Make Silicon Devices | The Silicon Engine | Computer History Museum*, www.computerhistory.org/siliconengine/photolithography-techniques-are-used-to-make-silicon-devices/.
8. "The Story of the Intel® 4004." *Intel*, www.intel.com/content/www/us/en/history/museum-story-of-intel-4004.html.
9. "Computer History Museum." *Intel's Microprocessor*, www.computerhistory.org/revolution/story/285.
10. Courtland, Rachel. "Intel Now Packs 100 Million Transistors in Each Square Millimeter." *IEEE Spectrum: Technology, Engineering, and Science News*, 30 Mar. 2017, spectrum.ieee.org/nanoclast/semiconductors/processors/intel-now-packs-100-million-transistors-in-each-square-millimeter.
11. "1958: All Semiconductor 'Solid Circuit' Is Demonstrated." *1958: All Semiconductor "Solid Circuit" Is Demonstrated | The Silicon Engine | Computer History Museum*, www.computerhistory.org/siliconengine/all-semiconductor-solid-circuit-is-demonstrated/.
12. "1959: Practical Monolithic Integrated Circuit Concept Patented." *1959: Practical Monolithic Integrated Circuit Concept Patented | The Silicon Engine | Computer History Museum*,

www.computerhistory.org/siliconengine/practical-monolithic-integrated-circuit-concept-patented/.

Internet

1. "The Birth of the Web." *CERN*, home.cern/science/computing/birth-web.
2. Leiner, Barry M., et al. "Brief History of the Internet." *Internet Society*, 14 Aug. 2020, www.internetsociety.org/internet/history-internet/brief-history-internet/.
3. "Hobbes' Internet Timeline." *IETF*, Nov. 1997, tools.ietf.org/html/rfc2235.
4. "World Wide Web." *The World Wide Web Project*, 20 Dec. 1990, info.cern.ch/hypertext/WWW/TheProject.html.
5. Grimes, William. "Raymond Tomlinson, Who Put the @ Sign in Email, Is Dead at 74." *The New York Times*, The New York Times, 7 Mar. 2016, www.nytimes.com/2016/03/08/technology/raymond-tomlinson-email-obituary.html.
6. Hafner, Katie. "Frank Heart, Who Linked Computers Before the Internet, Dies at 89." *The New York Times*, The New York Times, 25 June 2018, www.nytimes.com/2018/06/25/technology/frank-heart-who-linked-computers-before-the-internet-dies-at-89.html.
7. "E-Commerce Payments Trends: United States." *J P Morgan*, 2019, www.jpmorgan.com/merchant-services/insights/reports/united-states.
8. "How America Banks: Household Use of Banking and Financial Services, 2019 FDIC Survey." *FDIC*, 19 Oct. 2020, www.fdic.gov/analysis/household-survey/index.html.
9. "Railways Launch Upgraded IRCTC Website, Call It a 'New Year Gift' for Travellers." *Hindustan Times*, 31 Dec. 2020, www.hindustantimes.com/india-news/railways-launch-upgraded-irctc-website-call-it-a-new-year-gift-for-travellers/story-cysAvQQ3O1uCuqeG9w49fP.html.
10. Varghese, George. "An Interview with Leonard Kleinrock." *Communications of the ACM*, 1 Oct. 2019, dl.acm.org/doi/pdf/10.1145/3363183.

Compass

1. "Nova Reperta: Invention of the Compass (Lapis Polaris Magnes) - National Maritime Museum." *Royal Museum of Greenwich*, collections.rmg.co.uk/collections/objects/101930.html.
2. "THE COMPASS OF COLUMBUS." The American Catholic Historical Researches, vol. 7, no. 1, 1911, pp. 1–7. JSTOR, www.jstor.org/stable/44374849.
3. "Mariner's Compass - National Maritime Museum." *Royal Museum Greenwich*, collections.rmg.co.uk/collections/objects/42488.html.
4. Neckam, Alexander. *Alexandri Neckam De naturis rerum libri duo, with the poem of the same author, De laudibus divinae sapientiae.* United Kingdom: Longman, Green, Longman, Roberts, and Green, 1863, https://www.google.com/books/edition/Alexandri_Neckam_De_naturis_rerum_libro/POAIAAAAIAAJ.
5. "Earth Fact Sheet." *NASA*, NASA, nssdc.gsfc.nasa.gov/planetary/factsheet/earthfact.html.
6. Abulafia, David. The Boundless Sea: A Human History of the Oceans. United States, Oxford University Press, 2019, pp. 471.

7. "How It All Began." *Distinctive Collections Spotlights*, libraries.mit.edu/collections/vail-collection/topics/masterworks/how-it-all-began/.
8. Schmidl, P. S. "Two Early Arabic Sources on the Magnetic Compass". Journal of Arabic and Islamic Studies, vol. 1, Mar. 2017, pp. 81-132, https://doi.org/10.5617/jais.4547.
9. Harris, Karen. "Who Invented the Magnetic Compass?" *History Daily*, 5 Nov. 2018, historydaily.org/who-invented-the-magnetic-compass.
10. "Dead Reckoning: Time and Navigation." *Time and Navigation - The Untold Story of Getting from Here to There.*, timeandnavigation.si.edu/navigating-at-sea/navigating-without-a-clock/dead-reckoning.
11. "A Compass in Every Smartphone." *IEEE Spectrum: Technology, Engineering, and Science News*, 29 Jan. 2010, spectrum.ieee.org/semiconductors/devices/a-compass-in-every-smartphone.
12. Gasser, Achilles Pirmin. De magnete, seu rota perpetui motus, libellus. N.p., Ulhart d. Ä., 1558, https://www.google.com/books/edition/De_magnete_seu_rota_perpetui_motus_libel/ucUNwEpgk60C?hl=en.

Firearm

1. "Survival Rates Similar for Gunshot, Stabbing Victims Whether Brought to the Hospital by Police or EMS, Penn Medicine Study Finds." *Penn Medicine News*, University Of Pennsylvania, 2 Jan. 2014, www.pennmedicine.org/news/news-releases/2014/january/survival-rates-similar-for-gun.
2. Coupland, R. M., & Meddings, D. R. (1999). Mortality associated with use of weapons in armed conflicts, wartime atrocities, and civilian mass shootings: literature review. BMJ (Clinical research ed.), 319(7207), 407–410. https://doi.org/10.1136/bmj.319.7207.407.
3. Project, Borgen. "Top 12 Most Deadliest Wars in History." *The Borgen Project*, 22 Nov. 2019, borgenproject.org/top-12-deadliest-wars-in-history/.
4. Hall, Michelle. "By the Numbers: World War II's Atomic Bombs." *CNN*, 6 Aug. 2013, www.cnn.com/2013/08/06/world/asia/btn-atomic-bombs/index.html.
5. "Table 20 - Murder by State, Types of Weapons, 2019." *FBI*, FBI, 29 Aug. 2019, ucr.fbi.gov/crime-in-the-u.s/2019/crime-in-the-u.s.-2019/tables/table-20.
6. Asher, Jeff, et al. "Why Does Louisiana Consistently Lead the Nation in Murders?" *The New York Times*, The New York Times, 15 Feb. 2021, www.nytimes.com/2021/02/15/upshot/why-does-louisiana-consistently-lead-the-nation-in-murders.html.
7. "The Chemistry of Gunpowder." *Compound Interest*, 3 May 2015, www.compoundchem.com/2014/07/02/the-chemistry-of-gunpowder/.
8. "Facts About Sodium Azide." *Centers for Disease Control and Prevention*, Centers for Disease Control and Prevention, 4 Apr. 2018, emergency.cdc.gov/agent/sodiumazide/basics/facts.asp.
9. Mitchell, Gareth. "How Far Can a Bullet Fired from a Handgun Travel?" *BBC Science Focus Magazine*, www.sciencefocus.com/science/how-far-can-a-bullet-fired-from-a-handgun-travel/.

10. "A Canadian Sniper Breaks the Record for the Longest Confirmed Kill Shot - but How?" *BBC News*, BBC, 23 June 2017, www.bbc.com/news/newsbeat-40381047.
11. Ingraham, Christopher. "There Are More Guns than People in the United States, According to a New Study of Global Firearm Ownership." *The Washington Post*, WP Company, 27 Apr. 2019, www.washingtonpost.com/news/wonk/wp/2018/06/19/there-are-more-guns-than-people-in-the-united-states-according-to-a-new-study-of-global-firearm-ownership/.
12. Baker, Mike, and Lucy Tompkins. "Boulder's Pain Is Deepened by a Lost Fight for Gun Control." *The New York Times*, The New York Times, 25 Mar. 2021, www.nytimes.com/2021/03/24/us/boulder-shooting-gun-control.html.
13. GATLING, RICHARD J. "US36836A - Improvement in Revolving Battery-Guns." *United States Patent*, 4 Nov. 1862, patents.google.com/patent/US36836A/en.
14. MAXIM, HIRAM S. "US319596A - Machine-Gun." *United States Patent*, 9 June 1885, patents.google.com/patent/US319596A/en.
15. COLT, SAMUEL. "USX9430I1 - Improvement in Fire-Arms." *United States Patent*, 25 Feb. 1836, patents.google.com/patent/USX9430I1/en.
16. Killicoat, Phillip. "Weaponomics : The Global Market for Assault Rifles." *Open Knowledge Repository*, World Bank, Washington, DC, 1 Apr. 2007, openknowledge.worldbank.org/handle/10986/7024.
17. Chivers, C.J. "The AK-47: 'The Gun' That Changed The Battlefield." *NPR*, NPR, 12 Oct. 2010, www.npr.org/templates/story/story.php?storyId=130493013.
18. "History." *GLOCK Inc.*, us.glock.com/en/LEARN/Brand/History.
19. Lowrey, Annie. "America's 25-Year Love Affair with Glocks." *NBCNews.com*, NBCUniversal News Group, 13 Jan. 2011, www.nbcnews.com/id/wbna41046787.
20. Francescani, Chris. "US Secret Service Switching to 9mm Glock Pistols." *ABC News*, ABC News Network, 1 Aug. 2019, abcnews.go.com/US/us-secret-service-switching-9mm-glock-pistols/story?id=64719349.
21. "9mm By The Numbers - The History of the 9mm Cartridge." *Fenix Ammunition*, fenixammo.com/pages/a-brief-history-of-the-9mm-cartridge.
22. "Gatling Gun." *History.com*, A&E Television Networks, 4 Mar. 2010, www.history.com/topics/american-civil-war/gatling-gun.
23. LUGER, GEORG. "US753414A - RECOIL-LOADING SMALL-ARM." *United States Patent*, 1 Mar. 1904, patents.google.com/patent/US753414/en.

Nuclear weapon

1. Reif, Kingston, and Shannon Bugos. "Issue Briefs." *Surging U.S. Nuclear Weapons Budget a Growing Danger | Arms Control Association*, Arms Control Association, 19 Mar. 2020, www.armscontrol.org/issue-briefs/2020-03/surging-us-nuclear-weapons-budget-growing-danger.
2. *Congressional Budget Justification Department of State, Foreign Operations, and Related Programs*. US State Department, 11 Mar. 2019, www.state.gov/wp-content/uploads/2019/05/FY-2020-CBJ-FINAL.pdf.
3. "The Hidden Costs of Our Nuclear Arsenal: Overview of Project Findings." *Brookings*, 14 Apr. 2017,

www.brookings.edu/the-hidden-costs-of-our-nuclear-arsenal-overview-of-project-findings/.
4. "Budget Act for 2020–21: Information." *Budget Act for 2020–21: Information - Education Budget (CA Dept of Education)*, 30 Sept. 2020, www.cde.ca.gov/fg/fr/eb/yr20ltr0929.asp.
5. "Nuclear Notebook: United States Nuclear Weapons, 2021." *Bulletin of the Atomic Scientists*, 29 Jan. 2021, thebulletin.org/premium/2021-01/nuclear-notebook-united-states-nuclear-weapons-2021/.
6. "Nuclear Notebook: How Many Nuclear Weapons Does Russia Have in 2021?" *Bulletin of the Atomic Scientists*, 24 Mar. 2021, thebulletin.org/premium/2021-03/nuclear-notebook-russian-nuclear-weapons-2021/.
7. "Nuclear Notebook." *Bulletin of the Atomic Scientists*, 7 Dec. 2020, thebulletin.org/nuclear-notebook/.
8. "Worldwide Deployments of Nuclear Weapons, 2017." *Bulletin of the Atomic Scientists*, 22 Oct. 2020, thebulletin.org/2017/09/worldwide-deployments-of-nuclear-weapons-2017/.
9. "FIRST ATOMIC BOMB DROPPED ON JAPAN." *The New York Times*, The New York Times, 7 Aug. 1945, timesmachine.nytimes.com/timesmachine/1945/08/07/issue.html.
10. "Nuclear Notebook: Indian Nuclear Forces, 2020." *Bulletin of the Atomic Scientists*, 29 Jan. 2021, thebulletin.org/premium/2020-07/nuclear-notebook-indian-nuclear-forces-2020/.
11. "Nuclear Notebook: Chinese Nuclear Forces, 2020." *Bulletin of the Atomic Scientists*, 14 Dec. 2020, thebulletin.org/premium/2020-12/nuclear-notebook-chinese-nuclear-forces-2020/.
12. Broad, William J., and David E. Sanger. "As U.S. Modernizes Nuclear Weapons, 'Smaller' Leaves Some Uneasy." *The New York Times*, The New York Times, 12 Jan. 2016, www.nytimes.com/2016/01/12/science/as-us-modernizes-nuclear-weapons-smaller-leaves-some-uneasy.html.
13. "Counting the Dead at Hiroshima and Nagasaki." *Bulletin of the Atomic Scientists*, 17 Sept. 2020, thebulletin.org/2020/08/counting-the-dead-at-hiroshima-and-nagasaki/.
14. "TOLL AT HIROSHIMA PLACED AT 210,000; Mayor of First Atomic Target in Japan Reports Casualties Far Above Earlier Totals." *The New York Times*, The New York Times, 29 Aug. 1949, timesmachine.nytimes.com/timesmachine/1949/08/29/86781406.html?pageNumber=2.

Eye glasses

1. Lienhard, John H. *How Invention Begins: Echoes of Old Voices in the Rise of New Machines*. Oxford Univ. Press, 2008, https://www.google.com/books/edition/How_Invention_Begins/EQZoS9CRvs0C?hl=en.
2. "About Us." *The Worshipful Company of Spectacle Makers*, 10 Feb. 2021, www.spectaclemakers.com/about-us/.

3. *What Percentage of the Population Wears Glasses?*,
   www.glassescrafter.com/information/percentage-population-wears-glasses.html.
4. Modiano, Mario S. "How Archimedes Stole Sun to Burn Foe's Fleet; Far Better Conditions' Large Number of Mirrors." *The New York Times*, The New York Times, 11 Nov. 1973,
   timesmachine.nytimes.com/timesmachine/1973/11/11/91030219.html?pageNumber=16.
5. Handley, Neil. "The Invention of Spectacles." *College of Optometrists*,
   www.college-optometrists.org/the-college/museum/online-exhibitions/virtual-spectacles-gallery/the-invention-of-spectacles.html.
6. Knight, Sam. "The Spectacular Power of Big Lens | The Long Read." *The Guardian*, Guardian News and Media, 10 May 2018,
   www.theguardian.com/news/2018/may/10/the-invisible-power-of-big-glasses-eyewear-industry-essilor-luxottica.
7. American Journal of Physics. United States, American Association of Physics Teachers, 1975, pp. 776,
   https://www.google.com/books/edition/American_Journal_of_Physics/DhpVAAAAMAAJ.
8. Bordsen, John. *The Seattle Times*, The Seattle Times Company, 9 Aug. 1995,
   archive.seattletimes.com/archive/?date=19950809&slug=2135585.
9. Shekoski, Mimi. "20/20 Vision: How to Convert 20/20 Based Measure to Diopters." *Happy Eyesight*, 22 Mar. 2018, www.happyeyesight.com/get-20-20-vision/.
10. "Global Prescription Lens Market Outlook to 2027 - Featuring Prive Revaux, ZEISS International & Rodenstock Among Others." *Yahoo! Finance*, Yahoo!, 13 Oct. 2020,
    finance.yahoo.com/news/global-prescription-lens-market-outlook-101400707.html.
11. Efron, Nathan. "Centenary Celebration of Fick's Eine Contactbrille." *Archives of Ophthalmology*, JAMA Network, 1 Oct. 1988,
    jamanetwork.com/journals/jamaophthalmology/article-abstract/637484.
12. "The Thirteenth Century." *A History of Western Philosophy: Vol. 2: Philosophy from St. Augustine to Ockham*, by Ralph M. McInerny, vol. 2, Regnery, 1970,
    https://maritain.nd.edu/jmc/etext/hwp218.htm.
13. Bacon, Roger, and Bridges, J. H.. The 'Opus Majus' of Roger Bacon, Ed. United Kingdom, Williams & Norgate, 1900,
    https://www.google.com/books/edition/The_Opus_Majus_of_Roger_Bacon_Ed/6F0XAQAAMAAJ?hl=en.

Microscope

1. Poppick, Laura. "Let Us Now Praise the Invention of the Microscope." *Smithsonian.com*, Smithsonian Institution, 30 Mar. 2017,
   www.smithsonianmag.com/science-nature/what-we-owe-to-the-invention-microscope-180962725/.
2. Leeuwenhoek, Antoni Van. "1677 Observations, communicated to the publisher by Mr. Antony van Leewenhoeck, in a dutch letter of the 9th Octob. 1676. here English'd: concerning little animals by him observed in rain-well-sea- and snow water; as also in water wherein pepper had lain infused", Phil. *Trans. R. Soc.* 12821–831, doi.org/10.1098/rstl.1677.0003

3. "The Microscope." *Science Museum*, 19 Aug. 2019, www.sciencemuseum.org.uk/objects-and-stories/medicine/microscope.
4. "Electron Microscopes." *Oxford University Museum of Natural History*, oumnh.ox.ac.uk/electron-microscopes.
5. The World Book Encyclopedia. Volume 13, M, World Book, Inc., 2015, pp. 517-520.
6. "Field-Ion Micrograph Of Atoms Of Iridium by Prof. Erwin Mueller/Science Photo Library." *Fine Art America*, fineartamerica.com/featured/field-ion-micrograph-of-atoms-of-iridium-prof-erwin-muellerscience-photo-library.html.
7. "ZEISS Group." *ZEISS International*, www.zeiss.com/corporate/int/about-zeiss/history.html.
8. "ANNUAL REPORT 2019/20." *ZEISS Group*, 9 Dec. 2020, www.zeiss.com/content/dam/corporate-new/annualreport/2019-2020/download/annual_report_2019-20_zeiss-group.pdf.
9. "Designing Advanced Scanning Probe Microscopy Instruments." *NIST*, 30 Jan. 2020, www.nist.gov/programs-projects/designing-advanced-scanning-probe-microscopy-instruments.
10. Grady, Denise. "50 Years Later, Rosalind Franklin's X-Ray Fuels Debate." *The New York Times*, The New York Times, 25 Feb. 2003, www.nytimes.com/2003/02/25/science/a-revolution-at-50-50-years-later-rosalind-franklin-s-x-ray-fuels-debate.html.
11. "Atomic Imaging Turns 50." *Chemical and Engineering News*, 28 Nov. 2005, cen.acs.org/articles/83/i48/Atomic-Imaging-Turns-50.html.
12. Smyth MS, Martin JH. x ray crystallography. Mol Pathol. 2000;53(1):8-14. https://doi.org/10.1136/mp.53.1.8.
13. Hooke, Robert. Micrographia: Or, Some Physiological Descriptions of Minute Bodies Made by Magnifying Glasses, with Observations and Inquiries Thereupon. United States, Science Heritage Limited, 1665, https://www.google.com/books/edition/Micrographia/LsbBada4VVYC?hl=en.
14. "X-Ray Crystallography: Revealing Our Molecular World." *Science Museum*, 26 Sept. 2019, www.sciencemuseum.org.uk/objects-and-stories/chemistry/x-ray-crystallography-revealing-our-molecular-world.

Petroleum

1. "Fact #676: May 23, 2011 U.S. Refiners Produce about 19 Gallons of Gasoline from a Barrel of Oil." *Energy.gov*, 23 May 2011, www.energy.gov/eere/vehicles/fact-676-may-23-2011-us-refiners-produce-about-19-gallons-gasoline-barrel-oil.
2. "Differences Between Diesel and Petrol." *European Automobile Manufacturers Association*, 24 Sept. 2016, www.acea.be/news/article/differences-between-diesel-and-petrol.
3. "Top Five Countries with Biggest Oil Refining Capacities." *NS Energy*, 22 Feb. 2019, www.nsenergybusiness.com/news/top-countries-oil-refining-capacities/.
4. "Statistical Review of World Energy 2020 | 69th Edition." *British Petroleum*, 17 June 2020,

www.bp.com/content/dam/bp/business-sites/en/global/corporate/pdfs/energy-economics/statistical-review/bp-stats-review-2020-full-report.pdf.
5. "ICSC 1561 - DIESEL FUEL No. 2." *International Programme on Chemical Safety*, Oct. 2004, www.inchem.org/documents/icsc/icsc/eics1561.htm.
6. Swinton, W E. "Physician contributions to nonmedical science: Abraham Gesner, inventor of kerosene." Canadian Medical Association journal vol. 115,11 (1976): 1126-33, https://www.ncbi.nlm.nih.gov/pmc/articles/PMC1878918/.
7. GESNER, ABRAHAM. "US12612A - IMPROVEMENT IN PROCESSES FOR MAKING KEROSENE." *United States Patent*, 27 Mar. 1855, patents.google.com/patent/US12612A/en.
8. YOUNG, JAMES. "US8833A - IMPROVEMENT IN MAKING PARAFFINE-OIL." *United States Patent*, 23 Mar. 1852, patents.google.com/patent/US8833A/en.
9. "Founder of the Modern Oil Industry to Be Honoured." *BBC News*, BBC, 8 Nov. 2011, www.bbc.com/news/uk-scotland-15648088.
10. "Bathgate Oil Works." *Scottish Shale*, www.scottishshale.co.uk/places/oil-works/bathgate-chemical-works/.
11. Waszak, Stanislaw. "World's First Oil Well Still Bubbling up Black Gold in Poland." *Yahoo! Finance*, Yahoo!, 16 Nov. 2014, finance.yahoo.com/news/worlds-first-oil-well-still-bubbling-black-gold-045512403.html.
12. "Who Made America? | Innovators | Edwin Drake." *PBS*, Public Broadcasting Service, www.pbs.org/wgbh/theymadeamerica/whomade/drake_hi.html.
13. Curl, Herbert, and Kevin O'Donnell . "CHEMICAL AND PHYSICAL PROPERTIES OF REFINED PETROLEUM PRODUCTS." *NATIONAL OCEANIC AND ATMOSPHERIC ADMINISTRATION I Environmental Research Laboratories* , Oct. 1977, repository.library.noaa.gov/view/noaa/11031/noaa_11031_DS1.pdf.
14. Sorkhabi, Rasoul, and Bruce Blanche. "The Standard Oil Story III: The Rise, Fall and Rise of The Standard Oil Company." *GEO ExPro*, 21 Jan. 2014, www.geoexpro.com/articles/2011/11/the-standard-oil-story-iii-the-rise-fall-and-rise-of-the-standard-oil-company.
15. Burclaff, Natalie. "Rockefeller: Making of a Billionaire." *Inside Adams: Science, Technology & Business*, 14 Jan. 2020, blogs.loc.gov/inside_adams/2020/01/rockefeller-billionaire/.
16. "Gross Domestic Product, 4th Quarter and Year 2020 (Advance Estimate) ." *U.S. Bureau of Economic Analysis (BEA)*, 28 Jan. 2021, www.bea.gov/news/2021/gross-domestic-product-4th-quarter-and-year-2020-advance-estimate.

Plastic

1. Ritchie, Hannah, and Max Roser. "Plastic Pollution." *Our World in Data*, 1 Sept. 2018, ourworldindata.org/plastic-pollution.
2. Biemiller, Amy. "Home." *MIT Engineering*, 20 Mar. 2013, engineering.mit.edu/engage/ask-an-engineer/can-we-safely-burn-used-plastic-objects-in-a-domestic-fireplace/.
3. "International Year of Glass 2022 (IYOG2022) The Global Glass Economy and Its Wider Social Consequences." *International Year of Glass in 2022*, 17 Nov. 2020, www.iyog2022.org/images/files/77-economicsiyog-200925.pdf.

4. Beckman, Eric. "The World of Plastics, in Numbers." *The Conversation*, 8 July 2019, theconversation.com/the-world-of-plastics-in-numbers-100291.
5. "U.S. and World Population Clock." *Population Clock*, 15 Mar. 2021, www.census.gov/popclock/.
6. "History and Future of Plastics." *Science History Institute*, 20 Nov. 2019, www.sciencehistory.org/the-history-and-future-of-plastics.
7. Hopkins, Jared S., and Drew Hinshaw. "Coronavirus Vaccine Makers Are Hunting for Vital Equipment: Glass Vials." *The Wall Street Journal*, Dow Jones & Company, 16 June 2020, www.wsj.com/articles/coronavirus-vaccine-makers-are-hunting-for-vital-equipment-glass-vials-11592317525.
8. Gómez, Fernando J., and Simonetta Rima. "Setting the Facts Straight on Plastics." *World Economic Forum*, 4 Oct. 2019, www.weforum.org/agenda/2019/10/plastics-what-are-they-explainer/.
9. Cantonwine, David E et al. "Bisphenol A and Human Reproductive Health." Expert review of obstetrics & gynecology vol. 8,4 (2013): 10.1586/17474108.2013.811939. https://doi.org/10.1586/17474108.2013.811939
10. "Joseph Priestley, Discoverer of Oxygen National Historic Chemical Landmark." *American Chemical Society*, www.acs.org/content/acs/en/education/whatischemistry/landmarks/josephpriestleyoxygen.html.
11. "Fashion's Tiny Hidden Secret." *UN Environment Programme*, 13 Mar. 2019, www.unep.org/news-and-stories/story/fashions-tiny-hidden-secret.
12. "Preferred Fiber & Materials Market Report 2019." *Textile Exchange*, 2019, store.textileexchange.org/wp-content/uploads/woocommerce_uploads/2019/11/Textile-Exchange_Preferred-Fiber-Material-Market-Report_2019.pdf.

Rubber

1. Harford, Tim. "The Horrific Consequences of Rubber's Toxic Past." *BBC News*, BBC, 23 July 2019, www.bbc.com/news/business-48533964.
2. "History." *Continental USA*, www.continental.com/en-us/company/history/history.
3. "Largest Tire Manufacturer." *Statista*, 19 Mar. 2021, www.statista.com/statistics/225677/revenue-of-the-leading-tire-producers-worldwide/.
4. Wagner, Neil. "Why the Prices of Natural and Synthetic Rubber Do Not Always Bounce Together : Beyond the Numbers." *U.S. Bureau of Labor Statistics*, U.S. Bureau of Labor Statistics, 14 Feb. 2020, www.bls.gov/opub/btn/volume-9/mobile/why-the-prices-of-natural-and-synthetic-rubber-do-not-always-bounce-together.htm.
5. "Natural Rubber Leading Producers Worldwide 2019." *Statista*, 27 Jan. 2021, www.statista.com/statistics/275397/caoutchouc-production-in-leading-countries/.
6. "EPDM RUBBER OR ETHYLENE PROPYLENE DIENE MONOMER RUBBER-M CLASS." *Stern Rubber Company*, 5 Mar. 2015, sternrubber.com/blog/epdm-rubber-or-ethylene-propylene-diene-monomer-rubber-m-class.
7. De Guzman, Doris. "History of the Synthetic Rubber Industry." *ICIS Explore*, 7 May 2008,

www.icis.com/explore/resources/news/2008/05/12/9122056/history-of-the-synthetic-rubber-industry/.
8. "The ROMANCE of RUBBER." *The New York Times*, The New York Times, 23 Sept. 1906, timesmachine.nytimes.com/timesmachine/1906/09/23/101799104.html?pageNumber=33.
9. "History." *Bridgestone Corporation*, www.bridgestone.com/corporate/history/index.html.
10. Isalska, Anita. "Travel - The Ingenious Story behind Michelin Stars." *BBC*, BBC, 25 Oct. 2018, www.bbc.com/travel/story/20181024-the-ingenious-story-behind-michelin-stars.

Lock

1. Pearce, Tom. "A History of Locks: Precision Locksmiths." *Precision Locksmiths*, 11 Apr. 2020, www.precision-locksmiths.co.uk/a-history-of-locks/.
2. Fessenden, Marissa. "Medieval Chastity Belts Are a Myth." *Smithsonian.com*, Smithsonian Institution, 20 Aug. 2015, www.smithsonianmag.com/smart-news/medieval-chastity-belts-are-myth-180956341/.
3. "Locks & Keys." *National Postal Museum*, postalmuseum.si.edu/exhibition/about-postal-operations-mail-processing/locks-keys.
4. Walmsley, Roy. "World Prison Population List: Eleventh Edition." *National Institute of Corrections*, 19 Jan. 2021, nicic.gov/world-prison-population-listeleventh-edition.
5. Abloy, Assa. "About Yale." *Us.yalehome.com*, us.yalehome.com/en/about-yale/.
6. Tucker, Lindsay. "The Evolution of the Lock." *Boston Magazine*, Boston Magazine, 30 Nov. 2016, www.bostonmagazine.com/2011/06/27/the-evolution-of-the-lock/.
7. *The World Book Encyclopedia. Volume 12, L,* World Book, Inc., 2015, pp. 410-412.
8. Radner, Karen, *GATEKEEPERS AND LOCK MASTERS: THE CONTROL OF ACCESS IN ASSYRIAN PALACES,* Your Praise is Sweet: A Memorial Volume for Jeremy Black from Students, Colleagues and Friends. United Kingdom, British Institute for the Study of Iraq, 2010.
9. "Bramah Locks." *Bramah UK: British Lock Manufacturer, Locksmiths, and Alarm Engineers*, www.bramah.co.uk/default.asp?lnc=bramah_locks.
10. "Key Figures & Financial Data." *ASSA ABLOY*, 2021, www.assaabloy.com/group/en/investors/key-figures.

Battery

1. "The Functions of Lead Batteries." *Battery Council*, aboutbatteries.batterycouncil.org/What-is-a-lead-battery.
2. "Lithium-Ion Battery." *Clean Energy Institute*, 25 Sept. 2020, www.cei.washington.edu/education/science-of-solar/battery-technology/.
3. Araujo, Keith. "Battery Cell Comparison." *Epec Engineered Technologies - Build to Print Electronics*, www.epectec.com/batteries/cell-comparison.html.
4. "Tesla, Inc. (TSLA) Stock Historical Prices & Data." *Yahoo! Finance*, Yahoo!, 26 Mar. 2021, finance.yahoo.com/quote/TSLA/history/.

5. Boudette, Neal E., and Coral Davenport. "G.M. Will Sell Only Zero-Emission Vehicles by 2035." *The New York Times*, The New York Times, 28 Jan. 2021, www.nytimes.com/2021/01/28/business/gm-zero-emission-vehicles.html.
6. Ewing, Jack. "Volvo Plans to Sell Only Electric Cars by 2030." *The New York Times*, The New York Times, 2 Mar. 2021, www.nytimes.com/2021/03/02/business/volvo-electric-cars.html.
7. "Battery: Voltaic Pile." *National Museum of American History*, americanhistory.si.edu/collections/search/object/nmah_703289.
8. Alarco, Jose, and Peter Talbot. "The History and Development of Batteries." *Phys.org*, 30 Apr. 2015, phys.org/news/2015-04-history-batteries.html.
9. *Gross Domestic Product, 4th Quarter and Year 2020 (Advance Estimate)*. U.S. Bureau of Economic Analysis (BEA), 28 Jan. 2021, www.bea.gov/news/2021/gross-domestic-product-4th-quarter-and-year-2020-advance-estimate.
10. GASSNER, CARL. "US373064A - GALVANIC BATTERY." *United States Patent*, 15 Nov. 1887, patents.google.com/patent/US373064A/en.
11. EDISON, THOMAS A. "US678722A - REVERSIBLE GALVANIC BATTERY." *United States Patent*, 16 July 1901, patents.google.com/patent/US678722A/en.
12. Yoshino, Akira, et al. "US4668595A - SECONDARY BATTERY." *United States Patent*, 26 May 1987, patents.google.com/patent/US4668595A/en.
13. "Pioneers of Innovation: The Battery That Changed the World." *Energy Factor*, Exxon Mobil, 15 Dec. 2020, energyfactor.exxonmobil.com/energy-innovation/rd/the-history-of-batteries-lithium-ion-batteries-energy-factor/.

Electric Motor

1. DAVENPORT, THOMAS. "US132A - Improvement in Propelling Machinery by Magnetism and Electro-Magnetism." *United States Patent*, 25 Feb. 1837, patents.google.com/patent/US132A/en.
2. Davenport, Thomas. "Davenport Electric Motor, US Patent #132." *Smithsonian Institution*, 1837, www.si.edu/object/nmah_703302.
3. "Electric Motor for Alternating Current." *National Museum of American History*, 1888, americanhistory.si.edu/collections/search/object/nmah_739995.
4. Henry, Joseph. "ON A RECIPROCATING MOTION PRODUCED BY MAGNETIC ATTRACTION AND REPULSION." *Silliman's American Journal of Science*, XX, July 1831, pp. 340–343, commons.princeton.edu/josephhenry/electric-motor/
5. Kinnaird, Clark. "How Many Electric Motors Are in Your Car?" *Automotive - Technical Articles - TI E2E Support Forums*, 16 Oct. 2020, e2e.ti.com/blogs_/b/behind_the_wheel/posts/how-many-electric-motors-are-in-your-car.
6. HURLEY, SELINA. "The Surprisingly Old Story Of London's First Ever Electric Taxi." *Science Museum Blog*, 27 Jan. 2020, blog.sciencemuseum.org.uk/the-surprisingly-old-story-of-londons-first-ever-electric-taxi/.
7. T. G. Wilson and P. H. Trickey, "D-C machine with solid-state commutation," in Electrical Engineering, vol. 81, no. 11, pp. 879-884, Nov. 1962, doi: 10.1109/EE.1962.6446586.

8. TESLA, NIKOLA. "US381968A - Electro-Magnetic Motor." *United States Patent*, 1 May 1888, patents.google.com/patent/US381968/en.
9. Arnett, Ryan. "The History of DC Motors." *Parvalux Electric Motors*, 10 Aug. 2020, www.parvalux.com/news/history-of-dc-motors/.
10. Dittler, Sabine. "A Detour to Success – The World's First Electric Streetcar." *Siemens.com*, new.siemens.com/global/en/company/about/history/stories/first-electric-streetcar.html.

Photography

1. "The Fire This Time." *The New York Times*, The New York Times, 11 June 1972, timesmachine.nytimes.com/timesmachine/1972/06/11/91333708.html?pageNumber=215.
2. Tong, Traci. "How the Vietnam War's Napalm Girl Found Hope after Tragedy." *The World from PRX*, 21 Feb. 2018, www.pri.org/stories/2018-02-21/how-vietnam-wars-napalm-girl-found-hope-after-tragedy.
3. Estrin, James. "Kodak's First Digital Moment." *The New York Times*, The New York Times, 12 Aug. 2015, lens.blogs.nytimes.com/2015/08/12/kodaks-first-digital-moment/.
4. *US4131919A - Electronic Still Camera*. United States Patent, 26 Dec. 1978, patents.google.com/patent/US4131919.
5. This 23-year-old Czech victim of dysentery in Nazi camp at Flossenburg, Germany, was found by 97th Division of U.S. Army / Signal Corps U.S. Army. May 4. Photograph. www.loc.gov/item/2001696922/.
6. "Niépce and the Invention of Photography." *Nicéphore Niépce's House Museum*, 17 Feb. 2021, photo-museum.org/niepce-invention-photography/.
7. "Niépce's Catalog of Works." *Nicéphore Niépce's House Museum*, 17 Feb. 2021, photo-museum.org/catalog-works-niepce/.
8. "View from the Window at Le Gras | 100 Photographs | The Most Influential Images of All Time." *Time*, Time, 100photos.time.com/photos/joseph-niepce-first-photograph-window-le-gras.
9. *Daguerreobase*, Nederlands Fotomuseum, www.daguerreobase.org/en/collections.
10. Prokudin-Gorskiĭ, Sergeĭ Mikhaĭlovich. *Ėmir Bukharskiĭ. Bukhara*. Library of Congress, 1911, www.loc.gov/pictures/item/2018680327/.

Pasteurization

1. "CFR - Code of Federal Regulations Title 21." *Food and Drug Administration*, 1 Apr.
2. 2020, www.accessdata.fda.gov/scripts/cdrh/cfdocs/cfcfr/cfrsearch.cfm?fr=1240.61.
3. Roman, Gabrielle. "A History of Pasteurization." *Culture Magazine*, 18 Apr. 2015, culturecheesemag.com/cheese-bites/history-pasteurization/.
4. "Raw Milk Questions and Answers." *Centers for Disease Control and Prevention*, Centers for Disease Control and Prevention, 15 June 2017, www.cdc.gov/foodsafety/rawmilk/raw-milk-questions-and-answers.html.
5. "Louis Pasteur." *Lemelson-MIT*, lemelson.mit.edu/resources/louis-pasteur.

6. "BRITISH PHYSICIANS NOW INDORSE STRAUS; Former Skepticism as to Merits of Pasteurized Milk Is Disappearing. SOME AMAZING RESULTS Demonstrations In Germany and In Parts of London Show Great Reduction In Infantile Mortality." *The New York Times*, The New York Times, 23 May 1909, timesmachine.nytimes.com/timesmachine/1909/05/23/106718278.html?pageNumber=23.
7. Janina. "The History of Raw Milk and Pasteurization." *Food Policy For Thought*, 6 May 2014, foodpolicyforthought.com/2014/05/06/the-history-of-raw-milk-and-pasteurization/.

Television

1. "TELEVISOR LETS RADIO FANS 'LOOK IN' AS WELL AS LISTEN; John L. Baird of Scotland Demonstrates Apparatus Which Sends and Detects Pictures of Moving Objects." *The New York Times*, The New York Times, 26 Apr. 1926, timesmachine.nytimes.com/timesmachine/1926/04/25/100069222.html?pageNumber=227.
2. "Baird Works London-to-Glasgow Television As Preliminary to Transatlantic Test Soon." *The New York Times*, The New York Times, 27 Mar. 1927, timesmachine.nytimes.com/timesmachine/1927/05/27/118644122.html?pageNumber=3.
3. Hsu, Tiffany. "Wary Companies Hesitate on Super Bowl Commercials, Citing Pandemic." *The New York Times*, The New York Times, 15 Dec. 2020, www.nytimes.com/2020/12/15/business/media/super-bowl-2021-ads-coronavirus.html.
4. "More than Half the World Watched Record-Breaking 2018 World Cup." *FIFA*, 21 Dec. 2018, www.fifa.com/worldcup/news/more-than-half-the-world-watched-record-breaking-2018-world-cup.
5. "TIME CAPSULE Vol.21." *SONY*, www.sony.net/SonyInfo/CorporateInfo/History/capsule/21/.
6. The Iconoscope for Television. Nature 132, 962–963 (1933). https://doi.org/10.1038/132962e0
7. "How It All Began - History of the BBC." *BBC News*, BBC, www.bbc.com/historyofthebbc/research/story-of-bbc-television/how-it-all-began.
8. "The BBC Steps in: 1929-1935 - History of the BBC." *BBC News*, BBC, www.bbc.com/historyofthebbc/100-voices/birth-of-tv/the-bbc-steps-in.
9. "Television to Start in Brazil." *The New York Times*, The New York Times, 17 Sept. 1950, timesmachine.nytimes.com/timesmachine/1950/09/17/87052567.html?pageNumber=106.
10. Gramlich, John. *5 Facts about Fox News*. Pew Research Center, 18 Aug. 2020, www.pewresearch.org/fact-tank/2020/04/08/five-facts-about-fox-news/.
11. Koblin, John. "How Much Do We Love TV? Let Us Count the Ways." *The New York Times*, 30 June 2016, www.nytimes.com/2016/07/01/business/media/nielsen-survey-media-viewing.html.

12. "Table 11A. Time Spent in Leisure and Sports Activities for the Civilian Population by Selected Characteristics, Averages per Day, 2019 Annual Averages." *U.S. Bureau of Labor Statistics*, 25 June 2020, www.bls.gov/news.release/atus.t11A.htm.

Refrigerator

1. *The World Book Encyclopedia. Volume 16, R*, World Book, Inc., 2015, pp. 200-203.
2. Eschner, Kat. "One Man Invented Two of the Deadliest Substances of the 20th Century." *Smithsonian.com*, Smithsonian Institution, 18 May 2017, www.smithsonianmag.com/smart-news/one-man-two-deadly-substances-20th-century-180963269/.
3. Eschner, Kat. "Leaded Gas Was a Known Poison the Day It Was Invented." *Smithsonian.com*, Smithsonian Institution, 9 Dec. 2016, www.smithsonianmag.com/smart-news/leaded-gas-poison-invented-180961368/.
4. "Ice Cream Market Size to Reach USD 91.90 Billion by 2027." *GlobeNewswire News Room*, Fortune Business Insights, 22 Feb. 2021, www.globenewswire.com/news-release/2021/02/22/2179178/0/en/Ice-Cream-Market-Size-to-Reach-USD-91-90-Billion-by-2027-Premiumization-of-Ice-creams-to-Emerge-as-a-Prominent-Trends-states-Fortune-Business-Insights.html.
5. "1,1,1,2-Tetrafluoroethane." *National Center for Biotechnology Information. PubChem Compound Database*, U.S. National Library of Medicine, pubchem.ncbi.nlm.nih.gov/compound/1_1_1_2-Tetrafluoroethane#section=Use-and-Manufacturing.
6. Long, Heather. "23% Of American Homes Have 2 (or More) Fridges." *CNNMoney*, Cable News Network, 27 May 2016, money.cnn.com/2016/05/27/news/economy/23-percent-of-american-homes-have-2-fridges/index.html.
7. "Equipment of Households with Electrical Household Appliances and Others (Germany)." *Federal Statistical Office*, 29 Oct. 2019, www.destatis.de/EN/Themes/Society-Environment/Income-Consumption-Living-Conditions/Equipment-Consumer-Durables/Tables/liste-equipment-households-electrical--household-appliance-others-germany.html.
8. "The Hidden Truth behind India's Low Refrigerator Ownership." *Economic Times Blog*, 24 Mar. 2017, economictimes.indiatimes.com/blogs/et-commentary/the-hidden-truth-behind-indias-low-refrigerator-ownership/.
9. "Royal Society Names Refrigeration, Pasteurisation and Canning as Greatest Three Inventions in the History of Food and Drink: Royal Society." *Royal Society*, 13 Sept. 2012, royalsociety.org/news/2012/top-20-food-innovations/.
10. Donnelly, Jim. "History - British History in Depth: The Irish Famine." *BBC*, BBC, 17 Feb. 2011, www.bbc.co.uk/history/british/victorians/famine_01.shtml.
11. "How Do Refrigerated Trailers Work?" *ABCO Transportation*, 5 May 2017, www.shipabco.com/refrigerated-trailers-work.
12. "A Look at Ammonia Refrigeration Through the Years." *ACHR News RSS*, ACHR News, 27 Nov. 2018, www.achrnews.com/articles/84047-a-look-at-ammonia-refrigeration-through-the-years.

13. "The Montreal Protocol on Substances That Deplete the Ozone Layer - United States Department of State." *U.S. Department of State*, U.S. Department of State, 5 Jan. 2021, www.state.gov/key-topics-office-of-environmental-quality-and-transboundary-issues/the-montreal-protocol-on-substances-that-deplete-the-ozone-layer/.
14. "Are Refrigerators like Air Conditioners?" *Boston.com*, The Boston Globe, 1 Aug. 2011, archive.boston.com/news/science/articles/2011/08/01/are_refrigerators_like_air_conditioners/.
15. Varrasi, John. "Global Cooling: The History of Air Conditioning." *The American Society of Mechanical Engineers*, 6 June 2011, www.asme.org/topics-resources/content/global-cooling-the-history-of-air-conditioning.
16. "Carrier Reports First Quarter 2020 Earnings." *Carrier Corporate*, 8 May 2020, www.corporate.carrier.com/news/news-articles/carrier_reports_first_quarter_2020_earnings.html.
17. "Refrigerant Transition & Environmental Impacts." *EPA*, Environmental Protection Agency, 18 Nov. 2020, www.epa.gov/mvac/refrigerant-transition-environmental-impacts.
18. "Substitutes in Household Refrigerators and Freezers." *EPA*, Environmental Protection Agency, 26 June 2019, www.epa.gov/snap/substitutes-household-refrigerators-and-freezers.
19. "Regulation (EU) No 517/2014 of the European Parliament and of the Council of 16 April 2014 on Fluorinated Greenhouse Gases and Repealing Regulation (EC) No 842/2006." *European Environment Agency*, 20 Apr. 2016, www.eea.europa.eu/policy-documents/regulation-eu-no-517-2014.

Airplane

1. Boyne, Walter James, Bilstein, Roger E. and Crouch, Tom D.. *"History of flight"*. Encyclopedia Britannica, 12 Nov. 2020, https://www.britannica.com/technology/history-of-flight.
2. Enlightenment Now the Case for Reason, Science, Humanism and Progress, by Steven Pinker, Penguin Books, 2019, pp. 262-263; https://books.google.com/books?id=J6grDwAAQBAJ
3. "1903 Wright Flyer." *National Air and Space Museum*, 27 Feb. 2017, airandspace.si.edu/media/NASM-A19610048000-NASM2018-10793.
4. "The World's First Scheduled Airline." *Smithsonian National Air and Space Museum*, airandspace.si.edu/exhibitions/america-by-air/online/early_years/early_years01.cfm.
5. *The World of Air Transport in 2018*, www.icao.int/annual-report-2018/Pages/the-world-of-air-transport-in-2018.aspx.
6. "Theory of Flight." *MIT*, 16 Mar. 1997, web.mit.edu/16.00/www/aec/flight.html.
7. McCartney, Scott. "A Prius With Wings vs. a Guzzler in the Clouds." *The Wall Street Journal*, Dow Jones & Company, 12 Aug. 2010, www.wsj.com/articles/SB10001424052748704901104575423261677748380.
8. Bland, Alastair. "How Bad Is Air Travel for the Environment?" *Smithsonian.com*, Smithsonian Institution, 26 Sept. 2012, www.smithsonianmag.com/travel/how-bad-is-air-travel-for-the-environment-51166834/.

9. "The A380: Speed, Power and Efficiency in One Package." *Engine Alliance*, 11 June 2019, www.enginealliance.com/a380/.
10. "The Anaesthetic of Familiarity." *Unweaving the Rainbow: Science, Delusion, and the Appetite for Wonder*, by Richard Dawkins, HMH, 2000, pp. 1–7.
11. "Airbus Topples Boeing as Biggest Plane Maker: DW: 02.01.2020." Deutsche Welle, 2 Jan. 2020, www.dw.com/en/airbus-topples-boeing-as-biggest-plane-maker/a-49536539.
12. Airbus. "Airbus Reports Full-Year (FY) 2019 Results, Delivers on Guidance." *Airbus*, Airbus, 29 July 2020, www.airbus.com/newsroom/press-releases/en/2020/02/airbus-reports-full-year-2019-results.html.
13. Chang, Kenneth. "What Does Keep Them Up There?" *The New York Times*, The New York Times, 9 Dec. 2003, www.nytimes.com/2003/12/09/news/staying-aloft-what-does-keep-them-up-there.html.

Paint

1. "FIND 'MONA LISA,' ARREST ROBBER; Picture Recovered in Florence Officially Identified as Da Vinci Masterpiece. REVENGE IS THIEF'S PLEA Wanted to Compensate Italy for Napoleon's Looting of Art Treasures. BUT OFFERED IT FOR SALE Italian Prisoner a Former Employe at the Louvre ;- Picture to Go Back to France at Once. FIND 'MONA LISA,' ARREST ROBBER." *The New York Times*, The New York Times, 13 Dec. 1913, timesmachine.nytimes.com/timesmachine/1913/12/13/100413284.html?pageNumber=1.
2. "Prussian Blue." *American Chemical Society*, 23 Jan. 2017, www.acs.org/content/acs/en/molecule-of-the-week/archive/p/prussian-blue.html.
3. "Highest Insurance Valuation for a Painting." *Guinness World Records*, www.guinnessworldrecords.com/world-records/highest-insurance-valuation-for-a-painting.
4. "Nebamun Tomb-Painting ." *The British Museum*, www.britishmuseum.org/collection/object/Y_EA37977.
5. "Lead in Paint." *Sandberg*, 16 June 2020, www.sandberg.co.uk/laboratories/chemistry/lead-in-paint/.
6. "Analysis of the Materials Used in the 'Earlier Mona Lisa'." *The Mona Lisa Foundation*, 9 Aug. 2018, monalisa.org/2012/09/08/leonardos-materials-the-canvas-the-paint/.
7. *How Prussian Blue Changed Everything: Christie's*. Christies, 14 Dec. 2017, www.christies.com/features/The-transformational-powers-of-Prussian-Blue-8758-1.aspx.
8. "Health Effects of Lead Exposure." *Centers for Disease Control and Prevention*, Centers for Disease Control and Prevention, 7 Jan. 2020, www.cdc.gov/nceh/lead/prevention/health-effects.htm.
9. "About Sherwin-Williams." *About Us*, oem.sherwin-williams.com/ap/en/aboutus.
10. "Company History." *PPG*, corporate.ppg.com/our-company/company-history.aspx.

11. Garside, M. "Leading Paint and Coatings Companies Globally by Revenue 2020." *Statista*, 28 Jan. 2021, www.statista.com/statistics/543989/leading-paint-companies-worldwide-by-revenue/.
12. Mueller, Allie. "A Very Brief History of Acrylic Paint." *Kensington Art Supply*, 19 Sept. 2016, kensingtonartsupply.com/kas/a-very-brief-history-of-acrylic-paint/.

Tea and Coffee

1. "Global Tea Market Revenue 2012-2025." *Statista*, 26 Mar. 2021, www.statista.com/statistics/326384/global-tea-beverage-market-size/.
2. "The History of Coffee." *National Coffee Association*, www.ncausa.org/about-coffee/history-of-coffee.
3. "Global $465.9 Billion Coffee Market (Value, Volume) Analysis and Forecast to 2026." *GlobeNewswire News Room*, Research and Markets, 9 Feb. 2021, www.globenewswire.com/news-release/2021/02/09/2172014/0/en/Global-465-9-Billion-Coffee-Market-Value-Volume-Analysis-and-Forecast-to-2026.html.
4. Gardner, E., Ruxton, C. & Leeds, A. "Black tea – helpful or harmful? A review of the evidence." Eur J Clin Nutr 61, 3–18 (2007). https://doi.org/10.1038/sj.ejcn.1602489
5. Woodall, Bernie. "Coroner Says South Carolina Teenager Died after Drinking Caffeine Quickly." *Reuters*, Thomson Reuters, 16 May 2017, www.reuters.com/article/us-south-carolina-death-caffeine-idUSKCN18C05O.
6. Stromberg, Joseph. "This Is How Your Brain Becomes Addicted to Caffeine." *Smithsonian.com*, Smithsonian Institution, 9 Aug. 2013, www.smithsonianmag.com/science-nature/this-is-how-your-brain-becomes-addicted-to-caffeine-26861037/.
7. Lewis, Danny. "Archaeologists Find World's Oldest Tea in the Tomb of a Han Dynasty Emperor." *Smithsonian.com*, Smithsonian Institution, 13 Jan. 2016, www.smithsonianmag.com/smart-news/archaeologists-find-worlds-oldest-tea-tomb-han-dynasty-emperor-180957790/.
8. Lu, H., Zhang, J., Yang, Y. et al. "Earliest tea as evidence for one branch of the Silk Road across the Tibetan Plateau." Sci Rep 6, 18955 (2016). https://doi.org/10.1038/srep18955
9. Lu, Yu. The Classic of Tea: Origins & Rituals. United Kingdom, Ecco Press, 1974, https://www.google.com/books/edition/The_Classic_of_Tea/6qogPwAACAAJ
10. Leung LK, Su Y, Chen R, Zhang Z, Huang Y, Chen ZY. "Theaflavins in black tea and catechins in green tea are equally effective antioxidants." J Nutr. 2001 Sep;131(9):2248-51. doi: 10.1093/jn/131.9.2248.
11. Wack, Margaret. "Black Tea vs. Green Tea: What's the Difference?" *ArtfulTea*, ArtfulTea, 6 Feb. 2020, www.artfultea.com/tea-wisdom-1/black-tea-vs-green-tea-whats-the-difference.
12. Shahbandeh, M. "Top Coffee Producing Countries 2020." *Statista*, 23 Mar. 2021, www.statista.com/statistics/277137/world-coffee-production-by-leading-countries/.
13. "History of Twinings." *Twinings UK & Ireland*, www.twinings.co.uk/our-communi-tea/about-twinings/history-of-twinings.
14. "Quality from Dallmayr Since 1700." *Alois Dallmayr – Find out More about Our Company*, www.dallmayr.com/deen/company/.
15. "Tea." *EST: Tea*, www.fao.org/economic/est/est-commodities/tea/en/.

16. Chang, Kaison. *World Tea Production and Trade: Current and Future Development.* FOOD AND AGRICULTURE ORGANIZATION OF THE UNITED NATIONS, 2015, www.fao.org/3/i4480e/i4480e.pdf.
17. Willson, Cyril. "The clinical toxicology of caffeine: A review and case study." Toxicology reports vol. 5 1140-1152. 3 Nov. 2018, doi:10.1016/j.toxrep.2018.11.002
18. Avey, Tori. "The Caffeinated History of Coffee." *PBS*, Public Broadcasting Service, 8 Apr. 2013, www.pbs.org/food/the-history-kitchen/history-coffee/.
19. "About Tea and Coffee." *The New York Times*, The New York Times, 12 Nov. 1899, timesmachine.nytimes.com/timesmachine/1899/11/12/117934476.html?pageNumber=27.
20. "GREEN TEA AND BLACK." *The New York Times*, The New York Times, 27 Feb. 1886, timesmachine.nytimes.com/timesmachine/1886/02/27/103099125.html?pageNumber=3.

Cinema

1. "The Lumière Brothers: Pioneers of Cinema and Photography: National Science and Media Museum." *National Science and Media Museum Blog*, 8 Jan. 2021, blog.scienceandmediamuseum.org.uk/the-lumiere-brothers-pioneers-of-cinema-and-colour-photography/.
2. "The Story of the Kelly Gang (1906): United Nations Educational, Scientific and Cultural Organization." *United Nations Educational, Scientific and Cultural Organization*, www.unesco.org/new/en/communication-and-information/memory-of-the-world/register/full-list-of-registered-heritage/registered-heritage-page-8/the-story-of-the-kelly-gang-1906/.
3. "Media and Entertainment Spotlight." *SelectUSA.gov*, www.selectusa.gov/media-entertainment-industry-united-states.
4. "The Jazz Singer | AFI CATALOG OF FEATURE FILMS." *AFI*, catalog.afi.com/Catalog/moviedetails/1535.
5. *The Role of Film in Society.* Thought Economics, 3 June 2015, thoughteconomics.com/the-role-of-film-in-society/.
6. "MIDDLEBURY SHOWS HISTORY OF MOVIES; Students Throng College Play-House to See Films Dating From 1893." *The New York Times*, The New York Times, 21 Nov. 1937, timesmachine.nytimes.com/timesmachine/1937/11/21/94463252.html?pageNumber=44.
7. "The Great Train Robbery." *The Library of Congress*, 1903, www.loc.gov/item/00694220.
8. Green, Richard. "The Great Train Robbery." *The Unwritten Record*, National Archives and Records Administration, 3 Dec. 2013, unwritten-record.blogs.archives.gov/2013/12/03/the-great-train-robbery/.
9. Youngs, Ian. "Louis Le Prince, Who Shot the World's First Film in Leeds." *BBC News*, BBC, 22 June 2015, www.bbc.com/news/entertainment-arts-33198686.
10. "Glass Copy Negative of Roundhay Garden Scene: Science Museum Group Collection." *Science Museum* , 1888,

collection.sciencemuseumgroup.org.uk/objects/co8646638/glass-copy-negative-of-ro undhay-garden-scene-by-louis-le-prince-glass-negative.
11. "Technicolor Camera." *National Museum of American History*, americanhistory.si.edu/collections/search/object/nmah_759495.
12. "Technicolor Technology." *George Eastman Museum*, www.eastman.org/technicolor/technology.
13. Ramachandran, T. "In Charts: How the Indian Film Industry Compares with the Rest of the World." *The Hindu*, The Hindu, 13 Nov. 2017, www.thehindu.com/entertainment/movies/indian-film-industry-goes-places/article203 80070.ece.

Phonograph

1. EDISON, THOMAS A. "US200521A - Improvement in Phonograph or Speaking Machines." *United States Patent*, 19 Feb. 1878, patents.google.com/patent/US200521A/en.
2. EDISON, THOMAS A. "US227679A - Phonograph." *United States Patent*, 18 May 1880, patents.google.com/patent/US227679A/en.
3. "History of the Cylinder Phonograph." *The Library of Congress*, www.loc.gov/collections/edison-company-motion-pictures-and-sound-recordings/articl es-and-essays/history-of-edison-sound-recordings/history-of-the-cylinder-phonograp h/.
4. Murphy, Mary Jo. "Remembering the Phonograph." *The New York Times*, The New York Times, 3 July 2015, www.nytimes.com/2015/07/03/arts/music/remembering-the-phonograph.html.
5. Editorial. "THE PHONOGRAPH." *The New York Times*, The New York Times, 7 Nov. 1877, timesmachine.nytimes.com/timesmachine/1877/11/07/80668334.html?pageNumber= 4.
6. BERLINER, EMILE. "US372786A - Gramophone." *United States Patent*, 8 Nov. 1887, patents.google.com/patent/US372786A/en.
7. "A Brief Guide to Microphones - What a Microphone Does." *Audio-Technica*, www.audio-technica.com/en-us/support/a-brief-guide-to-microphones-what-a-microp hone-does/.
8. "NEW RECORD AIMED TO BOLSTER SALES; Columbia Offers 45-Minute Disk While Philco and Magnavox Show Two-Speed Players OTHERS MAY FOLLOW SUIT Whole Album of Music Played on Single Vinylite Platter With Slower Turntable." *The New York Times*, The New York Times, 21 June 1948, timesmachine.nytimes.com/timesmachine/1948/06/21/86744870.html?pageNumber= 29.
9. "Berliner Gramophone: Science Museum Group Collection." *Science Museum Group*, collection.sciencemuseumgroup.org.uk/objects/co117385/berliner-gramophone-gram ophone.
10. "Record for Gramophone by HMV, 1897.: Science Museum Group Collection." *Science Museum Group*, collection.sciencemuseumgroup.org.uk/objects/co8191368/record-for-gramophone-b y-hmv-1897-gramophone-record.

11. "Record Pressing - 1897-1904." *VictorRecords.com*, www.victorrecords.com/18971904.
12. "Berliner Gramophone Record." *National Museum of American History*, americanhistory.si.edu/collections/search/object/nmah_852763.

Rocket

1. "Tuesday January 13, 1920." A Severe Strain on Credulity, The New York Times, 13 Jan. 1920, timesmachine.nytimes.com/timesmachine/1920/01/13/issue.html.
2. Dunbar, Brian. *July 20, 1969: One Giant Leap For Mankind*. NASA, 19 Feb. 2015, www.nasa.gov/mission_pages/apollo/apollo11.html.
3. "Sergei Korolev: Father of the Soviet Union's Success in Space." *ESA*, 9 Mar. 2007, www.esa.int/About_Us/ESA_history/50_years_of_humans_in_space/Sergei_Korolev_Father_of_the_Soviet_Union_s_success_in_space.
4. Harbaugh, Jennifer. "Biography of Wernher Von Braun." *NASA*, 18 Feb. 2016, www.nasa.gov/centers/marshall/history/vonbraun/bio.html.
5. Wilford, John Noble. *Wernher Von Braun, Rocket Pioneer, Dies*. The New York Times, 18 June 1977, www.nytimes.com/1977/06/18/archives/wernher-von-braun-rocket-pioneer-dies-wernher-von-braun-pioneer-in.html.
6. "A Man – and an Equation." *ESA*, 14 Oct. 2012, blogs.esa.int/rocketscience/2012/10/14/a-man-and-an-equation/.
7. Dunbar, Brian. "Saturn V." *NASA*, NASA, www.nasa.gov/centers/johnson/rocketpark/saturn_v.html.
8. "The Mayflower." *History.com*, A&E Television Networks, 4 Mar. 2010, www.history.com/topics/colonial-america/mayflower.
9. WILFORD, JOHN NOBLE. "ASTRONAUTS LAND ON PLAIN; COLLECT ROCKS, PLANT FLAG; A Powdery Surface Is Closely Explored." *The New York Times*, The New York Times, 21 July 1969, timesmachine.nytimes.com/timesmachine/1969/07/21/90114388.html?pageNumber=1.
10. *The World Book Encyclopedia. Volume 16, R*, World Book, Inc., 2015, pp. 384–391.

Satellite

1. "SOVIET FIRES EARTH SATELLITE INTO SPACE." *The New York Times*, The New York Times, 5 Oct. 1957, timesmachine.nytimes.com/timesmachine/1957/10/05/issue.html.
2. Wood, Therese. "Who Owns Our Orbit: Just How Many Satellites Are There in Space?" *World Economic Forum*, 23 Oct. 2020, www.weforum.org/agenda/2020/10/visualizing-easrth-satellites-sapce-spacex/.
3. Garcia, Mark. "International Space Station Facts and Figures." *NASA*, NASA, 28 Apr. 2016, www.nasa.gov/feature/facts-and-figures.
4. Wattles, Jackie. "Elon Musk's SpaceX Now Owns about a Third of All Active Satellites in the Sky." *CNN*, Cable News Network, 11 Feb. 2021, www.cnn.com/2021/02/11/tech/spacex-starlink-satellites-1000-scn/index.html.
5. "World Cup United: How Satellites Make Football a Truly Global Game." *Euronews*, 12 June 2014,

www.euronews.com/2014/06/12/world-cup-united-how-satellites-make-football-a-truly-global-game.
6. "Celebrating 60 Years of the World's First Weather Satellite." *NOAA National Environmental Satellite, Data, and Information Service (NESDIS)*, 31 Mar. 2020, www.nesdis.noaa.gov/content/celebrating-60-years-world%E2%80%99s-first-weather-satellite.
7. "Meteorological Satellite, TIROS." *Meteorological Satellite, TIROS | National Air and Space Museum*, airandspace.si.edu/collection-objects/tiros-meteorological-satellite/nasm_A19650289000.
8. "How High and Fast Does an Aircraft Fly?" *Schiphol Amsterdam Airport*, www.schiphol.nl/en/you-and-schiphol/page/five-questions-about-aircraft/.
9. "RAMMB/CIRA SLIDER: Satellite Loop Interactive Data Explorer in Real-Time with GOES-16 and Himawari-8 Satellite Imagery." *Colorado State University*, rammb-slider.cira.colostate.edu/.
10. Watkins, Derek. "Japan's New Satellite Captures an Image of Earth Every 10 Minutes." *The New York Times*, The New York Times, 10 July 2015, www.nytimes.com/interactive/2015/07/10/science/An-Image-of-Earth-Every-Ten-Minutes.html.
11. Benson, Michael. "Watching Earth Burn." *The New York Times*, The New York Times, 28 Dec. 2020, www.nytimes.com/2020/12/28/opinion/climate-change-earth.html.
12. "How Much Does It Cost?" *ESA*, www.esa.int/Science_Exploration/Human_and_Robotic_Exploration/International_Space_Station/How_much_does_it_cost.
13. Chang, Kenneth. "Want to Buy a Ticket to the Space Station? NASA Says Soon You Can." *The New York Times*, The New York Times, 7 June 2019, www.nytimes.com/2019/06/07/science/space-station-nasa.html.
14. "DOD Releases Fiscal Year 2021 Budget Proposal." *U.S. DEPARTMENT OF DEFENSE*, 10 Feb. 2020, www.defense.gov/Newsroom/Releases/Release/Article/2079489/dod-releases-fiscal-year-2021-budget-proposal/.
15. Leary, Warren E. "HUBBLE TELESCOPE LOSES LARGE PART OF OPTICAL ABILITY." *The New York Times*, The New York Times, 28 June 1990, www.nytimes.com/1990/06/28/us/hubble-telescope-loses-large-part-of-optical-ability.html.
16. Wilford, John Noble. "Astronomers Say Hubble Repairs Were Successful." *The New York Times*, The New York Times, 13 Jan. 1994, www.nytimes.com/1994/01/13/us/astronomers-say-hubble-repairs-were-successful.html.
17. "SATELLITE IN ORBIT BEAMS TV FROM U.S. TO EUROPE; PICTURES CLEAR IN FRANCE." *The New York Times*, The New York Times, 11 July 1962, timesmachine.nytimes.com/timesmachine/1962/07/11/80409440.html?pageNumber=1.
18. Halloran, Richard. "U.S. PLANS BIG SPENDING INCREASE FOR MILITARY OPERATIONS IN SPACE." *The New York Times*, The New York Times, 17 Oct. 1982, www.nytimes.com/1982/10/17/us/us-plans-big-spending-increase-for-military-operations-in-space.html.

Mobile phone

1. Anjarwalla, Tas. "Inventor of Cell Phone: We Knew Someday Everybody Would Have One." *CNN*, Cable News Network, 9 July 2010, www.cnn.com/2010/TECH/mobile/07/09/cooper.cell.phone.inventor/index.html.
2. Ha, Peter. "All-TIME 100 Gadgets." *Time*, Time Inc., 25 Oct. 2010, content.time.com/time/specials/packages/article/0,28804,2023689_2023708_2023656,00.html.
3. *The Mobile Economy*, 21 Feb. 2021, www.gsma.com/mobileeconomy/.
4. Belson, Ken. "NTT DoCoMo Introduces New Generation Cellphone Service." *The New York Times*, The New York Times, 1 Oct. 2001, www.nytimes.com/2001/10/01/business/ntt-docomo-introduces-new-generation-cellphone-service.html.
5. Monaghan, Angela. "Nokia: the Rise and Fall of a Mobile Phone Giant." *The Guardian*, Guardian News and Media, 3 Sept. 2013, www.theguardian.com/technology/2013/sep/03/nokia-rise-fall-mobile-phone-giant.
6. "Smartphone Market Share - OS." *IDC*, 15 Dec. 2020, www.idc.com/promo/smartphone-market-share/os.
7. "TeliaSonera First in the World with 4G Services." *Telia Company*, 14 Dec. 2009, www.teliacompany.com/en/news/press-releases/2009/12/teliasonera-first-in-the-world-with-4g-services/.
8. "Net Proceeds in Key U.S. Spectrum Auction Tops $80 Billion." *Reuters*, Thomson Reuters, 15 Jan. 2021, www.reuters.com/article/us-usa-spectrum/net-proceeds-in-key-u-s-spectrum-auction-tops-80-billion-idUSKBN29K2EG.
9. Li, Kenneth, and Ju-min Park. "Who Was First to Launch 5G? Depends Who You Ask." *Reuters*, Thomson Reuters, 5 Apr. 2019, www.reuters.com/article/us-telecoms-5g/who-was-first-to-launch-5g-depends-who-you-ask-idUSKCN1RH1V1.
10. "5G Speed Is Data Transmission in Real Time." *Deutsche Telekom*, 8 Oct. 2018, www.telekom.com/en/company/details/5g-speed-is-data-transmission-in-real-time-544498.
11. "The 8 Attributes of 5G Network Performance." *Verizon Business*, www.verizon.com/business/resources/5g/8-currencies-5g-network-performance/.
12. "Nokia 3310." *Nokia 3310 - Full Phone Specifications*, www.gsmarena.com/nokia_3310-192.php.
13. "Nokia 1110." *Nokia 1110 - Full Phone Specifications, Price*, www.fonearena.com/nokia-1110_253.html.
14. Silver, Laura. "Smartphone Ownership Is Growing Rapidly Around the World, but Not Always Equally." *Pew Research Center's Global Attitudes Project*, Pew Research Center, 25 Aug. 2020, www.pewresearch.org/global/2019/02/05/smartphone-ownership-is-growing-rapidly-around-the-world-but-not-always-equally/.

Soap

1. "Sodium Stearate." *American Chemical Society*, 9 Nov. 2020, www.acs.org/content/acs/en/molecule-of-the-week/archive/s/sodium-stearate.html.

2. "Procter & Gamble." *Procter & Gamble - Ohio History Central*, ohiohistorycentral.org/w/Procter_%26_Gamble.
3. "History: Colgate-Palmolive." *Colgate*, www.colgatepalmolive.com/en-us/about/history.
4. "Our History." *Unilever Global Company Website*, www.unilever.com/our-history.html.
5. Tolsma, Marisa. "What Is Tallow?: Benefits of Beef Tallow." *Bumblebee Apothecary*, 18 June 2020, bumblebeeapothecary.com/what-is-tallow-benefits-of-beef-tallow/.
6. "Development of Tide Synthetic Detergent - National Historic Chemical Landmark." *American Chemical Society*, 25 Oct. 2006, www.acs.org/content/acs/en/education/whatischemistry/landmarks/tidedetergent.html.
7. "NEW WASH DETERGENT IS INTRODUCED HERE." *The New York Times*, The New York Times, 15 Oct. 1948, timesmachine.nytimes.com/timesmachine/1948/10/15/97154416.html?pageNumber=19.
8. "Lifebuoy with Puralin Plus Bath Size Soap." *Smithsonian Institution*, www.si.edu/object/lifebuoy-puralin-plus-bath-size-soap:nmah_1448558.
9. "What Sodium Lauryl Sulfate & Is It Safe?", *Colgate*, www.colgate.com/en-us/oral-health/selecting-dental-products/what-sodium-lauryl-sulfate-is-it-safe.
10. Jabr, Ferris. "Why Soap Works." *The New York Times*, The New York Times, 13 Mar. 2020, www.nytimes.com/2020/03/13/health/soap-coronavirus-handwashing-germs.html.
11. "Soaps & Detergents History." *The American Cleaning Institute (ACI)*, www.cleaninginstitute.org/understanding-products/why-clean/soaps-detergents-history.

Pain killer

1. Trevor Brown, Alan Dronsfield. "Pain Relief: from Coal Tar to Paracetamol." *RSC Education*, 1 July 2005, edu.rsc.org/feature/pain-relief-from-coal-tar-to-paracetamol/2020140.article.
2. Commissioner, Office of the. "A Guide to Safe Use of Pain Medicine." *U.S. Food and Drug Administration*, FDA, 9 Feb. 2009, www.fda.gov/consumers/consumer-updates/guide-safe-use-pain-medicine.
3. Yardley, William. "Jack Fishman Dies at 83; Saved Many From Overdose." *The New York Times*, The New York Times, 14 Dec. 2013, www.nytimes.com/2013/12/15/business/jack-fishman-who-helped-develop-a-drug-to-treat-overdoses-dies-at-83.html.
4. Sinha, Shreeya, et al. "Heroin Addiction Explained: How Opioids Hijack the Brain." *The New York Times*, The New York Times, 19 Dec. 2018, www.nytimes.com/interactive/2018/us/addiction-heroin-opioids.html.
5. "What Is an Epidural." *American Pregnancy Association*, 2 Oct. 2020, americanpregnancy.org/healthy-pregnancy/labor-and-birth/what-is-an-epidural-782/.
6. "Upfront: The Birth of the Epidural." *The Royal College of Midwives*, 11 Mar. 2014, www.rcm.org.uk/news-views/rcm-opinion/upfront-the-birth-of-the-epidural/.
7. Team, EBI Web. "Sufentanil." *European Bioinformatics Institute*, 22 Feb. 2017, www.ebi.ac.uk/chebi/searchId.do?chebiId=CHEBI:9316.

8. Rapaport, Lisa. "Pain Relief during Labor Varies across U.S." *Reuters*, Thomson Reuters, 11 Jan. 2019, www.reuters.com/article/us-health-childbirth-anesthesia/pain-relief-during-labor-varies-across-u-s-idUSKCN1P529Y.
9. "Drug Overdose Deaths." *Centers for Disease Control and Prevention*, Centers for Disease Control and Prevention, 19 Mar. 2020, www.cdc.gov/drugoverdose/data/statedeaths.html.
10. Nemo, Leslie. "How Did Opium Poppies Get Their Painkilling Properties?" *LiveScience*, Purch, 30 Aug. 2018, www.livescience.com/63471-opium-poppy-painkiller-genome.html.
11. Guo, Li, et al. "The Opium Poppy Genome and Morphinan Production." *Science*, American Association for the Advancement of Science, 19 Oct. 2018, science.sciencemag.org/content/362/6412/343.
12. "How Opioid Drugs Activate Receptors." *National Institutes of Health*, U.S. Department of Health and Human Services, 5 June 2018, www.nih.gov/news-events/nih-research-matters/how-opioid-drugs-activate-receptors.
13. Trickey, Erick. "Inside the Story of America's 19th-Century Opiate Addiction." *Smithsonian.com*, Smithsonian Institution, 4 Jan. 2018, www.smithsonianmag.com/history/inside-story-americas-19th-century-opiate-addiction-180967673/.
14. National Institute on Drug Abuse. "Naloxone." *National Institute on Drug Abuse*, 21 Oct. 2020, www.drugabuse.gov/publications/drugfacts/naloxone.
15. "How Aspirin Works." *UChicago Medicine*, UChicago Medicine, 31 July 1995, www.uchicagomedicine.org/forefront/news/how-aspirin-works.

Telescope

1. "Reflecting Telescope: Royal Society Picture Library." *Royal Society Picture Library*, pictures.royalsociety.org/image-rs-8462.
2. Hall A. Rupert and Simpson A. D. C.1996 *An account of the Royal Society's Newton telescope Notes Rec. R. Soc. Lond.*501–11, doi.org/10.1098/rsnr.1996.0001
3. *Mirrors Webb/NASA.* NASA, jwst.nasa.gov/content/observatory/ote/mirrors/index.html.
4. Tyndall, Amy. "Hubble's Greatest Discoveries: the Age of the Universe." *BBC Science Focus Magazine*, 10 Aug. 2020, www.sciencefocus.com/space/hubble-space-telescope-the-age-of-the-universe/.
5. O'Connell, Patrick M. "Closed for Months, Historic Space Observatory Once Again Has a Bright Future after U. of C. Donation." *Chicagotribune.com*, Chicago Tribune, 30 Apr. 2020, www.chicagotribune.com/news/breaking/ct-yerkes-observatory-university-of-chicago-new-life-20200501-xmdjrdwau5hqvdb63ls6q2jauq-story.html.
6. "Introducing the Gran Telescopio CANARIAS." *Gran Telescopio CANARIAS*, www.gtc.iac.es/gtc/gtc.php.
7. Houlton, Susan. "Spain Inaugurates One of the World's Most Powerful Telescopes." *Deutsche Welle*, 24 July 2009, www.dw.com/en/spain-inaugurates-one-of-the-worlds-most-powerful-telescopes/a-4516752.

8. Cox, Lauren. "Who Invented the Telescope?" *Space.com*, Space, 21 Dec. 2017, www.space.com/21950-who-invented-the-telescope.html.
9. "About Chandra :: Telescope System." *Chandra*, chandra.harvard.edu/about/telescope_system.html.
10. "Imagine the Universe!" *NASA*, NASA, imagine.gsfc.nasa.gov/observatories/technology/xray_telescopes2.html.
11. "About the Compton Gamma Ray Observatory." *NASA*, NASA, heasarc.gsfc.nasa.gov/docs/cgro/cgro/.
12. "Key Facts - Webb/NASA." *NASA*, NASA, www.jwst.nasa.gov/content/about/faqs/facts.html.
13. Garner, Rob. "Discoveries." *NASA*, NASA, 27 Jan. 2017, www.nasa.gov/content/goddard/2017/highlights-of-hubble-s-exploration-of-the-universe.

Alcohol

1. "Table 20. Use of Selected Substances in the Past Month among Persons Aged 12 Years and over, by Age, Sex, Race, and Hispanic Origin: United States, Selected Years 2002–2017." *Center for Disease Control*, 2018, www.cdc.gov/nchs/data/hus/2018/020.pdf.
2. "Underage Drinking." *National Institute on Alcohol Abuse and Alcoholism*, U.S. Department of Health and Human Services, Oct. 2020, www.niaaa.nih.gov/publications/brochures-and-fact-sheets/underage-drinking.
3. *Global Status Report on Alcohol and Health 2018*. World Health Organization, 21 Aug. 2019, www.who.int/substance_abuse/publications/global_alcohol_report/gsr_2018/en/.
4. Barclay, Eliza. *Our Ability To Digest Alcohol May Have Been Key To Our Survival*. NPR, 3 Dec. 2014, www.npr.org/sections/thesalt/2014/12/03/368044880/our-ability-to-digest-alcohol-may-have-been-key-to-our-survival.
5. McGovern, Patrick et.al. "The Earliest Alcoholic Beverage in the World." *The Earliest Alcoholic Beverage in the World | Research - Penn Museum*, www.penn.museum/research/project.php?pid=12.
6. Krasowski, Matthew D. "Toxic Alcohols." *AACC*, www.aacc.org/cln/articles/2012/february/toxic-alcohols.
7. Manning, Aaliyah. "Drinking Rubbing Alcohol Can Bring Deadly Results." *The Recovery Village Drug and Alcohol Rehab*, The Recovery Village Drug and Alcohol Rehab, 10 Nov. 2020, www.therecoveryvillage.com/alcohol-abuse/rubbing-alcohol-dangerous/.
8. Oreskovich MR, Kaups KL, Balch CM, et al. Prevalence of Alcohol Use Disorders Among American Surgeons. Arch Surg. 2012;147(2):168–174. doi:10.1001/archsurg.2011.1481
9. "Alcohol Poisoning." *Alcohol Poisoning | Alcohol.org.nz*, www.alcohol.org.nz/alcohol-its-effects/health-effects/alcohol-poisoning.
10. "Global Drinking Demographics." *Alcohol.org*, www.alcohol.org/guides/global-drinking-demographics/.
11. Rudnik, Alesia. "Who Benefits from the Alcoholisation of Belarus?" *Belarus Digest*, 11 May 2017, belarusdigest.com/story/who-benefits-from-the-alcoholisation-of-belarus/.

12. Daley, Jason. "2,100-Year-Old Roman Tavern Unearthed, Empty Cups and All." *Smithsonian.com*, Smithsonian Institution, 14 Mar. 2016, www.smithsonianmag.com/smart-news/2100-year-old-roman-tavern-uncovered-empty-cups-and-all-180958395/.
13. Boissoneault, Lorraine. "Ancient Humans Liked Getting Tipsy, Too." *Smithsonian.com*, Smithsonian Institution, 10 July 2017, www.smithsonianmag.com/history/ancient-humans-liked-getting-tipsy-too-180963990/.
14. "Prehistoric China." *Penn Museum*, www.penn.museum/sites/biomoleculararchaeology/?page_id=247.
15. McGovern, Patrick E. "Intoxicating: The Science of Alcohol." *Penn Museum*, 1 Apr. 2015, www.penn.museum/sites/biomoleculararchaeology2/2015/04/01/intoxicating-the-science-of-alcohol/.
16. "Alcohol Consumption Dips in Kerala, but Govt Rakes in the Revenue." *Malayala Manorama*, 30 July 2020, www.onmanorama.com/kerala/top-news/2020/07/30/alcohol-consumption-kerala-beverages-corporation.html.
17. "Economy." *Department of Economics and Statistics*, Government of Kerala, 20 Feb. 2020, www.ecostat.kerala.gov.in/index.php/economy.
18. "Kerala Budget Analysis 2019-20." *PRS Legislative Research*, 10 Apr. 2021, prsindia.org/budgets/states/kerala-budget-analysis-2019-20.

Tobacco

1. Lal, Pranay. "Bidi – A Short History ." *CURRENT SCIENCE, VOL. 96, NO. 10, 25 MAY 2009*, 25 Mar. 2009, pp. 1335–1337, www.indiaenvironmentportal.org.in/files/Bidi.pdf.
2. "Tobacco in India." *World Health Organization*, World Health Organization, www.who.int/india/health-topics/tobacco.
3. "Tobacco." *World Health Organization*, World Health Organization, 27 May 2020, www.who.int/news-room/fact-sheets/detail/tobacco.
4. Mishra, Shanu, and M B Mishra. "Tobacco: Its historical, cultural, oral, and periodontal health association." Journal of International Society of Preventive & Community Dentistry vol. 3,1 (2013): 12-8. doi:10.4103/2231-0762.115708
5. D'Souza, Manoranjan S, and Athina Markou. "Neuronal mechanisms underlying development of nicotine dependence: implications for novel smoking-cessation treatments." Addiction science & clinical practice vol. 6,1 (2011): 4-16, www.ncbi.nlm.nih.gov/pmc/articles/PMC3188825/.
6. "Fast Facts." *Centers for Disease Control and Prevention*, Centers for Disease Control and Prevention, 21 May 2020, www.cdc.gov/tobacco/data_statistics/fact_sheets/fast_facts/index.htm.
7. *British American Tobacco - The Global Market*, www.bat.com/group/sites/UK__9D9KCY.nsf/vwPagesWebLive/DO9DCKFM.
8. "Tobacco Economics." *Philip Morris International*, www.pmi.com/who-we-are/tobacco-facts/tobacco-economics.
9. Rubin, Rebecca. "Global Entertainment Industry Surpasses $100 Billion for the First Time Ever." *Variety*, Variety, 12 Mar. 2020,

variety.com/2020/film/news/global-entertainment-industry-surpasses-100-billion-for-the-first-time-ever-1203529990/.
10. Turak, Natasha. "Oil and Gas Companies Set to Lose $1 Trillion in Revenues This Year." *CNBC*, CNBC, 30 Apr. 2020, www.cnbc.com/2020/04/30/coronavirus-creating-1-trillion-revenue-loss-for-oil-and-gas-companies.html.
11. "Our History." *Imperial*, www.imperialbrandsplc.com/about-us/our-history.html.
12. "The Toll of Tobacco in China." *Campaign for Tobacco-Free Kids*, 1 Oct. 2020, www.tobaccofreekids.org/problem/toll-global/asia/china.
13. "Tobacco History." *CNN*, Cable News Network, edition.cnn.com/US/9705/tobacco/history/.
14. "Key Milestones." *Philip Morris International*, Philip Morris International, 31 July 2019, www.pmi.com/who-we-are/key-milestones.
15. "British American Tobacco - Our History – a Timeline.", *British American Tobacco*, www.bat.com/history.

Bridge

1. *Alcántara Bridge - History of Puente Trajan at Alcantara.* BridgesDB, www.bridgesdb.com/bridge-list/alcantara-bridge/.
2. Anderson, Chris. "24 Of the World's Most Amazing Bridges." *CNN*, Cable News Network, 12 July 2017, www.cnn.com/travel/article/most-amazing-bridges/index.html.
3. *Port Authority of New York and New Jersey George Washington Bridge*, www.panynj.gov/bridges-tunnels/en/george-washington-bridge.html.
4. "THE GOLDEN GATE BRIDGE." *The New York Times*, The New York Times, 30 May 1937, timesmachine.nytimes.com/timesmachine/1937/05/30/101009420.html?pageNumber=40.
5. Buckley, Chris. "China's New Bridges: Rising High, but Buried in Debt." *The New York Times*, The New York Times, 10 June 2017, www.nytimes.com/2017/06/10/world/asia/china-bridges-infrastructure.html.
6. "Why Bogibeel Bridge - World's Longest Steel Bridge - Is an Engineering Marvel: Latest News & Updates at DNAIndia.com." *DNA India*, DNA India, 26 Dec. 2018, www.dnaindia.com/india/photo-gallery-why-bogibeel-bridge-world-s-longest-steel-bridge-is-an-engineering-marvel-2700304.
7. "An Ancient Work of Art." *An Ancient Work of Art | Site Du Pont Du Gard*, www.pontdugard.fr/en/ancient-work-art.
8. "History of Iron Bridge." *English Heritage*, www.english-heritage.org.uk/visit/places/iron-bridge/history/.
9. "Arkadiko Bridge." *Atlas Obscura*, Atlas Obscura, 24 Sept. 2018, www.atlasobscura.com/places/arkadiko-bridge.
10. "Design & Construction Stats." *Golden Gate Bridge, Highway and Transportation District*, www.goldengate.org/bridge/history-research/statistics-data/design-construction-stats/.

Dam

1. Yardley, Jim. "Chinese Dam Projects Criticized for Their Human Costs." *The New York Times*, The New York Times, 19 Nov. 2007, www.nytimes.com/2007/11/19/world/asia/19dam.html.
2. "Hydropower Explained." *U.S. Energy Information Administration (EIA)*, 8 Apr. 2021, www.eia.gov/energyexplained/hydropower/.
3. "TEXANS PERISH IN DISASTROUS FLOODS; Raging Colorado River Engulfs Fifty People at Austin. GREAT DAM BROKEN AWAY Raging Wall of Water Rushes Through the City. Stream a Mile Wide Sweeps Away Electric Power House Leaving City in Darkness -- Over 100 Buildings Destroyed." *The New York Times*, The New York Times, 8 Apr. 1900, timesmachine.nytimes.com/timesmachine/1900/04/08/102583348.html?pageNumber=1.
4. "Hydropower Program." *Bureau of Reclamation*, 3 Feb. 2016, www.usbr.gov/power/edu/history.html.
5. "AUSTIN'S BIG REGATTA.; Oarsmen Competing in Celebration of the Completion of the Colorado Dam." *The New York Times*, The New York Times, 9 June 1893, timesmachine.nytimes.com/timesmachine/1893/06/09/109699865.html?pageNumber=3.
6. Ahmed, Niaz. "கரிகாலன் கட்டிய கல்லணை: தமிழர்களின் நீர் மேலாண்மை குறித்து நீங்கள் அறிவீர்களா?" *BBC News தமிழ்*, BBC, 19 Aug. 2020, www.bbc.com/tamil/india-53826135.
7. Perera, Duminda, et al. "Ageing Water Storage Infrastructure: An Emerging Global Risk." *United Nations University Institute for Water, Environment and Health, Hamilton, Canada.*, 25 Jan. 2021, inweh.unu.edu/ageing-water-storage-infrastructure-an-emerging-global-risk/.
8. "The Columbia River Basin Provides More than 40% of Total U.S. Hydroelectric Generation." *Today in Energy - U.S. Energy Information Administration (EIA)*, 27 June 2014, www.eia.gov/todayinenergy/detail.php?id=16891.
9. "THE COLUMBIA RIVER SYSTEM INSIDE STORY." *Bonneville Power Administration*, Apr. 2001, www.bpa.gov/p/Generation/Hydro/hydro/columbia_river_inside_story.pdf.
10. "Typhoon Nina–Banqiao Dam Failure." *Encyclopædia Britannica*, Encyclopædia Britannica, Inc., www.britannica.com/event/Typhoon-Nina-Banqiao-dam-failure.

Canal

1. "How the Panama Canal Took a Huge Toll On the Contract Workers Who Built It." *Smithsonian.com*, Smithsonian Institution, 18 Apr. 2018, www.smithsonianmag.com/history/how-panama-canal-took-huge-toll-on-contract-workers-who-built-it-180968822/.
2. Tuttle, Robert. "Analysis | Why a Canal Built in 1869 Is More Important Than Ever." *The Washington Post*, WP Company, 29 Mar. 2021, www.washingtonpost.com/business/energy/why-a-canal-built-in-1009-is-more-important-than-ever/2021/03/26/2aef3bb8-8dfe-11eb-a33e-da28941cb9ac_story.html.
3. *About ACP*. Panama Canal Authority, www.pancanal.com/eng/general/canal-faqs/physical.html.
4. "About Suez Canal." *Suez Canal Authority*, www.suezcanal.gov.eg/English/About/SuezCanal/Pages/AboutSuezCanal.aspx.

5. Russon, Mary-Ann. "The Cost of the Suez Canal Blockage." *BBC News*, BBC, 29 Mar. 2021, www.bbc.com/news/business-56559073.
6. "The New Suez Canal." *NASA*, NASA, 29 Apr. 2016, www.earthobservatory.nasa.gov/images/87948/the-new-suez-canal.
7. "French Panama Canal Failure (1881-1889)." *University of Kansas Medical Center*, 5 Mar. 2020, www.kumc.edu/school-of-medicine/history-and-philosophy-of-medicine/panama-canal/french-panama-canal-failure.html.
8. Mouynes, Rene. *Panama Canal History - End of the Construction*, www.pancanal.com/eng/history/history/end.html.
9. "The Enlargement of the Erie Canal a National Necessity." *The New York Times*, The New York Times, 14 Jan. 1863, timesmachine.nytimes.com/timesmachine/1863/01/14/80269467.html?pageNumber=4.
10. "The Grand Canal." *UNESCO World Heritage Centre*, whc.unesco.org/en/list/1443/.
11. "Erie Canal." *History.com*, A&E Television Networks, 15 Mar. 2018, www.history.com/topics/landmarks/erie-canal.

Bicycle

1. "Rover 'Safety' Bicycle, 1885: Science Museum Group Collection." *Science Museum Group*, collection.sciencemuseumgroup.org.uk/objects/co25833/rover-safety-bicycle-1885-bicycle.
2. Sonuparlak, Itir. "Bicycles as a Source of Income in Africa." *TheCityFix*, 1 June 2011, thecityfix.com/blog/bicycles-as-a-source-of-income-in-africa/.
3. Correspondent, Special. *6.44 Lakh Students to Get Free Bicycles in TN*. The Hindu, 27 June 2014, www.thehindu.com/news/national/tamil-nadu/644-lakh-students-to-get-free-bicycles-in-tn/article6155098.ece.
4. Bhattacharya, Pramit. "One in Three Households in India Owns a Two-Wheeler." *Mint*, 11 Dec. 2016, www.livemint.com/Politics/Yd2EAFIupVHDX0EbUdecsO/One-in-three-households-in-India-owns-a-twowheeler.html.
5. Long, Tony. "Aug. 30, 1885: Daimler Gives World First 'True' Motorcycle." *Wired*, 3 June 2017, www.wired.com/2011/08/0830daimler-first-true-motorcycle/.
6. Dawson, Louise. "How the Bicycle Became a Symbol of Women's Emancipation." *The Guardian*, Guardian News and Media, 4 Nov. 2011, www.theguardian.com/environment/bike-blog/2011/nov/04/bicycle-symbol-womens-emancipation.
7. "The Development of the Velocipede." *Smithsonian Institution*, www.si.edu/spotlight/si-bikes/si-bikes-velocipede.
8. LALLEMENT, PERRE. *US59915A - Improvement in Velocipedes*. United States Patent, 20 Nov. 1866, patents.google.com/patent/US59915A/en.
9. "Velocipede, 1868." *Smithsonian Institution*, www.si.edu/object/velocipede-1868:nmah_843079.

10. "Statistics of the United States, (Including Mortality, Property, &c.,) in 1860 : Comp. from the Original Returns and Being the Final Exhibit of the Eighth ..." *HathiTrust*, babel.hathitrust.org/cgi/pt?id=chi.12697213&view=1up&seq=586&size=125.
11. "U.S. Census Bureau QuickFacts: United States." *Census Bureau QuickFacts*, www.census.gov/quickfacts/fact/table/US/SEX255219.
12. Dunlop, John Boyd. *US435995A - John Boyd Dunlop*. United States Patent, 9 Sept. 1890, patents.google.com/patent/US435995A/en.
13. "Columbia Bicycles Heritage - America's First Bicycle." *Columbia Bicycles*, 12 Apr. 2018, columbiabicycles.com/heritage/.
14. *Motorcyclist Safety*. Governors Highway Safety Association, www.ghsa.org/issues/motorcycle-safety.
15. "Royal Enfield Journey Since 1901: Royal Enfield." *Royal Enfield Journey Since 1901| Royal Enfield*, www.royalenfield.com/in/en/our-world/since-1901/.
16. "American Drivers Have Bicyclists to Thank for a Smooth Ride to Work." *Smithsonian.com*, Smithsonian Institution, 12 Sept. 2016, www.smithsonianmag.com/travel/american-drivers-thank-bicyclists-180960399/.
17. Santora, Marc. "Woman Hit by a Bicyclist in Central Park Dies." *The New York Times*, The New York Times, 22 Sept. 2014, www.nytimes.com/2014/09/23/nyregion/woman-dies-after-being-struck-by-central-park-cyclist.html.
18. "Honda Engines: About Us." *Honda*, engines.honda.com/company/about-us.

X-ray

1. "X-Rays, CT Scans and MRIs." *The American Academy of Orthopaedic Surgeons*, orthoinfo.aaos.org/en/treatment/x-rays-ct-scans-and-mris/.
2. "History of Medicine: Dr. Roentgen's Accidental X-Rays." *Columbia University Department of Surgery*, columbiasurgery.org/news/2015/09/17/history-medicine-dr-roentgen-s-accidental-x-rays.
3. "To X-Ray or Not to X-Ray?" *World Health Organization*, World Health Organization, 14 Apr. 2016, www.who.int/news-room/feature-stories/detail/to-x-ray-or-not-to-x-ray-.
4. "Featured History: Magnetic Resonance Imaging." *UW Radiology*, 21 July 2016, rad.washington.edu/blog/featured-history-magnetic-resonance-imaging/.
5. "Ultrasound." *National Institute of Biomedical Imaging and Bioengineering*, U.S. Department of Health and Human Services, July 2016, www.nibib.nih.gov/science-education/science-topics/ultrasound.
6. "Computed Tomography (CT)." *National Institute of Biomedical Imaging and Bioengineering*, U.S. Department of Health and Human Services, www.nibib.nih.gov/science-education/science-topics/computed-tomography-ct.
7. "Magnetic Resonance Imaging (MRI)." *National Institute of Biomedical Imaging and Bioengineering*, U.S. Department of Health and Human Services, www.nibib.nih.gov/science-education/science-topics/magnetic-resonance-imaging-mri.
8. Campbell, S. "A short history of sonography in obstetrics and gynaecology." Facts, views & vision in ObGyn vol. 5,3 (2013): 213-29, www.ncbi.nlm.nih.gov/pubmed/24753947.

9. Rivera-Ruiz, Moises et al. "Einthoven's string galvanometer: the first electrocardiograph." Texas Heart Institute journal vol. 35,2 (2008): 174-8, www.ncbi.nlm.nih.gov/pubmed/18612490.
10. Markel, Howard. "'I Have Seen My Death': How the World Discovered the X-Ray." *PBS*, Public Broadcasting Service, 20 Dec. 2012, www.pbs.org/newshour/health/i-have-seen-my-death-how-the-world-discovered-the-x-ray.
11. "X-RAY LOCATES BULLET." *The New York Times*, The New York Times, 18 Oct. 1900, timesmachine.nytimes.com/timesmachine/1900/10/18/105754176.html?pageNumber=1.

Cement

1. "The Foundation of Civilization." *Cement*, Portland Cement Association, www.cement.org/civilization.
2. Garside, M. "U.S. and World Cement Production 2020." *Statista*, 15 Feb. 2021, www.statista.com/statistics/219343/cement-production-worldwide/.
3. Keskeys, Paul. "Redemption: The 'World's Ugliest Building' Just Won a Major Architecture Award - Architizer Journal." *Architizer Journal*, 13 Aug. 2019, architizer.com/blog/inspiration/stories/worlds-ugliest-building-wins-major-architecture-award/.
4. Rodgers, Lucy. "Climate Change: The Massive CO2 Emitter You May Not Know About." BBC News, BBC, 17 Dec. 2018, www.bbc.com/news/science-environment-46455844.
5. Watts, Jonathan. "Concrete: the Most Destructive Material on Earth." *The Guardian*, Guardian News and Media, 25 Feb. 2019, www.theguardian.com/cities/2019/feb/25/concrete-the-most-destructive-material-on-earth.
6. Van Mead, Nick. "A Brief History of Concrete: from 10,000BC to 3D Printed Houses." *The Guardian*, Guardian News and Media, 25 Feb. 2019, www.theguardian.com/cities/2019/feb/25/a-brief-history-of-concrete-from-10000bc-to-3d-printed-houses.
7. "Cement and Concrete." *The World Book Encyclopedia. Volume 3, C, World Book, Inc.*, 2015, pp. 343-345.
8. "How Concrete Is Made." *Portland Cement Association*, www.cement.org/cement-concrete/how-concrete-is-made.

Elevator

1. Prisco, Jacopo. *A Short History of the Elevator*. Cable News Network, 8 Feb. 2019, www.cnn.com/style/article/short-history-of-the-elevator/index.html.
2. Kiuntke, Florian. *Lift Me up – Werner Von Siemens Presents the World's First Electric Elevator*. Siemens, 5 Dec. 2020, new.siemens.com/global/en/company/about/history/stories/electric-elevator.html.
3. "INSTALLING ELEVATORS IN A BIG SKYSCRAPER; Commodious Electric Cars for The Times's New Building. A Speed of 500 Feet a Minute -- Safety Appliances to Bar Accidents -- The Newest Signaling System." *The New York Times*, The New York

Times, 8 Nov. 1903, timesmachine.nytimes.com/timesmachine/1903/11/08/105065527.html?pageNumber=24.
4. "HOTEL LIGHTED BY ELECTRICITY." *The New York Times*, The New York Times, 8 June 1880, timesmachine.nytimes.com/timesmachine/1880/06/08/98903805.html?pageNumber=2.
5. "WHAT'S NEW IN ELEVATORS; ELEVATORS AS TRAFFIC COPS." *The New York Times*, The New York Times, 13 Apr. 1986, timesmachine.nytimes.com/timesmachine/1986/04/13/645586.html?pageNumber=265.
6. TUFTS, OTIS. "US25061A - Elevator or Hoisting Apparatus for Hotels." *United States Patent*, 9 Aug. 1859, patents.google.com/patent/US25061A/en.
7. "Elevator And Escalator Fun Fact." *National Elevator Industry, Inc.*, 2007, www.neii.org/presskit/pressmaster.cfm?link=7.
8. "Investors." *Otis Elevator Company*, www.otisinvestors.com/investors.

Skyscraper

1. "Productivity Growth." *The Age of Diminished Expectations*, by Paul R. Krugman, MIT, 1997, pp. 11, https://www.google.com/books/edition/The_Age_of_Diminished_Expectations/awA0yp1V8c8C?hl=en.
2. "EMPIRE STATE TOWER, TALLEST IN WORLD, IS OPENED BY HOOVER; THE HIGHEST STRUCTURE RAISED BY THE HAND OF MAN." *The New York Times*, The New York Times, 2 May 1931, timesmachine.nytimes.com/timesmachine/1931/05/02/102231255.html?pageNumber=1.
3. "Europe's Tallest Skyscraper Costs More Than Dubai's Burj Khalifa, Media Reports." *The Moscow Times*, The Moscow Times, 15 Nov. 2018, www.themoscowtimes.com/2018/11/15/europe-tallest-skyscraper-costs-more-than-dubais-burj-khalifa-media-reports-a63505.
4. "Europe's Tallest Skyscraper Costs More Than Dubai's Burj Khalifa, Media Reports." *The Moscow Times*, The Moscow Times, 2 Apr. 2021, www.themoscowtimes.com/2018/11/15/europe-tallest-skyscraper-costs-more-than-dubais-burj-khalifa-media-reports-a63505.
5. "BUILDING BIG: Databank: Empire State Building." *PBS*, Public Broadcasting Service, www.pbs.org/wgbh/buildingbig/wonder/structure/empire_state.html.
6. "Home Insurance Building." *History.com*, A&E Television Networks, 22 Apr. 2010, www.history.com/topics/landmarks/home-insurance-building.
7. Poon, Linda. *Charting the Booms and Busts of NYC's Skyscraper History*. Bloomberg, 27 Oct. 2015, www.bloomberg.com/news/articles/2015-10-27/new-timeline-traces-the-booms-and-busts-of-new-york-city-s-skyscrapers.
8. "Park Row Building." *The Skyscraper Center*, www.skyscrapercenter.com/new-york-city/park-row-building/9142/.
9. "The Stories: The Towers." *Petronas Twin Towers*, www.petronastwintowers.com.my/the-stories/.

10. "Tallest Buildings." *100 Tallest Completed Buildings in the World - The Skyscraper Center*, www.skyscrapercenter.com/buildings.
11. Hollister, Nathaniel, et al. "Tall Buildings in Numbers." *CTBUH Journal*, vol. 2011, no. 3, 2011, pp. 54–55, https://global.ctbuh.org/resources/papers/153-Journal2011_IssueIII_TBIN.pdf.
12. "Facts & Figures." *Facts & Figures | Empire State Building*, www.esbnyc.com/about/facts-figures.
13. "Makkah Royal Clock Tower ." *Makkah Royal Clock Tower - The Skyscraper Center*, www.skyscrapercenter.com/building/makkah-royal-clock-tower/84.
14. "The 2020 CTBUH Awards Announced the Lakhta Center a Five-Time Winner, Receiving Awards of Excellence." *Lakhta Center*, 6 Dec. 2019, lakhta.center/en/article/?id=1350.
15. "QuickFacts, New York County (Manhattan Borough), New York." *United States Census Bureau*, www.census.gov/quickfacts/fact/table/newyorkcountymanhattanboroughnewyork/LND110210#LND110210.
16. Ritchie, Hannah. "Which Countries Are Most Densely Populated?" *Our World in Data*, 6 Sept. 2019, ourworldindata.org/most-densely-populated-countries.

Dynamite

1. "Alfred Nobel's Patents." *NobelPrize.org*, www.nobelprize.org/alfred-nobel/list-of-alfred-nobels-patents.
2. Lieffers, Caroline. "How the Panama Canal Took a Huge Toll On the Contract Workers Who Built It." *Smithsonian.com*, Smithsonian Institution, 18 Apr. 2018, www.smithsonianmag.com/history/how-panama-canal-took-huge-toll-on-contract-workers-who-built-it-180968822/.
3. "HOOVER DAM." *Bureau of Reclamation*, 13 Mar. 2015, www.usbr.gov/lc/hooverdam/educate/kidfacts.html.
4. "Highway History." *U.S. Department of Transportation/Federal Highway Administration*, 27 June 2017, www.fhwa.dot.gov/infrastructure/back0403.cfm.
5. Mangravite, Andrew. "Meeting the Miner's Friend." *Science History Institute*, 9 Dec. 2017, www.sciencehistory.org/distillations/magazine/meeting-the-miners-friend.
6. Nobel, Alfred. "US78317A - Improved Explosive Compound." *United States Patent*, 26 May 1868, patents.google.com/patent/US78317A/en.
7. Butterfield, Fox. "Ideas Abound, but Blocking Oklahoma-Type Bombs Is Seen as Unlikely." *The New York Times*, The New York Times, 3 May 1995, www.nytimes.com/1995/05/03/us/terror-oklahoma-bomb-ideas-abound-but-blocking-oklahoma-type-bombs-seen-unlikely.html.
8. "Alfred Nobel." *Science History Institute*, 13 Dec. 2017, www.sciencehistory.org/historical-profile/alfred-nobel.
9. "An Explosive Discovered in a Deadly Way." *The New York Times*, The New York Times, 3 May 1995, www.nytimes.com/1995/05/03/us/terror-in-oklahoma-an-explosive-discovered-in-a-deadly-way.html.
10. "PubChem Compound Summary for CID 4510, Nitroglycerin" PubChem, National Center for Biotechnology Information, https://pubchem.ncbi.nlm.nih.gov/compound/Nitroglycerin. Accessed 14 April, 2021.

11. "Ascanio Sobrero" NobelPrize.org, Nobel Media AB 2021, https://www.nobelprize.org/alfred-nobel/ascanio-sobrero. Accessed 14 April, 2021.
12. "PubChem Compound Summary for CID 8490, Cyclonite" PubChem, National Center for Biotechnology Information, https://pubchem.ncbi.nlm.nih.gov/compound/Cyclonite. Accessed 14 April, 2021.
13. Bacon, John U. "Opinion | How the Explosive Destruction of Halifax Holds Lessons - and Hope - for Beirut." *The Washington Post*, WP Company, 10 Aug. 2020, www.washingtonpost.com/opinions/2020/08/07/how-explosive-destruction-halifax-holds-lessons-hope-beirut/.

Pesticide

1. United States Census Office. 9Th Census, 1870, and Francis Amasa Walker. "Statistical atlas of the United States based on the results of the ninth census with contributions from many eminent men of science and several departments of the government." [*New York J. Bien, lith*, 1874] Map. Retrieved from the Library of Congress, www.loc.gov/item/05019329.
2. POPHAM, JOHN N. "REPORT PROGRESS IN MALARIA FIGHT; Authorities Tell Convention on Tropical Medicine of Effects From the Use of DDT." *The New York Times*, The New York Times, 7 Dec. 1948, timesmachine.nytimes.com/timesmachine/1948/12/07/96606296.html?pageNumber=27.
3. Rice, Robert. "DDT." *The New Yorker*, The New Yorker, 20 June 2017, www.newyorker.com/magazine/1954/07/17/ddt-2.
4. The Nobel Prize in Physiology or Medicine 1948. *NobelPrize.org*. Nobel Media AB 2021. www.nobelprize.org/prizes/medicine/1948/summary/
5. National Center for Biotechnology Information. "PubChem Compound Summary for CID 24572, Lead arsenate" *PubChem*, pubchem.ncbi.nlm.nih.gov/compound/Lead-arsenate.
6. National Center for Biotechnology Information. "PubChem Compound Summary for CID 72720419, Copper acetoarsenite" *PubChem*, pubchem.ncbi.nlm.nih.gov/compound/Copper-acetoarsenite.
7. "Chemically-Related Groups of Active Ingredients." *EPA*, Environmental Protection Agency, 3 Oct. 2016, www.epa.gov/ingredients-used-pesticide-products/chemically-related-groups-active-ingredients.
8. Callaway, Ewen. "Pathogen Genome Tracks Irish Potato Famine Back to Its Roots." *Nature News*, Nature Publishing Group, 21 May 2013, www.nature.com/news/pathogen-genome-tracks-irish-potato-famine-back-to-its-roots-1.13021.
9. Gianessi, Leonard, and Ashley Williams. *Repeat of Great Bengal Famine Unlikely Thanks to Fungicides*. CropLife Foundation, Nov. 2012, croplife.org/wp-content/uploads/pdf_files/Repeat-of-Great-Bengal-Famine-Unlikely-Thanks-to-Fungicides.pdf.
10. Institute of Medicine (US) Committee on the Economics of Antimalarial Drugs; Arrow KJ, Panosian C, Gelband H, editors. Saving Lives, Buying Time: Economics of Malaria Drugs in an Age of Resistance. Washington (DC): National Academies Press

(US); 2004. 5, A Brief History of Malaria. Available from: www.ncbi.nlm.nih.gov/books/NBK215638/.

Safety match

1. EMERY, STEUART M. "OUR LOWLY MATCH HAS ROUNDED OUT A CENTURY; It Is Only in the Last 100 Years That It Has Taken the Place of the Cumbersome Tinder Box -- Growth of The Industry Is Traced From Its Origin." *The New York Times*, The New York Times, 27 Nov. 1927, timesmachine.nytimes.com/timesmachine/1927/11/27/96681672.html?pageNumber=219.
2. "The History of Matches." *Swedish Match Industries AB*, www.swedishmatchindustries.com/en/the-fire-academy/the-history-of-matches/.
3. Wisniak, Jaime. "Matches-The Manufacture of Fire ." *Indian Journal of Chemical Technology* , vol. 12, May 2005, pp. 369–380, www.researchgate.net/publication/286993224_Matches-The_manufacture_of_fire
4. *A Brief History of Flame: The Story of the Everyday Lighter*. WILSONS & CO., sharrowmills.com/pages/the-story-of-the-everyday-lighter.
5. Eschner, Kat. "Friction Matches Were a Boon to Those Lighting Fires–Not So Much to Matchmakers." *Smithsonian.com*, Smithsonian Institution, 27 Nov. 2017, www.smithsonianmag.com/smart-news/friction-matches-were-boon-those-lighting-firesnot-so-much-matchmakers-180967318/.
6. "A History of the World - Object : John Walker's Friction Light." *BBC*, BBC, 2014, www.bbc.co.uk/ahistoryoftheworld/objects/hQR9oN5LTeCLcuKfPDMJ9A.
7. "White Phosphorus." *American Chemical Society*, 7 Sept. 2020, www.acs.org/content/acs/en/molecule-of-the-week/archive/w/white-phosphorus.html.
8. Chemistry of Phosphorus (Z=15). 21 Aug. 2020, chem.libretexts.org/@go/page/566.
9. Stinson, Andrew. *Smoke and Mirrors: Small Lighters Leave Big Impact - The: Corsair*. The Corsair, 2 Oct. 2015, www.thecorsaironline.com/corsair/culture/2015/10/02/smoke-and-mirrors-small-lighters-leave-big-impact.
10. "THE DEADLY WHITE PHOSPHORUS." *The New York Times*, The New York Times, 28 July 1910, timesmachine.nytimes.com/timesmachine/1910/07/28/104944978.html?pageNumber=6.

Kerosene lamp

1. Fry, Richard. "The Number of People in the Average U.S. Household Is Going up for the First Time in over 160 Years." *Pew Research Center*, Pew Research Center, 30 May 2020, www.pewresearch.org/fact-tank/2019/10/01/the-number-of-people-in-the-average-u-s-household-is-going-up-for-the-first-time-in-over-160-years/.
2. Webster, George. "Solar Lamps Replace Toxic Kerosene in Poorest Countries." *CNN*, Cable News Network, 1 Feb. 2012, www.cnn.com/2012/01/10/tech/innovation/solar-powered-led-lamps/index.html.
3. Melik, James. "Solutions Sought to End Use of Kerosene Lamps." *BBC News*, BBC, 26 Sept. 2012, www.bbc.com/news/business-18262217.

4. "THE FATAL KEROSENE LAMP." *The New York Times*, The New York Times, 1 Oct. 1876, timesmachine.nytimes.com/timesmachine/1879/10/29/81768364.html?pageNumber=3.
5. Ritchie, Hannah, and Max Roser. "Access to Energy." *Our World in Data*, 20 Sept. 2019, ourworldindata.org/energy-access#what-share-of-people-have-access-to-electricity.
6. "About Petromax: A German Brand with Tradition and History." *Petromax*, 28 Feb. 2019, www.petromax.com/petromax/about/.
7. "First Light." *Coleman Canada*, www.colemancanada.ca/en_US/FirstLight.html.
8. "Grace's Guide." *Tilley Lamp Co - Graces Guide*, www.gracesguide.co.uk/Tilley_Lamp_Co.
9. GHOSH, DEEPANNITA. *ILLUMINATING THE PAST: ARTIFICIAL LIGHTING IN AMERICA (1610-1930) AND A GUIDE TO LIGHTING HISTORIC HOUSE MUSEUMS*. University of Georgia, 2004, getd.libs.uga.edu/pdfs/ghosh_deepannita_200405_mhp.pdf.
10. Thomas, Bill. "Amish Are Hosts On Indiana Farm." *The New York Times*, The New York Times, 6 July 1969, timesmachine.nytimes.com/timesmachine/1969/07/06/89010796.html?pageNumber=270.

Vacuum Tube

1. Whelan , M., and W. Kornrumpf. *Vacuum Tubes (Valves)*, edisontechcenter.org/VacuumTubes.html.
2. JOSEPH, J. Applications of the Thermionic Valve. Nature 109, 522–523 (1922). https://doi.org/10.1038/109522a0.
3. FLEMING, JOHN AMBROSE. "US803684A - Instrument for Converting Alternating Electric Currents into Continuous Currents." *United States Patent*, 7 Nov. 1905, patents.google.com/patent/US803684/en.
4. DE FOREST, LEE. "US879532A - Space Telegraphy." *United States Patent*, 18 Feb. 1908, patents.google.com/patent/US879532/en.
5. EDISON, THOMAS A. "US307031A - ELECTRICAL INDICATOR." *United States Patent*, 21 Oct. 1884, patents.google.com/patent/US307031A/en.
6. Spencer, Percy L. "US2480679A - Prepared Food Article and Method of Preparing." *United States Patent*, 30 Aug. 1949, patents.google.com/patent/US2480679/en.
7. Whelan, M., et al. *Microwave Ovens*, edisontechcenter.org/Microwaves.html.
8. Detz, Remko J., and Bob van der Zwaan. "Surfing the Microwave Oven Learning Curve." *Journal of Cleaner Production*, Elsevier, 15 June 2020, www.sciencedirect.com/science/article/pii/S0959652620323258.

Washing Machine

1. "Housework in Late 19th Century America." *Digital History*, University of Houston, www.digitalhistory.uh.edu/topic_display.cfm?tcid=93.
2. "'Vowel' Washing Machine Made by Thomas Bradford & Company." *Museum of Applied Arts & Sciences*, 1880, collection.maas.museum/object/54294.

3. SHAW, WILLIAM. "Plea for the Washing Machine." *The New York Times*, The New York Times, 27 Aug. 1941, timesmachine.nytimes.com/timesmachine/1941/08/27/119447142.html?pageNumber=18.
4. Rosling, Hans. "The Magic Washing Machine." *TED*, Dec. 2010, www.ted.com/talks/hans_rosling_the_magic_washing_machine.
5. "Clothes Washing Machines." *Washing Machines*, edisontechcenter.org/WashingMachines.html.
6. "Whirlpool Corporation History & Heritage." *Whirlpool Corporation*, 19 Mar. 2021, www.whirlpoolcorp.com/history/.
7. "Financial History." *Electrolux Group*, www.electroluxgroup.com/en/financial-history-255/.
8. "Statistics Show When the Rot Really Started." *NZ Herald*, NZ Herald, 26 Aug. 2002, www.nzherald.co.nz/technology/statistics-show-when-the-rot-really-started/57TSK5T6M3HLNTLZB5YAKJULNA/.
9. Kiprop, Joseph. "Who Invented the Washing Machine?" *WorldAtlas*, WorldAtlas, 24 July 2018, www.worldatlas.com/articles/who-invented-the-washing-machine.html.
10. "Washing Machine." *Science Museum Group*, collection.sciencemuseumgroup.org.uk/objects/co8404998/washing-machine-washing-machine.
11. "USE OF WASHING MACHINES.; Report Shows 29 in Every 100 Homes Have Them." *The New York Times*, The New York Times, 17 Aug. 1925, timesmachine.nytimes.com/timesmachine/1925/08/17/98839259.html?pageNumber=27.
12. "India: Penetration Rate of Home Appliances 2018." *Statista*, 16 Apr. 2021, www.statista.com/statistics/370635/household-penetration-home-appliances-india/.
13. Koptyug, Evgenia. "Domestic Appliances: Household Penetration in Germany 2020." *Statista*, 11 Nov. 2020, www.statista.com/statistics/474528/domestic-appliances-household-penetration-germany/.
14. Figal Garone, Lucas & Olarte, Liliana & Peña, Ximena & León, Santiago. "The Acquisition of Home Durables among the Low-Income in Latin America and the Caribbean: Trends and Challenges.", 2019, https://idbinvest.org/en/download/7128.

GPS

1. Science Reference Section, Library of Congress. "What Is a GPS? How Does It Work?" The Library of Congress, 19 Nov. 2019, www.loc.gov/everyday-mysteries/item/what-is-gps-how-does-it-work/.
2. Murphy, Kate. "America Has a GPS Problem." *The New York Times*, The New York Times, 23 Jan. 2021, www.nytimes.com/2021/01/23/opinion/gps-vulnerable-alternatives-navigation-critical-infrastructure.html.
3. Anderson, Robert. "INDUSTRY IN ORBIT: HOW TO WIN IN SPACE." *The New York Times*, The New York Times, 7 Aug. 1983, www.nytimes.com/1983/08/07/business/business-forum-industry-in-orbit-how-to-win-in-space.html.

4. Halloran, Richard. "U.S. PLANS BIG SPENDING INCREASE FOR MILITARY OPERATIONS IN SPACE." *The New York Times*, The New York Times, 17 Oct. 1982, www.nytimes.com/1982/10/17/us/us-plans-big-spending-increase-for-military-operations-in-space.html.
5. Dembart, Lee, and International Herald Tribune. "U.S. Allows Precise Civilian Use : Satellite Locator Gets New Niche." *The New York Times*, The New York Times, 11 May 2000, www.nytimes.com/2000/05/11/business/worldbusiness/IHT-us-allows-precise-civilian-use-satellite-locator.html.
6. Lake, Matt. "Getting There With Help From Above." *The New York Times*, The New York Times, 29 Mar. 2001, www.nytimes.com/2001/03/29/technology/how-it-works-getting-there-with-help-from-above.html.
7. Janwalkar, Mayura. "India Becomes 4th Nation to Get IMO Nod for Navigation Satellite System." *The Indian Express*, 21 Nov. 2020, indianexpress.com/article/technology/science/india-becomes-4th-nation-to-get-imo-nod-for-navigation-satellite-system-7059589/.
8. "Magellan NAV 1000 GPS Receiver, 1988: Time and Navigation." *Time and Navigation - The Untold Story of Getting from Here to There.*, Smithsonian Institution, timeandnavigation.si.edu/multimedia-asset/magellan-nav-1000-gps-receiver-1988.
9. "Magellan NAV 1000 GPS Receiver, 1991: Science Museum Group Collection." *Science Museum Group*, collection.sciencemuseumgroup.org.uk/objects/co8365897/magellan-nav-1000-gps-receiver-1991-gps-receiver.
10. "1 Billion Now Use Galileo Smartphones." *GPS World*, 11 Sept. 2019, www.gpsworld.com/1-billion-now-use-galileo-smartphones/.
11. "Home." *Use Galileo*, 9 Apr. 2021, www.usegalileo.eu/.

Radar

1. "This Month in Physics History." *American Physical Society*, Apr. 2006, www.aps.org/publications/apsnews/200604/history.cfm.
2. "Air Traffic Control Comes of Age." *Smithsonian National Air and Space Museum*, airandspace.si.edu/exhibitions/america-by-air/online/heyday/heyday05.cfm.
3. "About Our WSR 88-D Radar." *National Weather Service*, NOAA's National Weather Service, 23 Mar. 2015, www.weather.gov/iwx/wsr_88d.
4. *2019 Airport Traffic Report*. PORT OF AUTHORITY OF NEW YORK, 18 May 2020, www.panynj.gov/content/dam/airports/statistics/statistics-general-info/annual-atr/ATR 2019.pdf.
5. "History of Radar." *Fraunhofer-Gesellschaft*, 8 Feb. 2018, www.blackvalue.de/en/radarbasics/radar-history.html.
6. "Role of Radar in the Weather and Climate Observing and Predicting System." *Weather Radar Technology beyond NEXRAD*, National Academy Press, 2002, pp. 8–9, www.nap.edu/read/10394/chapter/3#9
7. Zelnick, C. Robert. "A Harder Look at Capt. Rogers's Judgment." *The New York Times*, The New York Times, 17 July 1988,

timesmachine.nytimes.com/timesmachine/1988/07/17/328389.html?pageNumber=154.

RFID

1. Cardullo, Mario W., and William L. Parks. "US3713148A - Transponder Apparatus and System." *United States Patent*, 23 Jan. 1973, patents.google.com/patent/US3713148A/en.
2. Walton, Charles A. "US4384288A - Portable Radio Frequency Emitting Identifier." *United States Patent*, 17 May 1983, patents.google.com/patent/US4384288A/en.
3. "High-Tech Credit Fraud." *The New York Times*, The New York Times, 9 Oct. 1990, timesmachine.nytimes.com/timesmachine/1990/10/09/000690.html?pageNumber=91.
4. "Magnetic Stripe Technology." *IBM100*, IBM, www.ibm.com/ibm/history/ibm100/us/en/icons/magnetic/.
5. "What Is RFID? | The Beginner's Guide to RFID Systems." *Atlas RFID Store*, www.atlasrfidstore.com/rfid-beginners-guide/.
6. "The Rise of EMV and What It Means for the Magnetic Stripe." *NCR*, 5 Apr. 2021, www.ncr.com/blogs/payments/emv-magnetic-stripe.
7. Jones, Ashley. "The History of Contactless Payments." *Global Payments Integrated*, www.globalpaymentsintegrated.com/en-us/blog/2020/09/15/the-history-of-contactless-payments.
8. "EMV Chip Cards." *National Retail Foundation*, nrf.com/emv-chip-cards.
9. Harford, Tim. "The Cold War Spy Technology Which We All Use." *BBC News*, BBC, 20 Aug. 2019, www.bbc.com/news/business-48859331.
10. TELTSCH, KATHLEEN. "Lodge Tells U.N. Symbol Was Gift From Russians; RUSSIANS TAPPED U.S. EMBASSY SEAL." *The New York Times*, The New York Times, 27 May 1960, timesmachine.nytimes.com/timesmachine/1960/05/27/105437259.html?pageNumber=1.

Laser

1. Slade, Margot. "FIBER OPTICS: A SUPER FUTURE FOR A VARIETY OF SPECIALISTS." *The New York Times*, The New York Times, 17 Oct. 1982, www.nytimes.com/1982/10/17/jobs/fiber-optics-a-super-future-for-a-variety-of-specialists.html.
2. "Sea-Me-We 3." *Submarine Cable Map*, www.submarinecablemap.com/#/submarine-cable/seamewe-3.
3. "December 1958: Invention of the Laser." *American Physical Society*, Dec. 2003, www.aps.org/publications/apsnews/200312/history.cfm.
4. "Technology | Laser Diode | Products | Sony Semiconductor Solutions Group." *Sony Semiconductor Solutions Group*, www.sony-semicon.co.jp/e/products/laserdiode/technology.html.
5. "Nick Holonyak." *Lemelson*, lemelson.mit.edu/award-winners/nick-holonyak.
6. "World's Most Powerful Laser Developed by Thales and ELI-NP Achieves Record Power Level of 10 PW." *Thales Group*, 20 Mar. 2019,

www.thalesgroup.com/en/group/journalist/press-release/worlds-most-powerful-laser-developed-thales-and-eli-np-achieves.
7. Martin, Douglas. "Theodore Maiman, 79, Dies; Demonstrated First Laser." *The New York Times*, The New York Times, 11 May 2007, www.nytimes.com/2007/05/11/obituaries/11maiman.html.
8. "Frequently Asked Questions about Lasers." *U.S. Food and Drug Administration*, FDA, 7 Mar. 2018, www.fda.gov/radiation-emitting-products/laser-products-and-instruments/frequently-asked-questions-about-lasers.
9. "An ABC of Light and Lasers." *The New York Times*, The New York Times, 8 Sept. 1963, timesmachine.nytimes.com/timesmachine/1963/09/08/306232912.html?pageNumber=466.
10. "Laser-Beam Phone Calls With Glass Fibers Tried." *The New York Times*, The New York Times, 31 May 1976, timesmachine.nytimes.com/timesmachine/1976/05/31/86394621.html?pageNumber=28.

Traffic light

1. HOGE, JAMES B. "US1251666A - Municipal Traffic-Control System." *United States Patent*, 1 Jan. 1918, patents.google.com/patent/US1251666A/en.
2. ROBBINS, L.H. "AN UNASSUMING ROBOT COMMANDS 'STOP AND GO'; RULER OF THE TRAFFIC SIGNALS." *The New York Times*, The New York Times, 17 Mar. 1929, timesmachine.nytimes.com/timesmachine/1929/03/17/91764816.html?pageNumber=211.
3. "Japan to Greenlight 5G Base Stations on 200,000 Traffic Signals." *Nikkei Asia*, Nikkei Asia, 3 June 2019, asia.nikkei.com/Spotlight/5G-networks/Japan-to-greenlight-5G-base-stations-on-200-000-traffic-signals.
4. Bergal, Jenni. "States and Cities Try Smarter Signals to Reduce Red Lights." *The Pew Charitable Trusts*, 11 Mar. 2015, www.pewtrusts.org/en/research-and-analysis/blogs/stateline/2015/3/11/states-and-cities-try-smarter-signals-to-reduce-red-lights.
5. Willingham, AJ. "Commuters Waste an Average of 54 Hours a Year Stalled in Traffic, Study Says." *CNN*, Cable News Network, 22 Aug. 2019, www.cnn.com/2019/08/22/us/traffic-commute-gridlock-transportation-study-trnd.
6. "2019 Urban Mobility Report." *Mobility Division*, Texas A&M Transportation Institute, 9 Dec. 2019, static.tti.tamu.edu/tti.tamu.edu/documents/mobility-report-2019.pdf.
7. Friedman, Jordan. "These Are the Cities With the Worst Traffic Congestion." *U.S. News & World Report*, U.S. News & World Report, 23 Sept. 2020, www.usnews.com/news/cities/slideshows/cities-with-the-worst-traffic-in-the-world.
8. Eschner, Kat. "A Short History of the Crosswalk." *Smithsonian.com*, Smithsonian Institution, 31 Oct. 2017, www.smithsonianmag.com/smart-news/short-history-crosswalk-180965339/.
9. "Traffic Signals." *WSDOT*, 15 Nov. 2019, wsdot.wa.gov/Operations/Traffic/signals.htm.

10. *Fatality Facts 2019: Passenger Vehicle Occupants.* Insurance Institute for Highway Safety, www.iihs.org/topics/fatality-statistics/detail/passenger-vehicle-occupants.
11. Steele, Matt. "Does Minneapolis Have Too Many Stoplights?" *Streets.mn*, 22 July 2020, streets.mn/2013/12/11/too-many-stoplights/.
12. Wagner, Lise. "1868-2019: A Brief History of Traffic Lights." *Inclusive City Maker*, 12 Mar. 2021, www.inclusivecitymaker.com/1868-2019-a-brief-history-of-traffic-lights/.

Barcode

1. Greenman, Catherine. "There's More Than One Way to Scan a Bar Code." *The New York Times*, The New York Times, 20 Aug. 1998, www.nytimes.com/1998/08/20/technology/there-s-more-than-one-way-to-scan-a-bar-code.html.
2. Weightman, Gavin. "The History of the Bar Code." *Smithsonian.com*, Smithsonian Institution, 23 Sept. 2015, www.smithsonianmag.com/innovation/history-bar-code-180956704/.
3. Fox, Margalit. "N. Joseph Woodland, Inventor of the Bar Code, Dies at 91." *The New York Times*, The New York Times, 13 Dec. 2012, www.nytimes.com/2012/12/13/business/n-joseph-woodland-inventor-of-the-bar-code-dies-at-91.html.
4. Woodland, Norman J., and Bernard Silver. "US2612994A - Classifying Apparatus and Method." *United States Patent*, 7 Oct. 1952, patents.google.com/patent/US2612994A/en.
5. "Barcode 101: Guide to Barcode Symbologies." *GTIN INFO*, 16 Mar. 2021, www.gtin.info/barcode-101/.
6. "40 Years of the Barcode in Our Day-to-Day Life." *GS1UK*, www.gs1uk.org/get-a-barcode/40-years-of-the-barcode-in-the-uk.
7. Roberts, Sam. "George Laurer, Who Developed the Bar Code, Is Dead at 94." *The New York Times*, The New York Times, 11 Dec. 2019, www.nytimes.com/2019/12/11/technology/george-laurer-dead.html.

ATM

1. "A History of ATM Innovation." *NCR*, 20 Jan. 2021, www.ncr.com/blogs/banking/history-atm-innovation.
2. Brocklehurst, Steven. "The Man Who Really Invented the Cash Machine." *BBC News*, BBC, 27 June 2017, www.bbc.com/news/uk-scotland-40416025.
3. Oliveira Davies, Anthony Ivan, and James Goodfellow. "US3905461A - Access-Control Equipment." *United States Patent*, 16 Sept. 1975, patents.google.com/patent/US3905461A/en.
4. "The 2019 Federal Reserve Payments Study." *Board of Governors of the Federal Reserve System*, 6 Jan. 2020, www.federalreserve.gov/paymentsystems/2019-December-The-Federal-Reserve-Payments-Study.htm.
5. "How Diebold Nixdorf Is Leading the Banking Digitalisation Charge in Africa." *Company Report | Business Chief EMEA*, May 2019, www.businesschief.eu/company/how-diebold-nixdorf-leading-banking-digitalisation-charge-africa.

6. Gordon, Colin. "Worldwide Demand for Modern ATM Software and Hardware from NCR Continues to Grow." *NCR*, 21 Dec. 2018, www.ncr.com/company/blogs/financial/worldwide-demand-for-modern-atm-software-and-hardware-from-ncr-continues-to-grow.
7. King, Wayne, "Tellers Work 24-Hour Day, And Never Breathe a Word." *The New York Times*, The New York Times, 14 Mar. 1976, timesmachine.nytimes.com/timesmachine/1976/05/14/355464032.html?pageNumber=29.
8. White, Matt. "1967: First Cash Dispenser." *Guinness World Records*, Guinness World Records, 19 Aug. 2015, www.guinnessworldrecords.com/news/60at60/2015/8/1967-first-cash-dispenser-392981.
9. DAVIES ANTHONY, IVAN OLIVEIRA, and JAMES GOODFELLOW. "Improvements in or Relating to Customer-Operated Dispensing Systems ." *United Kingdom Patent*, 1 July 1970, worldwide.espacenet.com/publicationDetails/biblio?FT=D&date=19700701&DB=EPODOC&locale=en_EP&CC=GB&NR=1197183A&KC=A&ND=4.
10. Appelbaum, Binyamin, and Robert D. Hershey. "Paul A. Volcker, Fed Chairman Who Waged War on Inflation, Is Dead at 92." *The New York Times*, The New York Times, 9 Dec. 2019, www.nytimes.com/2019/12/09/business/paul-a-volcker-dead.html.
11. "The ATM Is 50 Years Old: How a Hole in the Wall Changed the World." *IOL*, 5 July 2017, www.iol.co.za/saturday-star/news/the-atm-is-50-years-old-how-a-hole-in-the-wall-changed-the-world-10141459.
12. Bátiz-Lazo, Bernardo. "A Brief History of the ATM." *The Atlantic*, Atlantic Media Company, 26 Mar. 2015, www.theatlantic.com/technology/archive/2015/03/a-brief-history-of-the-atm/388547/.
13. "It's Been 30 Years of ATMs in South Africa." *Standard Bank Community*, 11 Feb. 2012, community.standardbank.co.za/t5/Community-blog/It-s-been-30-years-of-ATMs-in-South-Africa/ba-p/1801.

Tank

1. "Tiananmen Square Protest Death Toll 'Was 10,000'." *BBC News*, BBC, 23 Dec. 2017, www.bbc.com/news/world-asia-china-42465516.
2. Larson, Caleb. "M103: Why Was the U.S. Army's Last Heavy Tank a Dud?" *The National Interest*, The Center for the National Interest, 1 Apr. 2021, nationalinterest.org/blog/buzz/m103-why-was-us-army%E2%80%99s-last-heavy-tank-dud-181688.
3. "List of Countries by Number of Tanks." *ArmedForces*, armedforces.eu/land_forces/ranking_tanks.
4. Hammer, Joshua. "A New View of the Battle of Gallipoli, One of the Bloodiest Conflicts of World War I." *Smithsonian.com*, Smithsonian Institution, 1 Feb. 2015, www.smithsonianmag.com/history/new-view-battle-gallipoli-one-bloodiest-conflicts-world-war-i-180953975/.
5. Heffer, Simon. *Simon Heffer: Why It's Time to Debunk the Churchill Myth*. New Statesman, 15 Jan. 2015,

www.newstatesman.com/politics/2015/01/simon-heffer-why-it-s-time-debunk-churchill-myth.
6. Harris, Peter. "Winston Churchill: Simply Imperfect." *The National Interest*, The Center for the National Interest, 2 Feb. 2015, nationalinterest.org/feature/winston-churchill-simply-imperfect-12162.
7. Wheelan, Charles. Naked Money: A Revealing Look at Our Financial System. United States, W. W. Norton, 2016, pp. 135-138, www.google.com/books/edition/Naked_Money_A_Revealing_Look_at_Our_Fina/12suCgAAQBAJ?hl=en
8. "Voices of the First World War: Tanks On The Somme." *Imperial War Museums*, www.iwm.org.uk/history/voices-of-the-first-world-war-tanks-on-the-somme.
9. Mitnick, Joshua. "Mighty Merkavas Fail In War Gone Awry: 'Boom, Flames and Smoke'." *Observer*, Observer, 21 Aug. 2006, observer.com/2006/08/mighty-merkavas-fail-in-war-gone-awry-boom-flames-and-smoke/.
10. Seaton, Andrew. "From Caution to Celebration: The NHS at 70." *History of Government*, 5 July 2018, history.blog.gov.uk/2018/07/05/from-caution-to-celebration-the-nhs-at-70/.

Zipper

1. JUDSON, WHITCOMB L. "US504037A - Shoe-Fastening." *United States Patent*, 29 Aug. 1893, patents.google.com/patent/US504037A/en.
2. JUDSON, WHITCOMB L. "US504038A - Clasp Locker or Unlocker for Shoes." *United States Patent*, 29 Aug. 1893, patents.google.com/patent/US504038A/en.
3. SUNDBACK, GIDEON. "US1236783A - Separable Fastener." *United States Patent*, 14 Aug. 1917, patents.google.com/patent/US1236783A/en.
4. "WHO INVENTED THE ZIPPER?" *The New York Times*, The New York Times, 21 Feb. 1947, timesmachine.nytimes.com/timesmachine/1947/02/21/88759515.html?pageNumber=18.
5. DAVIS, JACOB W. "US139121A - Improvement in Fastening Pocket-Openings." *United States Patent*, 20 May 1873, patents.google.com/patent/US139121A/en.
6. "CHANGE OF STRATEGY FOR TALON ZIPPERS." *The New York Times*, The New York Times, 7 Dec. 1981, www.nytimes.com/1981/12/07/business/change-of-strategy-for-talon-zippers.html.
7. "Levis History." *Levi Strauss & Co*, 30 Sept. 2019, www.levistrauss.com/levis-history/.
8. "Look down: Which Side of the Zipper Wars Do Your Jeans Belong to?" *ABC News*, ABC News, 1 Apr. 2019, www.abc.net.au/news/2019-04-01/zipper-manufacturers-are-preparing-for-a-market-battle/10957854.
9. "This Is YKK 2019." *YKK CORPORATION*, Sept. 2019, www.ykk.com/english/shared/pdf/corporate/csr/eco/report/2019/This_is_YKK_2019_all_en.pdf.
10. Shahbandeh, M. "Topic: Denim Market Worldwide." *Statista*, 2 Dec. 2020, www.statista.com/topics/5959/denim-market-worldwide/.

Principles

1. Fleming, John, et al. "New Index Shows Least-, Most-Accepting Countries for Migrants." *Gallup.com*, Gallup, 11 Apr. 2021, news.gallup.com/poll/216377/new-index-shows-least-accepting-countries-migrants.aspx.
2. Kristof, Nicholas, and Sheryl Wudunn. "The Women's Crusade." *The New York Times*, The New York Times, 17 Aug. 2009, www.nytimes.com/2009/08/23/magazine/23Women-t.html.
3. Mishel, Lawrence, and Julia Wolfe. "CEO Compensation Has Grown 940% since 1978: Typical Worker Compensation Has Risen Only 12% during That Time." *Economic Policy Institute*, 14 Aug. 2019, www.epi.org/publication/ceo-compensation-2018/.
4. "U.S. Census Bureau Releases New Educational Attainment Data." *The United States Census Bureau*, 30 Mar. 2020, www.census.gov/newsroom/press-releases/2020/educational-attainment.html.
5. Frenkel, Sheera, et al. "Delay, Deny and Deflect: How Facebook's Leaders Fought Through Crisis." *The New York Times*, The New York Times, 14 Nov. 2018, www.nytimes.com/2018/11/14/technology/facebook-data-russia-election-racism.html.
6. Nicas, Jack, et al. "Censorship, Surveillance and Profits: A Hard Bargain for Apple in China." *The New York Times*, The New York Times, 17 May 2021, www.nytimes.com/2021/05/17/technology/apple-china-censorship-data.html.
7. Friedman, Milton. "A Friedman Doctrine-- The Social Responsibility Of Business Is to Increase Its Profits." *The New York Times*, The New York Times, 13 Sept. 1970, www.nytimes.com/1970/09/13/archives/a-friedman-doctrine-the-social-responsibility-of-business-is-to.html.
8. Dowling, Stephen. "The Cheap Pen That Changed Writing Forever." *BBC Future*, BBC, 29 Oct. 2020, www.bbc.com/future/article/20201028-history-of-the-ballpoint-pen.
9. "In Our Time - The Poor Laws - BBC Sounds." *BBC News*, BBC, 20 Dec. 2018, www.bbc.co.uk/sounds/play/m0001m73.

INDEX

2-Methylbuta-1,3-diene, 196
6-aminopenicillanic acid, 130
1,2,3-trinitroxypropane, 286
2,2,4-trimethylpentane, 191
42-line bible, 41, 43

A

Abbott, Wallace, 160
Abbott Laboratories, 160
Abraj-al-bait, 284
AC. See alternating current
Academy of Motion Picture Arts, 230
Acetaminophen, 247
Acetylcholine, 259
acorn, 199
Acrylic fiber, 193
Actinophor, 296
Adenine, 227
adenosine, 227
Adidas, 89
Adolf Gygi, 277
adrenal glands, 227
Adriatic Sea, 157
Afghanistan, 223
AFP. See Agence France-Presse
Africa, 22, 25, 58, 62, 78, 295–96
Africans, 90
age
    bronze, 66
    ice, 32
    median, 265
Agence France-Presse (AFP), 154
agent
    anti-knocking, 216
    cleaning, 244
Agent Orange, 289

Agni, 26
agriculture, 29–32, 68, 70–71, 224, 226, 289–90, 338
airbus, 220
airbus consortium, 220
air conditioning, 218
aircrafts, 137, 220, 307
airfoil, 220
air-infused foam, 89
airlines, 220–21
airline tickets, 310
air molecules, 220
airplanes, 28–29, 107, 110, 191–92, 219–21, 270, 272
air pollution, 112
airports, small, 308
air pressure, 221
air traffic control, 306
Akira Yoshino, 203
AkzoNobel, 224
Alabama, 289
Alaca Höyük, 63
Alan Turing, 163
Alaska, 95
Alcántara, 261
Alcántara Bridge, 261–62
alcohol, 227, 246, 255–57
    isopropyl, 256
Aldrin, Edwin, 235
Aleksandr Prokhorov, 313
Aleppo Codex, 79
Alexander Fleming, 130
Alexander Graham Bell, 141–42, 231
Aliph, 35
alkanes, 190
    cyclic, 190
alkylbenzene sulfonates, 244
Älmhult, 74
Alois Dallmayr, 226

Aloysius Lilius, 49
alphabet, 34, 44
Altamira Cave, 223
alternating current (AC), 116, 136, 205–6, 298
alumina, 276
aluminium, 65, 71
aluminium cookware, 71
Aluminium Hydroxide, 245
Aluminium Oxide, 276
AlvordLake Bridge, 277
Amazon rainforest, 76
Amazon river, 266
American aircraft manufacturers, 220
American cities, 210
American civil war, 140
American colonies, 79, 259
American Express, 310
American independence, 44
American inventors, 15
American Revolution, 226
Americans, native, 29, 60, 236
American supermarkets, 318
Americas, 58, 60, 62, 103, 106, 124
Ameritech, 240
Amish, 295
Ammonia, 121–22, 216–17
Ammonium Chloride, 202
Ammonium Nitrate, 287
Ammonium Nitrate/Fuel Oil. See ANFO
amœbic dysentery, 134
amorphous solid, 157
Amoxicillin, 130
Ampicillin, 130
Amplitude Modulation, 136
AMPS, 241
Amsterdam, 153
Amur River, 132
analgesic, 246–48
analog, 241
Android phone, 242

Andrus, Jules, 167
anesthesia, 247
anesthetic, 12, 159–61, 217
    local, 160, 247
ANFO (Ammonium Nitrate/Fuel Oil), 287
Anhui, 278
Anhui Conch, 278
animal domestication, 31
animalhusbandry, 31
animal kingdom, 26
animals, pack, 31
animal skins, 23
animal triglycerides, 244
anode, 203, 298–99
antenna, 275
Anthology, 43
Anthony, David, 27
Anthony, Susan, 270
Anthropology, 64, 87
antibiotic resistance, 131
antibiotics, 129–31, 133–34, 331
antioxidants, 227
antiquity, 28, 64, 74, 82, 184
APCs. See ArmouredPersonnel Carrier
aperture, 45, 250
apes, great, 24
Apollo, 235
apparatus, 112, 141–42, 152
    photosensitive, 319
Apple, 311, 336
Apple iphone, 240
aqueduct, 261
Arabia, 31
archaeological site, 67
Archbishop's chair, 73
Arch bridges, 262, 265
Arch dam, 265
archeological artifacts, 157
Arctic Oil Company Works, 277
Areni-1 cave, 88

Argentina, 81
Arkadiko Bridge, 261
Arkwright, Richard, 149
armature, 142
ArmouredPersonnel Carrier (APCs), 325
Armstrong, Neil, 234
aromatics, 190
Arsenic, 289–90
AsahiKasei, 203
Al-Ashraf Umar, 173
Asia, 259, 265–66, 268, 314, 316
Asians, 333
askuttu, 198
Aspdin, Joseph, 277
asphalt, 94
AsphaltMagazine, 94
Aspirin, 247
Assam, 225–26
assault rifle, 178
assembly line, 10
assets, 52
    physical, 35
assistants, 145
    human, 150
AssociatedPress, 139, 154
association ofspectacle-frame makers, 185
Assyria, 198–99
Assyriologist, 199
AstraZeneca, 159
astronomy, 250, 307
Atalla, Mohamed, 166
Athens, 42, 73, 88
Atlanta, 127, 288, 308, 323
atlantic coast, 296
Atlantic Ocean, 268–69
ATM, 322–23
ATM cards, 311, 322
ATM machine, 322
atmosphere, 56, 104, 110, 117
atmospheric engine, 103

atomic nuclei, 275
atoms, 55, 305
AT&T, 143, 240
Audio-Technica, 233
Augenheilkunde, 184
Augsburg, 281
Augsburg Town Hall, 281
aurochs, wild, 31
Austin, 265
Australia, 32, 65, 92, 95, 179
australians, 29
Austria, 293
Austrian, 199
Austronesian islands, 60
Austronesians, 60, 62
authoritarian regime, 155
autobahns, 112
automobiles, 106–13, 197, 202–3, 270, 272, 329
autonomous vehicles, 308
awards Oscars, 230
axle, 27–28, 31
Azotobacter, 121

B

Baal, 34
Babbage, Charles, 163
babies, 127
Babylonian Calendar, 49
Babylonian Numerals, 37
babylonian period, 58
Babylonians, 34
Babylonian society, 33
Babylonian system, 37
BAC (Blood Alcohol Content), 257
Bacille Calmette-Guerin (BCG), 127
bacillus, 129
Bacon, Roger, 11
bacteria, 127, 130–31, 161, 186, 211
    nitrogen fixing, 121
bacterial cells, 131

bacterial gastroenteritis, 134
bacterium, 129, 133
BadischeAnilin undSodafabrik, 121
Baekeland, Leo, 193
Baghdad, 37, 58
Baird, John, 140, 213
Bakelite, 193
Bakhshali, 36
Bakhshalimanuscript, 36
Balcom, Homer, 283
Balki, 28
ball, cannon, 108
Ballbearings, 28
ballpoint pen, 81–83, 337
Baltimore, 105
Bandar Abbas, 306
bandgap, 157
Bangkok, 262
Bangladesh, 61
banking, 164
    mobile, 170
Banqiao Dam Failure, 266
Barcelona, 77, 316
barcode, 318, 320
    circular, 319
    standard, 319
barcode scanners, 314, 320
Bardeen, John, 166
Barnes, Thomas, 322
Barron, Robert, 199
Bascule bridges, 262
base-2, 38
Basel, 248, 289
BASF, 121–22
Bata, 89
Bathgate, 189
Bathgate Chemical Works, 189
batteries, 10, 104, 110, 151, 201–3
    nickel cadmium, 202
battery cars, 14

Bausch Health, 185
BayArea Rapid Transitsystem, 310
Bayer, 196, 246–47
BBC, 82, 295, 325
BBN (Bolt, Beranek and Newman), 168, 170
BCG. See Bacille Calmette-Guerin
beam bridges, 262
Beecham, 130
beedi, 260
Beijing, 269, 278, 308
Beirut, 78
Belarusian, 257
Belfast, 271
Belgium, 43, 133, 202
bell labs, 47, 166–67, 313
belly hill, 78
belts, chastity, 200
Benefon, 304
Benefon Esc, 304
Bengal Famine, 120
Bengalfamine of1943, 290, 324
Bengaluru, 61
Benz, Bertha, 112
Benz, Carl, 112
benzene, 191, 196, 244, 259
Benzylpenicillin, 130
Bergen, 129
Berkeley, 90
Berlin, 109, 116, 206, 214, 296
Bern, 68
Bernhardt, Sarah, 119
Bertha Benzmemorialroute, 112
Bessemer process, 64
betel leaf, 260
betel quid, 260
Bethlehem, 62
Bhagavad Gita, 180
Bible, 34, 80
Bibliothèque Nationale, 43
bicycles, 10, 12, 197, 200, 270–72

Bifenthrin, 289
binary data, 142
binary mode, 317
bipedalism, 26
BIPM (Bureau of Weights and Measures), 54
birch bark, 79
birds, 289
Birmingham, 247, 261, 270
Birmingham University, 92
birome, 81
BisphenolA (BPA), 194
bitcoins, 11, 14, 51
Bitumen of Judea, 208
Blackberry, 242
Black sea, 28
Blacksea, 73
black soot, 90
Blaisdell, George, 293
blake welts, 89
bleach, 132–33
Blitzkrieg, 325
Blueprints, 223
Blu-ray, 314
bobbin thread, 124
Bock, Walter, 196
Boéchat, Paul, 68
Boeing, 220–21, 238
Bogibeel Bridge, 263
Bogotá, 95, 317
Bökerwas, 67
Bollywood, 258
Boltzmann Constant, 100
Bombay, 92, 95, 140, 323
bombs, 26, 130, 180
    atomic, 177, 181
Bond, Walter, 167
bones, charred, 83
books, 9–13, 15–16, 34–35, 40–41, 43–44, 46–47, 83, 331–33, 337–38
boots, 88, 195, 247
Borin, Ernst, 130

Bosch, Carl, 121
Boston, 141, 143, 168, 170, 271, 275
Boston CityHall, 278
Boston SchoolforDeafMutes, 142
Boston Tea Party, 226
Botaisettlements, 31
Botswana, 155
BPA. See BisphenolA
Bradford, 293
Bragg, Lawrence, 187
Brahma Gupta, 36
Brahmaputra River, 262–63
bramah, 199
Bramah, Joseph, 199
Brattain, Walter, 166
Braun, 234
Brazil, 65, 215, 227, 259, 265
breast cancer, 274
Breast Tax, 23
brickphone, 240
bridges, 261–63, 265, 276–78
Bridgestone, 197
Brisbane, 58
Bristol, 260
Britain, 49, 61, 324
British American Tobacco, 260
BritishAmerican Tobacco, 260
british colonialism, 173
british economy, 324
British empire, 59, 61–62
british gas, 103
british government, 59
British India, 36, 105, 140
British Library, 34, 43
British Museum, 41, 49, 67, 72, 156–57
british troops, 296
broadcast, public, 215
Bronocice Pot, 27
Bronze knives, 66
bronze tachi, 67

Brooklyn, 263
Brooklyn Bridge, 261, 263
Brooklyn theatre, 119
Brown, Tom, 275
Brushless DC motors, 206
Brussels, 133, 202
brutalism, 278
Bryson, Bill, 216
Buddhism, 79
building
    public, 276, 279
    world's tallest, 283
Bukhara, 208
bulbs
    compact fluorescent, 118
    incandescent, 297
bullet, 108, 130, 178, 273, 275
    pistol, 179
Bulletin of the Atomic Scientists, 180
bupivacaine, 247
Bureau of Mines, 92
Burj Khalifa, 77, 281, 284
Burkina Faso, 295
Burma, 226
Burmah Shell oil company, 92
Burshane, 92
Burundi, 295
buses, public, 112
butadiene, 196
butane, 91, 293
Button, 208
buttress dam, 265
Byerly, David, 244

C

C-4, 287
cables, 262–63, 299, 314
    coaxial, 312
    diplomatic, 325
    inter-continental undersea, 313

cadmium, 202
Caesar, Augustus, 85
Caesar, Julius, 49
caesium-133, 47
Caffeine, 227
CaiLun, 40
Cairo, 22, 173
calcite, 223
calcium carbonate, 223, 245
calcium chloride, 133
calcium hydroxide, 223
calcium hypochlorite, 133–34
Calcium Oxide, 156
Calcium Sulfate Dihydrate, 276
calculators, electronic, 312
calculus, 46
Calcutta, 140, 153
calendar, 48–50
Calicut, 61
California, 172, 217, 220, 304, 310
California high speedrail, 107
Caliga, 88
call-sign, 136
calories, 24, 26
calotype, 208
Camels, 31
camera lenses, 158
cameras, 208–9, 229
Canada, 55, 238, 287, 296, 334
Canterbury Cathedral, 158
Canterbury Tales, 68
Cantilever bridges, 262
caoutchouc, 196
capacitor, 139
    electric, 142
Carbaryl, 289
Carbohydrate, 226
carbon, 64, 83, 130, 259, 331
carbon chains, 192
carbon granules, 233

card, smart, 311, 320
CardinalHugh ofProvence, 183
Cardullo, Mario, 311
Carl Auervon Welsbach, 293, 296
Carlstedt, Sven, 302
Carl Zeiss, 187
Carpathia, 136
Carribean, 259
Carrier, 217–18
Carrier, Willis, 217
carrier pigeons, 154
cars, 103, 106, 109, 111–12, 197, 200, 221
    electric, 104, 106, 110, 201, 203, 205–6
cartography, 59
cartridges, 178
Cartwright, Edmund, 146
cash machine. See ATM machine
cassettes, 233–34
castiron, 192
cast iron skillet, 192
Catch-22, 161
catechins, 227
cathedra, 73
cathedrals, 80, 158
cathode raytube, 214–15
Catholic church, 79
cats, 31, 151
cattles, 31, 316
cave paintings, 223
caves, 23, 87
CBS, 215
CCD, 209
CD, 233, 314
CDC, 128, 134, 211, 257
Celilo station, 116
cell, 201, 203
    photoconductive, 214
cell phones, 158, 209, 240
Cellulose, 193
cell walls, 24, 130

celsius, 99
Celsius, Anders, 99
cement, 276–78, 340
cement manufacturing, 278
Censorship, 336
Central America, 55, 268
Central Asia, 118
Central Park, 271, 283
century, first, 38, 40, 88
CEOs, 335
cephalosporins, 131
Cepheid variables, 250
cereal grain, 32
cereals, 30, 63
cerebral cortex, 160–61
cerebral hemispheres, 161
cerebral palsy, 290
Cerium, 293
CERN, 169
Cesium, 46
Cesium-133, 55
CFCs, 216–17
CGPM (General Conference on Weights and Measures), 55
CGPM conference, 55
Ch'a Ching, 226
Chad, 14, 76, 295
Chain Home, 307–8, 311
chains, hydrocarbon, 191
chairs, 72–73, 88, 192
chalice, 156
Chalon-sur-Saône, 208
ChancellorofExchequer, 324
Chandra observatory, 251
change, climate, 31, 106, 236, 339–40
channels, sodium, 160
charcoal, 179, 223
Charles James Napier, 61
Chaucer, Geoffrey, 68
checkout counter, 314
chemical bonds, 121

chemical elements, 203
chemical equation, 179
chemicals, dangerous, 194
chemistry, 83
Chengdu, 66, 226
Chennai, 116, 265
Chevron, 190
Chicago, 133–34, 283–84, 302, 304, 323, 328
ChicagoAirport, 310
chicken, 151
chicken cholera, 127
chicken domestication, 31
Chief Joseph, 266
childbirth, 129, 246–48
child mortality, 131
chimpanzees, 26
China, 31, 65–66, 105–7, 172, 226–27, 259, 265–66, 277–78, 325–26
China NationalBuildingMaterial, 278
chinese, 39–40, 154
chinese nationals, 309
chinese state, 278
chip cards, 310
chips, 167, 310
    micro, 166
chlorate ofpotash, 292
chlorination of water, 134
chlorine, 133–34
chlorine dioxide, 134
chloroform, 159
chloroprocaine, 247
chlorpyrifos, 289
cholera, 133–34
Cholera pandemic, 77, 133
Chomsky, Noam, 26
Christiaan Huygens, 11
Christianity, 79
Christian Ørsted, 205
Christmas, 231
Christmas Eve, 232
chromium, 64, 224

chromosomes, 188
Chronometer, 47, 59, 135, 304, 332
Chubb detectorlock, 199
churches, 80, 278, 333
Churchill, 325
church towers, 47
cigar, 260
cigarettes, 90, 258, 260
Cincinnati, 323
Cincinnati's Children'sHospital, 127
cinema, 15, 209, 230, 233, 332–33
Cinématographe, 229
cipher, 36
ciprofloxacin, 131
circumference, 225, 292
Cishan, 31
CityofBlades, 67
clay, 83
Clermont, 62
Cleveland, 57, 224, 316
clock, 15, 45, 47, 55, 167
    atomic, 305
    mechanical, 47
    pendulum, 47
    quartz, 47
    shadow, 46
clock cycle, 162
clock-maker, 45
clofazimine, 129
clothes, 13, 23, 122–24, 192, 194
clothing, 22–23, 26, 32, 35, 41
cloud computing, 218
CMOS sensors, 209
CNBM, 278
coal, 90–91, 103–5, 116, 157, 189
CoalBrookdale, 261
coal mines, 103, 105
coal oil, 189
coasts, 73, 118, 307
Cobalt Oxide, 158, 203

cocaine oropiates, 259
Coccleus, Johannes, 45
Cochliobolus miyabeanus, 290
cockroaches, 339
code, binary, 138–39, 142
codeine, 248
code monkeys, 336
code ofHammurabi, 33
code ofjustice, 33
codex, 43–44, 79
codexsinaiticus, 43
coffee, 225–27
coils, 206, 233
coke oven gas, 109
Coleman, William, 296
Coleman lanterns, 296
Colgate-Palmolive, 245
Colgate Toothpaste, 245
Collins, Arnold, 197
Cologne, 42, 109, 232, 246–47, 323
Colombia, 95, 227, 317
Colombo, Joe, 74
colonialism, 62, 179
Colorado, 285
Colorado river, 265
Colosseum, 38
Colt, Samuel, 178
ColtPaterson, 178
Columbia Bicycles, 271
Columbia records, 231–32
Columbia River, 264
Columbia University, 94, 274
Columbus, Christopher, 58, 171, 236, 323
Comic opera-house, 119
communications satellites, 238
communist china, 105
Compagnie Universelle, 268
companies
    dutch, 245
    livery, 184

railroad, 296
compass, 171–73
compound dyes, 83
compoundmicroscopes, 186
compounds, 161, 191
    fat-soluble, 161
Computable Numbers, 163
Computed tomography, 274
computer algorithms, 274
computer engineer, 14
computer memory, 166, 203
computer programming, 146
computers, 114–16, 128, 138–39, 143, 162–64, 167–69, 205–6, 242
    personal, 166
Concrete pavement, 278
condenser, 13, 102, 285, 329
condensing unit, 218
conductor, electrical, 116, 138
congenital anomalies, 290
Congo, 37
Congo rubber plant, 196
connective tissue, 24
consciousness, 338
conservation ofmomentum, 235
constant, fundamental, 100
contactlens, 184
contact lenses, 184
containers, 103, 268
    packaging, 192
    refrigerated, 218
Continental, 197
control, flood, 266, 269
Cooke, William, 139
cooking, 24–25, 90, 108
cookware, 71
Cooper, Martin, 334
Copenhagen, 74, 292
copper, 64, 82, 301, 316
copper(II) acetoarsenite, 289
copperbase, 71

copper pigments, 44
copper smelting, 64
copper wire, 312
copperwires, 138
Corn, 32
Cornelis Drebbel, 186
Corning glass, 71
coronavirus pandemic, 131, 255, 290
corporate lobbyists, 335
corrective eyeglass, 238
Correio-Mordo Reino, 86
cotton, 148, 151–52, 193–94, 296
cotton fibers, 144
cotton gin, 12–13, 15, 144, 149–52, 337
cotton lint, 151
countries, developing, 112, 116, 143, 218, 223, 272
countrymile, 12
countryside, 212
cousins, 71, 290
Coventry, 270
Coy, George, 143
Creighton, 224
Cristallo glass, 157
Crompton, Samuel, 149
crusades, 80
crystals, 187–88
CT Scans, 275
Cuba, 220, 259
Cullen, William, 217
cultural heritage, 80, 223
cultural norms, 77
Cumbria, 83
Cummings, Alexander, 76
Cuneiform, 34
cuneiform tablets, 118
Cyclopentane, 191
cyclotrimethylenetrinitramine, 287
cylinder, 102, 231
    wax, 232
Cyprus, 66

Czech republic, 307

D

daggers, 66
Daguerre, Louis, 208
daguerreotypes, 208
DaimlerReitwagen, 271
Dallas, 322
Damadian, Raymond, 275
Dame cathedral, 80
dams, 264–66, 278
    hydroelectric gravity, 264
dapsone, 129–30
Darien Gap, 95
Darlington, 105
DARPA, 169
Darwin, 26, 187–88
Davenport, Thomas, 205
David Curle Smith, 92
Davis, Jacob, 328
Dawkins, Richard, 219
Dawon Kahng, 166
Dayton, 323
DDT, 289
dead reckoning, 58, 171
Dead Sea, 30
debtor's prison, 196
decimal numbers, 38
Delft, 74
democracy, 107, 121, 143
denim market, 329
Denisovan cave, 22
Denmark, 74, 205, 212, 295
Denmark, Leila, 127
dental procedures, 160
depolarizer, 202
Deruta, 70
detergent, 16
    laundry, 245
Detroit, 111, 224, 316, 323

Deutsche Welle, 269
development, human, 41
devices
    electronic, 298
    household, 203
Devos, 85
diabetes, 255
DiacetylMorphine, 246
diamond, 157
Diamond, Jared, 31–32
diaphragm, 142, 231, 233
diatomaceous earth, 285
dichloro-diphenyl-trichloroethane, 289
Diclofenac, 248
Diebold Nixdorf, 323
diesel, 110, 190–91, 287
Diesel, Rudolf, 109
diesel engines, 62, 109–10, 178
diesel fuel, 109, 190
Digital cameras, 209
digital computers, 167
digital devices, 233
digits, 37, 319–20
diode, 298
diopters, 184–85
diphtheria, 127
disabilities, 131
diseases, viral, 134
disk drive, 14, 43
diversity, 335
DNA, 121, 187–88, 227
DNS (Domain Naming System), 169–70
Dobereiner'slamp, 293
Docutel, 322–23
dogs, 31, 127
Donald, Ian, 275
dopamine, 227, 259
doppler effect, 307
dopplereffect, 250
double coincidence of wants, 52

Dresden, 127, 187, 293
Dresden Codex, 14, 38
dried dung, 90
drinks
    standard, 256
    standard US, 256
driver'slicense, 209
drugs
    anti-inflammatory, 247
    synthetic, 248
Dubai, 284
dugout canoes, 60
Duisburg, 58
Dupont, 71, 193, 197
Duravit, 77
Durban, 22, 73
Dur-Sharrukin, 199
Dusseldorf, 58, 67
Duytslandt, 153
DVD, 241, 314
DVD players, 314
dynamic microphones, 233
dynamite, 285–86
DynaTec 8000x, 240

E

EAN-13 codes, 320
earth, 49–50, 63–65, 110, 137–38, 172, 237, 339
Eastern Roman Empire, 85
Eastman Kodak, 208, 228
East river, 283
eccentricity, 172
ECG, 275
echolocation, 275, 307
echoplanarimaging, 275
École Centrale Paris, 283
Economists, 338
economy, 37, 44, 50, 52–53, 338–39
    global, 268
Edinburgh, 44, 68, 86, 189

Edison, Thomas, 15, 116–17, 151, 229, 231–33, 332–33
Edison laboratory, 228
Édouard Michelin, 197
EduardMeier, 89
education, 80, 181, 257, 272
    public, 181
Egypt, 34–35, 40, 43, 66, 78, 82
Egyptian blue, 223
Egyptian locks, 199
Egyptian Museum, 67
Egyptian NationalLibrary, 173
Egyptians, 37, 60
Egyptian tomb, 82
Ehlrich, Paul, 339
Ehrich & Graetz, 296
Eiffel Tower, 187, 283
Eindhoven, 232
Einstein, Albert, 46, 334
Einthoven, Willem, 275
election, presidential, 107
electrical grids, 164
electricity, 59, 90–91, 104, 107, 113–16, 125, 203, 295, 333–34
    hydro, 264–65
electricity generation plants, 26, 104
electricity generators, 116
electricity grid, 128
electricity mains, 298
electric motors, 204–6
electric projects, 266
electric trams, 116
electrification, rural, 116, 119
Electrolux, 302
electrolysis, 203
electrolyte, 133, 201–2
    alkaline, 202
    dry, 202
electromagnetic equations. See Maxwell's equations
electromagnetic interference, 313
electromagnetic waves, 136, 187
electromagnetism, 205, 334

electronic appliances, 158
electronics, 192, 194
Electron Microscope, 101, 187–88
electrons, 136, 187, 313
electrum, 53
elephants, 188, 220, 235
elevators, 167, 205, 279–81, 283
ELI-NP, 314
Elizabeth I, 76
Elsener, Karl, 68
EMI, 124, 275
Êmir Bukharskiĭ, 208
Emitron, 215
Empire State Building, 65, 281, 283
EMV, 310–11
Endosulfan, 289–90
energy
    chemical, 114
    electrical, 233
    mechanical, 116
energy efficiency, 41
energy inefficiency, 137
energy levels, 46
energy source, 313
engines
    analytical, 163
    external combustion, 103–4, 109
    four-stroke, 112
    gas turbine, 62
England, 85–86, 135–36, 142–43, 145–46, 149, 184, 199, 260–61, 322
English Channel, 70, 277
English language, 62, 85
ENIAC, 38, 114, 162, 164, 167
Enkomi, 66–67
EnlightenmentNow, 118, 126, 128, 240
entropy, 100–101, 172
Entscheidungsproblem, 163
environmentalists, 339
EnvironmentalProtectionAgency, 217
enzymes, 121, 131, 247

EPDM (Ethylene Propylene Diene Monomer), 197
epidural injection, 246–47
Epiduralinjection, 247
Epistola, 173
equator, 55, 95, 106
equinoxes, 49–50
Equitable life building, 281
Erie Canal, 269
Eritrea, 76
Essen, Louis, 47
EssilorLuxottica, 185
ethanol, 255–56
ether, 160
Ethiopia, 295
Ethiopian plateau, 226
Ethylene Propylene Diene Monomer. See EPDM
ethylene vinylacetate, 89
Euclidean, 304
Eugen Langen, 109
EUMETSAT, 239
Euro, 53
Europay, 310
Europe, 37–38, 53–55, 79, 95, 184, 268, 314, 316–17
European Union, 61, 217
Euro zone, 206
EVA, 89
evolution, 22
explosion
    controlled, 178
    exothermic, 179
explosives, 122, 285, 287, 296
ExxonMobil, 190
Exxon Research, 203
eye glasses, 13, 157–58, 183–85
eyesight, 183

F
Faber, Kasper, 83
Faber-Castell, 82
Facebook, 154, 163, 336
Fahreinheitscale, 99
Fahrenheit, 99
Fahrenheit, Daniel, 99
Falklands war, 61
families, 121, 294, 338
    pea, 121
famines, 120, 218, 282, 290, 339
FAO, 227
Faraday, Michael, 205, 334
Farben, 193
farm jobs, 338
farsightedness, 185
Fasciculus Temporum, 42
fat, 24, 226, 244
fatal accidents, 113
fatal dose, 256
fatalityrate, 272
fatherofcapitalism, 52
fatherofgerm theory, 212
Fatimid Caliph, 82
FDA, 211
FDR (Franklin Roosevelt), 116, 119, 128, 180, 283
FederalReserve, 323
Fedex, 86
fentanyl, 247
Ferguson, 273–75
fermentation, 212, 227
Ferrocerium, 293
Fertile Crescent, 31
fertilizer, 65, 179, 282, 284, 287
    modern, 121
festivals, religious, 80
fetus, 275
f-gas'regulation, 217
Fiber optics, 312
fibers, 152
    natural, 148, 193–94
    optical, 158, 313
    textile, 158
fibres, optical, 115

filament, tantalum, 298
Finland, 241, 304
firearm, 68, 177–79, 286
fishing nets, 218
Fishman, Jack, 247
Fitzwater, Marlin, 318
flashpoint, 190
flint, 293
flint knives, 66
Florence, 222
Florey, Howard, 130
Florida, 289
fluorescent screen, 188
Fluoroquinolones, 131
flush toilet, 76
flying shuttle, 10, 12, 16, 144–46, 163
FM, 136
FM radio, 136
foci, 172, 185
football championship game, 215
Ford, 201
Ford, Henry, 53, 111, 151, 334
Ford, John, 224
FordMotorCompany, 111
Foreign Liquidations Commission, 67
forest, 76, 291, 298
formic acid, 256
Fort Rock, 87
FortRock Cave, 87
fossil fuels, 114
    dense, 109
fountain pen, 44, 81–83
four-stroke, 109
Foxnews, 215
franc, 229
France, 55, 95, 184–85, 202, 208, 270–71, 277
Francisco, San, 44, 87, 263, 277, 285
Francis Joseph Bassett, 148
Frankenstein Knickerbocker, 317
Frankfurt, 43

Franklin, Benjamin, 86
Franklin, James, 155
Franklin, Rosalind, 187
Franklin Roosevelt. See FDR
Frank Seiberling, 197
freedom, 137, 155, 325
freedom movements, 326
freezers, 218
French Academy, 55
French banks, 310
French Parliament, 55
Fresnoy-le-Grand, 71
fridge, 217
Friedrich Bayer, 247
frontal cortex, 160
Fry, Stephen, 293
Fuchs, Andreas, 198
fuel oil, 287
fuels, solid, 90, 92
Fujisawa, 203
fungus, 130, 290
furniture, 72–73, 88, 193, 338

G
Gaggenau Hausgeräte, 70
galaxies, 250
Galileo, 47, 99, 249, 304
Gallipoli, 324
GalliumArsenide, 313
gallium nitride, 314
Galvani, Luigi, 115
Galveston, 287
Galvin, Joseph, 240
gamma rays, 251
Gamma ray space telescopes, 251
Gandhi. See Mahatma Gandhi
Gansu province, 66
garderobe, 76
Garmin, 304
gas

444

greenhouse, 278
inflammable, 109
gas bill, 115
gas-burners, 280
gas industry, 258
gas mantle, 296
gasoline, 109, 190–91, 216, 235
gasoline engines, 109, 191
gas pipeline infrastructure, 92
gas pipelines, 91
gas stoves, 91–92
gas turbine powering, 108
Gates, Bill, 341
Gavà, 77
Gazprom, 284
GDP (Gross Domestic Product), 190, 257
GDP, annual, 240
Geigy, 289
General Conference on Weights and Measures. See CGPM
Geneva, 146, 163, 169, 185, 197
Genoa, 172
Genuine SwissArmyKnife, 68
Geography, 58
geomancy, 172
George Safford Parker, 82
George Washington, 263
Georg Horn, 77
Georgia, 23
Geostationary OperationalEnvironmentalSatellites, 239
geosynchronous orbit, 137, 238
German immigrants, 334
Germany, 42–44, 67, 89, 109, 153, 187, 196–97, 211, 296
germ theoryofdiseases, 127, 158, 186, 251
Gestapo, 324
Glasgow, 102, 134, 213
Glasgow University, 275
glass, 100, 117, 156–58, 185, 194
GlaxoSmithKline, 130
Glen, John, 159
Glines Canyon Dam, 266

globalization, 124
GlobalNavigation System Satellite, 305
Global Positioning System, 58–59, 173, 236, 239, 303–5
Global Temperatures, 110
Global Trade Item Number, 320
Glock, 178
Glocke, 47
GNSS, 305
Göbekli Tepe, 78
Goddard, Robert, 235
GOES-16, 239
GOES-17, 239
gold, 32, 51, 53, 65, 67
Golden Gate Bridge, 263
Golden Gate Park, 277
Goldman Sachs, 9
Goodenough, John, 203
Goodfellow, James, 322
Goodyear, 89, 197
Goodyear, Charles, 88, 192, 196
Goodyearwelt, 88
Goodyearwelt process, 88
Google, 9, 242, 311
Google books, 43
Google Maps, 59, 304
Google News, 163
Google Pixel 4a, 240
Gordian knot, 73
Gordion, 73
Gordios, 73
Gorges Dam, 264, 266, 278
Three, 264
Gothic architecture, 80
Gottlieb Daimler, 271
GPCR, 248
GPS satellites, 236
Graetz, Max, 296
grain mills, 206
grains, 263, 282
gram negative bacteria, 130

Gramophone, 232
Grand Canal, 269
Grand Coulee Dam, 264–66, 278
Gran Telescopio Canarias, 250
graphite, 83, 157, 203
gravestone, 28
Gray, Elisha, 142
Grazia Deruta, 70
grazing angle telescope, 251
Great Britain, 54
Great Buddhist Priests' Zen Teachings, 43
Greatleap forward, 325
Great Train Robbery, 229
Greco-Roman balanos lock, 199
Greece, 32, 34, 49–50, 53, 73, 76, 88
Greeland, 58
Green, Paris, 289
Greenwich, 59, 304
Greenwich Meridian, 239
Gregorian Calendar, 49
Griswold, 71
Griswold, Matthew, 70
grocery stores, 318, 335
Grohe, 77
Grosfillex, Jean, 73
Gross Domestic Product. See GDP
groups
    methyl, 196, 256
    nitrate, 286
GS1, 320
GSM, 241
GTC, 250–51
GTIN, 320
Guanine, 227
Guardian, 124, 154, 278
Gucci, 123
Guinea, 60
Guinness WorldRecords, 322
GulfofSuez, 268
guncotton, 296

gunpowder, 177, 179, 285
guns, 32, 177, 179
Gustave Eiffel, 283
Gutenberg, Johannes, 41, 43–44, 83
Gutenberg Bible. See 42-line bible
Gutenberg's press, 43
gynaecology, 275
gypsum, 276

H

Haber-Bosch process, 122
Haemophilus influenzae bacteria, 128
Haiti, 133
Halifax, 287, 296
Hammurabi, 34
hand lotions, 256
Hanover, 197
Hansen's disease, 129
Hansgrohe, 77
happiness, 160
    human, 337
Hargreaves, James, 149
Harley-Davidson, 272
Harrington, John, 76
Harris, Benjamin, 154
Harrison, John, 47, 59
Hart, Michael, 9
Havana, 220
Havas, 154
Hawaii, 60, 239, 250, 284
HB, 83
HDPE, 193
health care, universal, 106
health outcomes, 76
heart failure, 274
hearts, 84
heat, 24, 100–101, 104, 110, 157
heating coil, 166
Hebrew, 35
heddles, 145

Heinrich Böker, 67
Heinrich Hertz, 136, 307
helical column, 279
hemolytic uremic syndrome, 212
Henan Province, 266
Henckels, Peter, 67
Henlein, Peter, 45–46
Henry, Joseph, 204
Henry Bessemer's process, 282, 322, 331
hepatitis, 127, 134
Hephaestus, 26
heroin, 246, 248, 259
Hevea brasiliensis, 196
Hezbollah, 326
HFC-134a, 217
Hicky's Bengal Gazette, 153
Hidaka, 67
Hieroglyphs, 34
high school students, 272
highways, 95, 276, 278, 287
    interstate, 112
Hikkim, 86
Hilleman, Maurice, 127
Himachal Pradesh, 86
Himawari-8, 239
Hindu, 120
Hinduism, 79
Hingson, Robert, 247
Hipparchus, 50
Hiroshima, 180
Hiroshima bomb, 181
HIV/AIDS, 255
HMSArgus, 62
Hofmann, Fritz, 196
Hoge, James, 316
Hokkaido, 67
Holcim, 277
Holland, 226
Hollywood, 258
Holonyak, Nick, 314

home insurance building, 283
homo sapiens, 11, 22–23, 25, 78, 182
Honda, 272
Hong Kong, 309
Honjō Masamune, 67
hookah, 260
Hooke, Robert, 186
Hookless fastenercompany, 328
Hoover Dam, 265
Horologium Oscillatorium, 11
horses, 27, 29, 31, 111, 137–38, 142
horticulture, 277
hospitals, 211, 276
Hotel, Pierre, 243
hour clocks, 47
household appliances, 115, 167, 205
household refrigerators, 217
households, farm, 91
household washers, 300
houses, public coffee, 226
Houston, 289
Howe, Elias, 124
Hoya vision, 185
HSBC bank, 323
HTC dream, 242
Huai River basin, upper, 266
Hubble's constant, 238
Hubble space telescope, 238, 250
Hudson river, 283
HughesAircraft Company, 313
human art, 156
human beings, 11, 24
human burial, earliest, 78
human capital, 35, 52
human civilization, 29, 95, 101, 257, 276
human eyes, 251, 274
human hair, 251
human health, 77, 331
human impact, 224
human lifetime, 237

447

human productivity, 12, 143, 150, 282, 337
human societies, 23, 59, 62, 180
human suffering, 161
hunter-gatherers, 32
Hurricane Katrina, 308
Hyde Park, 283
hydraulic cement, 277
hydrocarbons, 190, 217
Hydrocarbon skeleton, 286
hydrochloric acid, 194
Hydrocodone, 248
hydro-electricity, 205, 264–65
Hydroelectric power plants, 266
hydrofluorocarbons, 217
hydrogen, 122, 202, 293
hydrophilic, 244
hydrophobic, 244
hydroxyl, 130
Hyperopia, 185
hypertext, 170
Hypertext Markup Language, 169
Hypertext Transfer Protocol, 169
hypoallergenic, 160

I

Ibach-Schwyz, 68
Iberian Peninsula, 251
IBM, 310
Ibuprofen, 247
ICAO, 221
ice, 55, 110
Iconoscope, 215
IG Farben, 196
ignition, 293
      spark, 191
IKEA, 74
Illinois, 261, 289, 301
images, digital, 209
imaging technique, 275
immigrants, 333, 339

immigration, 15, 333
IMP (Interface Message Processors), 169–70
Imperial ChemicalIndustries, 193
imperial units, 12
Improved Explosive Compound, 286
inauguration, 322
incandescence, 117, 297
incommensurability, 49
incompatibility, 49
incremental, 89
India, 61–62, 76–77, 88–92, 106–7, 116, 118–20, 259–60, 265–66, 272
      southern, 130, 272
India ink, 83
Indianapolis, 308
Indian Ocean, 73, 268
Indians, 86, 120, 125, 260, 268
indian subcontinent, 79
India Post, 86
India rubber, 196
Indo-Arabic numerals, 38
Indonesia, 60, 143, 197, 226, 259
indoor pollution, 114–15
Induction stoves, 92
industrial equipment, 63
industrial motors, 206
industrial production, 115
industrial revolution, 15–16, 40–41, 59, 61, 99, 101–4, 152
industrialrevolution, 301
industry, 32, 147, 218, 232, 335
Indus Valley Civilization, 76
infants, 212
infections, 129, 131, 212, 288
      bacterial, 129
      untreatable, 130
infinite number, 42
inflammation, 247
information content, 298
information technologies, 34, 218
infrared spectrum, 251
Ingvar Kamprad, 74

ink, 35, 83
    white, 83
innate grammar, 26
inoculation, 126
input tape, 164
inscriptions, 198
insecticide, arsenic-based, 289
installments, monthly, 124
institution, 10, 127, 143
instrument
    medical, 192
    musical, 28
    primitive, 82
integers, 49, 164
Intel, 167
intellectual property, 229, 332
intelligence, 23, 25
    artificial, 11, 148
IntelSat, 238
intercalary, 49
intercalated, 203
internal combustion engines, 101, 103–4, 108–11, 113–14, 203, 206, 270–71
International Bureau of Weights and Measures. See BIPM (Bureau of Weights and Measures)
internet, 140, 142, 168, 170, 241–42, 313–14
internet data, 158
Interplanetary, 236
inverted arch, 262
investment, 106, 181
Iowa, 303
IP (Internet Protocol), 170, 332
IPCC, 104
Iran, 67, 88, 182, 191, 265
Iranian military code, 306
Iraq, 37, 58, 80, 179, 182
Ireland, 153, 271, 290
IRNSS, 304
iron, 22, 26, 63–65, 122, 172
    cast, 64, 192
    crude, 64
    smelting, 64, 67

    solid, 64
iron(III) hexacyanoferrate(II), 223
iron age, 63, 65, 157
iron age knives, 67
iron alloys, 63–64
iron lungs, 128
iron metallurgy, 63, 65, 67, 157, 192
iron ore, 64
Iron Oxide, 158, 276
iron potts, 69
iron swords, 67
iron tires, 271
Iroquois theatre, 119
irrigation, 265–66
Ishango bone, 37
ISIS, 34, 80
Islam, 79
Islamic empire, 80
islamic prayer, 173
Islamic science, 38
Isle ofPortland, 277
ISMN, 320
isobutane, 217
isobutylphenyl propionic acid, 247
Isooctane, 191
isoprene, 192, 196
isoprene chains, 196
isoprene molecules, 196
Isopropylalcohol, 256
Israel, 43, 326
Israel-Lebanon war, 326
ISS, 238
Italian priest, 183
Italy, 37, 49, 66, 70, 183–85
Italy government officials, 334
Iwaka, 272

J

Jacquardloom, 10, 15, 150
Jacquard Loom, 12

Jail cigarettes, 52
Jainism, 79
jale, 56
Jamnagar Refinery, 191
Janesville, 82
Japan, 67, 70, 77, 80, 197, 265, 272
Japanese tourists, 57
Java, 226
Javel water, 134
Jawa Dam, 264
Jazz Singer, 230
Jean Baptiste Du Halde, 11, 40
Jeans, 328
Jena, 187
Jenner's vaccine, 127
Jeremia Chubb, 200
Jerusalem, 78
Jewett, Frank B., 328
Jewish immigrants, 334
Jews, 333–34
Jiahu, 257
Jiaozi, 53
Jikji, 43
Johann Abraham Wüsthof, 67
Johann Carolus, 153
Johann Diesbach, 223
Johann Döbereiner, 293
Johann Lauterjung, 68
John Boyd Dunlop, 271
John Day, 266
John Fleming, 298
John Pitcairne Jr, 224
Johnson, George, 133
Johnstown flood, 266
John Turton Randall, 92
Jōmon period, 70
Jona, 277
Jönköping, 292
Joseph-Auguste Pavin, 277
Joseph Marie Jacquard, 163

Joseph Monierat Paris, 277
Joseph Nicéphore Niépce, 208
Joule, 100
Judaism, 79
Judson, 328–29
Julian calendar, 49–50
Jura, 292

K

Kahn, Robert, 169
Kalambo, 25
Kalashnikov, Mikhail, 178
Kalgoorlie, 92
Kallanai, 265
Kamikaze, 80
Karecki, Marion, 322
Kariba Dam, 265
Karlsruhe, 270
Karolinska Institute, 292
Karpathos, 66
katana, 67
Kavak, 66
KaveriRiver, 265
Kay, John, 145
Kazakhstan, 31
KDKA, 136
Keck Observatory, 250
kelvin, 100
Kelvin & Hughes Scientific Instrument Company, 275
Kenya, 227
Kerala, 23, 257, 266, 290
kerosene, 90–91, 189–90, 296
kerosene lamps, 114, 118, 294–96
kerosene stove, 90–91
Khorsabad, 199
Khwarizmi, 37
kidney stones, 274
Kikuichi, 67
Kilby, Jack, 166
kinemacolor, 229

kinetoscope, 229
Kinetoscopic Record, 229
King, James, 301
kingdom of Lydia, 53
Kingdom of Travancore, 23
king Leopold II, 196
King's College, 139
King's College London, 187
Kintarō Hattori, 47
kitchen utensils, 158
Kleinrock, Leonard, 168, 334
klismos, 72–73, 88
klismos chair, 72
KLM (Koninklijke Luchtvaart Maatschappij), 220
klompen, 88
knife, 15, 66–68, 70–71, 74
    bronze, 66
Knoll, Max, 187
Kodachrome film, 208
Kodak, 208–9
Kodak cameras, 208
Kodak's film, 229
Koenig & Bauer, 44
Koninklijke Luchtvaart Maatschappij. See KLM
Korea, 43
Kozhikode, 61
Kraków, 27
kreteks, 260
Krosno, 190
Krugman, Paul, 282
Kuala Lumpur, 284

L

labor, human, 147
labor pain, 248
Labour party, 324
Labour Statistics, 125
LafargeHolcim, 277
La Gioconda, 222
La Joconde, 222

Lake, Summer, 87
Lake Edward, 37
Lake Erie, 224, 316
Lake Michigan, 272, 302
Lake Pontchartrain Causeway, 262
lakes, great, 37, 269
Lake Tanganyika, 25
Lakhta Center, 284
Lallement, Pierre, 271
Lamb, William, 283
lamps
    electric, 297
    electric arc, 119
    sesame-oil, 118
Lancet, 275
land masses, 58
Landolphia owariensis, 196
Landship committee, 325
language organ, 26
languages, dead, 232
Lanthanum-Nickel alloy, 202
lanthanum oxide, 296
laser, 13, 101, 233, 312–14
    near-infrared, 314
    red, 314
Laser Diode, 314
lasing material, 313
Latin, 43, 56
latitudes, 58–59, 135, 304
Laudanum, 246
Lausanne, 89
law enforcement, 179
laws, 10, 33, 100–101
    sumptuary, 23
LDPE (LowDensity PolyEthylene), 193
lead(II) chromate, 224
Lead(II) Sulfate, 202
Lead-based paints, 224
lead oxide, 202
lead pencils, 83

Leap Day, 49
Lebanon, 34
Leclanché, Georges, 202
Leclanché cell, 202
Leeds, 277
legumes, 121
Leh, 95
Lemoine, George, 292
Lenexa, 304
Leningrad Codex, 79
LensCrafters, 185
lenses, 185–86, 250
    corrective, 185
lepercolonies, 129
leprosy, 9, 129–30
letters, 85–86, 232
Levant, 31, 34
Lever Brothers, 245
Leverstarted, James, 245
lever-tumblerlock, 199
LEXINGTON, 273
libra, 56
libraryofAlexandria, 40
LibraryofCongress, 43, 58, 208, 229
libraryofPergamum, 40
Libya, 182
Lickobservatory, 250
LIDAR, 308
lidocaine, 160, 247
life, 131, 133, 211, 215, 302
    human, 257
Lifebuoy, 245
life expectancy, 131, 245, 335
Life Magazine, 134
life saver, 119
lifestyle, healthy, 90
light, electric, 15, 118–19, 280, 315
light bulb, 117, 119, 151, 158, 332–33
Light Emitting Diode, 314
lighter, 293

lighterfluid, 293
Li-ion, 202
Lilio, Luigi, 49
lime, 156, 223, 276
    slaked, 223
Lincoln, 133, 146
Lincoln, Abraham, 208, 289
Linde, 217
lingua franca, 62
Linus Yale Jr, 199
Linus Yale Sr, 199
lipid membranes, 130, 245
lipophilic, 244
Lippershay, 249
liquid oxygen, 235
literacy, 154
Lithium Cobalt Oxide, 203
Lithium compounds, 203
Lithium ion batteries, 203, 241
Lithium ions, 203
live attenuated virus, 127
Liverpool, 245
Live sports television, 215
live telecast, 239
localanesthetic, 246
locks, 124, 198–200, 268
    pin-tumbler, 199
    tumbler, 199
    unpickable, 200
lockstitch, 124
Locomotion, 105
Lodge, 70
Lodge, Joseph, 70
Löfgren, Nils, 160
Logi, 26
Lomb, Henry, 185
London, 77, 139, 205, 210–11, 213, 226, 260, 263, 316–17
longitude, 58–59, 135, 304
Long Term Evolution. See LTE
loom, 10, 144–46

452

lordofAdmiralty, 324
Lord ofAdmiralty, 325
Lorillard Tobacco, 260
LosAlamos NationalLaboratories, 311
Los Angeles, 107, 238, 313, 317
LosAngeles, 169, 308
los Muchachos, 250
Louisiana, 177, 289
Louis Le Prince, 229
Louvre Museum, 33, 38, 222
LowDensity PolyEthylene. See LDPE
LPG, 90–92
LPG stove, 91
LTE (Long Term Evolution), 241–42
Lucerne, 68
Luddites, 124, 338
Ludwig Boltzmann, 334
Ludwigshafen, 121
Lumière, Louis, 229
Lumière Brothers, 229
Lundström, Johan, 292
Lusaka, 265
Lusitania, 103
Luxor, 35, 46
Luxottica, 185
Lu Yu, 226
Lycurgus cup, 156
Lyon, 146, 163, 185, 208, 277

M

machine gun, 178
Macintosh, Charles, 134
Macrolides, 131
Madagascar, 60
Madras, 140
Magellan, 304
magnesium oxide, 296
magnetic attraction, 204
magnetic coating, 232
magnetic declination, 171

magnetic field, 173, 275, 299
magnetic strip cards, 10
Magnetic stripe card, 309–10
magnetism, 140, 205
magnetron, 299, 308
Mahatma Gandhi, 121, 148, 324
Maiman, Theodore, 313
Mainz, 43–44, 202
Majiayao Culture, 66
majolica, 70
Malaria, 289–90
Malaysia, 283–84
Malibu, 313
Mallet, Elizabeth, 153
Małopolskie, 27
Malthus, Thomas, 339
Mammography, 274
Manchester, 145–46, 149, 275, 277, 292
Manchuria, 132
manganese, 64, 202
Manganese Oxide, 157–58
Manhattan, 9, 116, 150, 283–84, 316
Mannheim, 111, 121
Mansfield, Peter, 275
mantle, 296
manufacturing, 314
manure crisis, great, 111
Mao, 325
maps, 40, 47, 57–59, 62, 113
Marathon, 190
Marble gravestone ofXanthippos, 73, 88
Marconi, 135
Marconi wireless telegraph company, 298
Marco Polo's book, 53
Margarine Unie, 245
Margery Brews, 85
Maricourt, 173
MarkAldrich, 91
mars, 56, 236, 308, 341
Mars Climate Orbiter, 56

Marseille, 277
Marsh Supermarket, 320
Maruishi, 160
Masaru Ibuka, 232
Maser, 313
Masjid-al-haram, 284
Massachusetts, 168
Massachusetts Institute of Technology, 168
mass murderer, 133
mass production, 166
MasterCard, 310
masterpiece, 261
matches, 291–93, 321
material
    raw, 194, 248
    war, 85
mathematics, 38, 47, 336
Mauna Kea, 250
Mauviel, 70
Maxim, Hiram, 178
Maxim Gun, 178
Maxwell, James, 208, 249, 334
Maxwell's equations, 47, 101, 208, 249
Mayan numerals, 38
Mayans, 138
Mayflower, 62, 236
Mayflowerpilgrims, 79
McAdam, John, 94
McAdam road, 94
McVicarat, John, 275
measles, 127–28, 194
Mecca, 48, 173, 284
mechanicalpress, 44
mechanical sensor, 173
Medina, 48
mediterranean, 40
mediterranian sea, 73, 184
memory, 159, 164, 311
    human, 34
Mendelian genetics, 47

Menlo Park, 116–17, 297
menstruating, 24
mental retardation, 131
Mercedes-Benz, 112
merchant receipts, 53
Merck, 127
Mercurius Politicus, 226
mercury, 99–100
mercury fumes, 208
Merkava tanks, 326
Mesoamerica, 34, 38
Mesopotamia, 31, 34, 157
metal conductors, 137
metallurgy, 26, 29, 63
Metaloxides, 158
metals, 64–65, 71, 83, 157, 297–98
    heavy, 194, 202
    precious, 53
    strategic, 312
meteorite iron, 63, 67
Meteosat-11, 239
methane, 91–92, 122
methanol, 256
Metropolitan Museum, 66, 73
Mexico, 195
Michelin, 197
Michelin Stars, 197
microbes, 134, 245
Micrographia, 187
microorganisms, 186
microphone, 233
microscope, 158, 185–88, 251
    compound, 186–87
microwave oscillations, 299
microwave ovens, 92, 299, 308
middle ages, 11, 13, 38, 47, 79–80
Middle East, 80
Middlesex, 321
Milan, 49, 115, 201
military arsenals, 181

military boot, 88
military conquests, 80
military explosives, 287
military resources, 326
milk, 159, 210–12
    pasteurized, 211
milky way galaxy, 239
Mills, Daniel, 88
Milwaukee, 272
Minamata, 290
minerals, 223, 235
    calcium, 276
    non-metallic, 64
mining, 285
    crypto coin, 218
mining asteroids, 235
Minoan civilization, 76
Miss Conway, 294
missiles, 273
    anti-tank, 326
Mississippi, 103, 289
Miss Juliet Corson, 91
Missouri, 261
Miss Sadie Conway, 294
MIT school of engineering, 194
MMR vaccine, 10, 127
mobile phones, 115, 137, 140, 167, 240–42, 249
Mobira Cityman, 240
Model T, 111
Modena, 183
Mona Lisa, 222
Monetaria moneta, 52
money, 34, 37, 50–53, 56, 322, 336
money cowry. See Monetaria moneta
Mongolia, 40, 118
Monobloc chair, 73
Mont-Blanc, 287
Moody's Investor Service, 267
moon, 47–49, 128, 137, 234–36, 249
Moore's law, 150, 167

morphine, 246–48
Morse, Samuel, 139
Morse code, 38, 115, 138, 313
Moscow, 22, 95, 311
MOSFET transistor, 166
mosques, 80, 158
Mosquitoes, 289
Mosul, 199
motherland, 78, 236
mothers, 44, 247–48
motion, rotary, 205
motion picture film stock, 228
Motorbikes, 10, 272
motorcycles, 272
Motorola, 240
motor vehicle, 29, 94, 113
Mount Everest, 187
Mount Hamilton, 250
movable types, 43–44, 331
movies, 13, 15, 115, 228, 230
mowing, 326
Mozes Lewenstein, 247
MRI, 10, 275
mRNA, 128
Muḥammad ibn Mūsā al-Khwārizmī, 37
Mullaperiyardam, 266
Mumbai. See Bombay
Munich, 89, 281
murder rate, 177
muscle power, 244
Museo Galileo, 58
Museum of Applied Arts & Sciences, 301
Museum of London, 88
Muslim, 155
Myanmar, 226, 326
Mycenaean, 66
Mycobacterium Leprae, 129

N

Nagasaki, 180, 197

Naked Economics, 35
NakedMoney, 51
naloxone, 247–48
nanoparticles, 156
nanoseconds, 166
napalm girl, 207
Naproxen, 247
Nara, 67
narcotics, 247–48
NASA, 238–39
Nasa earth observatory, 110
National Cash Register. See NCR
National Institute of Health, 257
NationalLibraryofFrance, 73
NationalLibraryofthe Netherlands, 42
National OceanicandAtmospheric Administration, 172
NationalPhysicalLaboratory, 47
Natufian culture, 31
natural disasters, 143
natural selection, 95
Nautilus, 62
navigational purposes, 172
navigation devices, 304
navigation services, 304
Nazareth, 78
Nazi concentration camps, 208–9
Nazi Germany, 326
Nazis, 325, 334
NBC, 215
NCR (National Cash Register), 320, 323
Neandarthals, 22
Nebamun, 223
Neckam, Alexander, 172
needle, 22, 173
negative effects, 289
Neiffies, Sarah, 127
neon lamp, 214
Neoprene, 197
nerve cells, 248
nerve damage, 131

nerve endings, 129
nervous systems, 129, 216, 226
Netherhampton, 67
Netherlands, 70, 74, 88, 95, 99, 185–86
Netherlands Fotomuseum, 208
network nodes, 170
networks
    cellular, 241
    digital, 241
    neural, 316
Neumann, John, 164
Newark, 94
Newcomen, Thomas, 102
Newcomen engine, 149
Newcomen's engine, 41, 103, 329
Newcomen's steam engine, 41, 331
New Deal, 300
Newhaven, 143
New Jersey, 228, 298
New Mexico Institute, 287
newspapers, 139, 143, 153–55, 209, 215
    dutch, 153
    indian, 120
Newton, Isaac, 250
Newton's laws, 47
New York, 210, 280, 284–85, 298, 302
New York City, 116–17, 261, 280, 284
New York StockExchange, 139
New York Times, 94, 213, 268–69, 281, 304–5, 328–29, 336
New York University, 328
New Zealand, 302
Ni-Cd, 202
nickel, 64
Nickel(III) Oxide-Hydroxide, 202
nickel coatings, 71
Nicotiana tabacum, 258
Nicotine, 259
Niger, 76
Nigeria, 15, 52, 295, 332
Nike, 89

Nikola Tesla, 116, 205
Nikolay Basov, 313
Ni-MH, 202
Nineveh, 199
Nipkow, Paul, 214
Nipkow disk, 214
Nipkowdisk, 214
Nippon Telegraph and Telephone, 241
Nitrocellulose, 192
nitrogen, 65, 121–22, 130, 286
nitrogen gas, 179
Nitroglycerin, 285–87
nitroglycerine, 286
NMT, 241
Noah'sArk, 80
Nobel, Alfred, 285–86
Nobel peace prize, 181
Nobel Prize, 289
Nokia, 240–41
Norimbergue, 46
Norman Joseph Woodland, 318
North Africa, 79
North America, 32, 58, 95, 259, 266
Northampton, 91
North Korea, 105, 182
north pole, 55, 171–72, 205
North RiverSteamboat, 62
North Stradbroke Island, 58
North Vietnamese, 207
Northwich, 193
Norway, 86, 129, 295
Norwegian Kroner, 86
Notre-Dame, 158
Nottingham, 247
Nottingham University, 275
Novartis, 248, 289
Novocain, 160
Novosibirsk, 22
NSAID, 247
NTT, 241

NTTDocomo, 241
Nuclear power, 104
Nuclear Weapon, 180–82
numbers
    binary, 146, 163
    personal-identification, 322
    random, 100
numerals, 38
Nuremberg, 15, 45–46, 83, 89, 274
nutritional value, 226
Nylons, 193
Nylon stockings, 193

O

objects
    metallic, 307
    moving, 307
Oceania, 60
Octane rating, 191
OECD average, 107
offices, 95, 200, 311
Ohio, 295, 323
Ohio railroad, 105
oil, 35, 223, 244, 278
    crude, 190
    linseed, 223
    palm, 244
oil boom, 190
oil paints, 223
Oklahoma, 296
Oklahoma City, 287
Ole Christensen Rømer, 249
Omura Shrine, 67
Oolong, 227
open defecation, 76
open heartsurgery, 159
operations
    control flow, 163
    logical, 165
Opiate, 248

opioid antagonist, 248
opioid medications, 247
opioidpainkillers, 246
opioid receptors, 248
opioids, 247-48
opium, 246
opium poppy, 246
Oppenheimer, Robert, 180
optical features, 156
optical fibercommunications, 314
optic nerve, 256
Opus Majus, 11, 183
oral polio virus, 127
orbit, 50, 56, 172, 251, 303
Oregon, 89
organelles, 188
organochlorines, 289
organophosphates, 289
Osaka, 67
Osmium, 122
Otis, 280-81
Otis, Elisha, 280
OurWorldin Data, 126
Oxford University, 36
oxidation, 117, 227
Oxide, 202
Oxycodone, 248
Oxygen, 10
Oyster Point, 57
ozone, 134

P

paan, 260
Pacific Ocean, 268
packet, 168-69
Paderborn, 246, 323
paduka, 88
painkiller, 15, 246, 248
paint, 222, 224
    acrylic, 223

painting, 183, 223
    egyptian, 223
    medieval, 223
Paisley, 87
Pakistan, 61, 76, 265
Palace ofparliament, 65
palaces, 39, 199
palimpsest, 79
Palma, 250
Palmolive, 245
Palmyra, 34
Panama, 95, 268
Panama canal, 268-69, 285, 287
Pantheon, 278
Panton, 74
Papaver somniferum, 246
paper, 38-41, 43, 79, 91, 275
paper cheques, 321
paper currency, 44
Papua New Guinea, 32
papyrus, 35, 40-41, 43, 79, 85
papyrus plant, 40
Parabellum, 178
Paracetamol, 247
paraffine-oil, 189
parchments, 40-41, 43, 85
Paris, 42-43, 54-55, 70-71, 73, 133-34, 222, 316-17
Parkesine, 192
ParkRowBuilding, 283
Parks, William, 311
Partial exemptions, 290
particles, 88, 187
Pasch, Gustav, 292
passengers, 106-7, 279, 308
passports, 62, 209
pasteurization, 15, 210-12, 329
pasteurization movement, 211
patent, 109, 111-12, 139, 141, 151, 189, 205
patent laws, 53
Patent Office, 249, 328

pattern
    circular, 319
    spiral, 214
peacekeepers, 133
pearl harbor attack, 62
pedals, 270
pediatrician, 127
Pelli, César, 284
pen, 10, 23, 81–83
pencils, 10, 82–83, 139
Penicillin, 130–31
Penicillium rubens, 130
Penn Museum, 64
Pennsylvania, 70, 295
Pennsylvania Museum of Archaeology, 64
Pennsylvania Quakers, 79
PennyBlack, 86
peptidoglycan, 130
Perraeaux, Louis, 272
Perry, Jack, 47
Persia, 157, 226
Persian mathematician, 37
Persson, Anders, 287
Perth, 92
pertussis, 127
Peshawar, 36
pesticides, 289–90
PET, 193
Petrie Museum, 22, 63
petrol, 190
petroleum, 189, 196, 216, 263, 295
petroleum fuels, 190
petroleum oil, 296
Petromax, 296
Petronas Towers, 283–84
Petrus Peregrinus, 173
Peyzae, 66
pharaohs, 28
pharmaceuticals, 311
phenol, 245

Philadelphia, 71, 89, 127, 318, 328
Philip Morris, 260
Phipps, James, 126
Phoenicians, 34
phones, smart, 41, 242
phonograph, 15, 151, 231–32, 332
phosphorus, 292
    red, 292
    white, 292
phosphorus atoms, 292
phosphorus sesquisulfide, 292
phossyjaw, 292
photodiode, 209
photographs, 209
photography, 185, 208–9, 230
photolithography, 166, 314
photons, 187, 313
photos, 9, 208–9
photosynthesis, 50
Phrygia, 73
physical brush, 206
physical carriers, 14, 141, 203
physical connections, 169
physics, 100
Phytophthora infestans, 290
pictures, 34, 207–8, 290
Pierce, Richard, 154
Pig, 31
pig iron, 63–64, 331
pigments, 223–24
pilgrims, 236
PIN code, 86
pin tumbler locks, 199–200
Pisa, 37
pistols, 178
piston, 102, 108, 178
Pit toilets, 77
Pittsburgh, 77, 136, 197
Pittsburgh plate glass company, 224
pixels, 209, 214

planets, 65, 110, 236, 249, 251
plant cell walls, 193
Planté, Gaston, 202
plants, 24, 50, 121, 123, 225
    industrial, 121
    opium, 247–48
plastic bags, 193
plastic cards, 321
plastics, 89, 157-58, 192–94, 196, 277
plastic syringes, single-use, 194
platinum, 293
platinum sponge, 293
plumbago, 83
Plunkettat, Roy, 71
Plutonium-239, 181
ply vessels, 263
pneumonia, 128, 274
point
    inflection, 301
    melting, 64, 156
    reference, 50
Poland, 94
police, 222, 271
polio, 127–28, 131, 194
polio vaccine, 128
Political freedom, 155
Polo, Marco, 11
polybutadiene, 197
polycarbonate, 194
polyester, 193, 232
PolyEthylene Terephthalate, 193
polyisoprene, 196
polymers, 193, 196
polypropylene, 73, 193
polytetrafluoroethylene. See PTFE
Poly Vinyl Chloride, 193
ponyexpress, 14, 137
Pope, Albert, 271
poppy, 246
population, 14, 133, 284, 295, 339–40

aging, 340
    human, 30, 35
Portland, 89, 323
Portland cement, 276–77
Portland CementAssociation, 276
Portugal, 86
positive feedback, 313
PostalService, 86
Postmaster General, 135
Potassium Hydroxide, 202
Potassium Nitrate, 179
pottery, 70
    ceramic, 70
    tin-glazed italian, 70
pottery fragments, 93
pound, 41, 56, 150, 337
Pound Sterling, 53
powder, 223
    bleaching, 134
    gun, 179
power, naval, 61
power accessories, 202
power loom, 338
PPG industries, 224
precipitation, 308
presbyopia, 185
president, 300, 318
press, 44, 208, 331
pressure, 110, 273
    atmospheric, 122
    high, 108, 179
Preston, 149
Priestley, Joseph, 196
Primus stove, 90
Princeton, 9
Principia Mathematica, 35
principles, scientific, 183
printers, 44
printing press, 13, 15, 38, 40–41, 43–44, 53, 331–32
prizes, 59, 199, 332

Procaine, 160
Procter & Gamble, 244
production, cereal, 63
productivity, 151–52, 282, 335–37, 341
profession, medical, 210
propane, 91, 217, 286
propofol, 159–61
prostaglandins, 247
proteins, 24, 121, 187, 226
protein sequence, 188
Proto-Canaanite, 34
protocol, 170
Prudhoe Bay, 95
Pseudomonas infections, 134
PTFE (polytetrafluoroethylene), 71
Public Affairs, 85
publictransit, 187
Puma, 68
pump, 159
punishments, 33
purification, 134
purine, 227
PVC, 193
pyrethrins/pyrethroids, 289
Pyridine, 259
Pyrrolidine, 259

Q
Qafzeh cave, 78
qahveh khaneh, 226
Qibla indicator, 173
quantum computing, 11
quantum mechanics, 100–101, 128, 187
Quartz watches, 47
Qur'an, 35, 79–80

R
R-134a/HFC-134a, 217
R-290, 217
R-600a, 217

rabies, 127
radar, 307–8
radar principle, 308
radiation, 55, 117
radio, 115, 135–37, 140, 155, 214–15, 242, 298–99, 307–8, 334
radioactive element, 181
radiocarbon, 22, 321
Radio Frequency ID, 310, 320
Radiolinja, 241
radio waves, 115, 136, 307–8, 310, 320
Radner, Karen, 199
railroads, 269
railroad tracks, 112
rails, 106
rail transit, 106
railways, 28, 103–7, 113–14, 270, 272
    indian, 107
Railway tracks, 65, 106
Rana Plaza, 123
randomness, 100
Ransome, 277
Ransome, Ernest, 277
rare-earth element, 293
Ray-ban, 185
Raytheon, 92, 299
RCA, 320
RCC (Reinforced Cement Concrete), 276, 278
RDX, 287
realnumbers, 46
Rebusprinciple, 34
receptors, adenosine, 227
Red Ochre, 223
Red Sea, 268
reference frame, 50
reference location, 59
reflecting telescopes, 250–51
reflectortelescope, 250
refracting telescopes, 250
refrigerant, 217
    non-flammable, 216

refrigeration, 101, 216, 218
refrigerators, 197, 206, 216–17
Regan, Bernard, 160
regions, 154
    backward, 289
    developing, 119
    tropical, 218
Reinforced Cement Concrete. See RCC
Reliance Industries, 191
religion, 10, 44, 53, 78–80, 257
religious ceremonies, 259
religious minorities, 79
religious persecution, 236
religious reconciliation, 168
religious texts, 80
renaissance, 44, 331
reproduction, human, 194
research, 12, 118, 156, 160, 242
reservoirs, 265–66
resistance, 137
    electrical, 233
Reuters, 154, 309
revenue, annual, 224
revolutions, 103, 297, 331
revolver, 178
Reynolds Pen Company, 81
RFID, 200, 309–11, 320
    passive, 311
RFID Systems, 311
RFID tags, 310–11
Rhine river, 121, 270
Rhodes, 66
rhumb lines, 59
rice, 31, 51, 290
Richard Gatling, 177
rifampicin, 129
Rigveda, 79
rivers, 104, 112, 225, 269, 285
RMS Carpathia, 136
RMS Titanic, 135

RNA, 227
roads, 58, 93–95, 106, 207, 272
    asphaltic, 95
    congested, 112
Roald Amundsen, 90
Roaring Springs Ranch, 87
Robertson, Robert, 283
robots, 11, 316
roca, 77
Rochester, 185, 208
Rockefeller, John, 295
rocket equation, 235
rockets, 65, 110, 234–36, 239, 305
    liquid-propellant, 235
Röhm, 223
Rolewinck, Werner, 42
Rolfe, John, 259
Roman aqueducts, 262
Romania, 65, 314
Roman numerals, 37–38
Romans, 56, 76, 94, 156
Roman soldiers, 88
Rome, 38, 70, 94, 222, 316–17
room temperature, 190, 244
Roversafetybicycle, 271
Royal Aviation Company, 220
RoyalEnfield, 272
Royal Mail, 86
RoyalMaternityHospital, 275
Royal TichelaarMakkum, 70
RPG-29, 326
rubber, 89, 192, 195–97, 316
    natural, 196–97
    polychloroprene, 197
rubbertrees, 196
rubbing alcohol, 256
rubycrystal, 313
ruler, 12, 54, 59, 61, 107
Rungpore, 225
rupees, 86

Ruska, Ernst, 187
Russia, 22, 104–7, 181–82, 191, 208, 238, 255
Russians, 215, 237

S

Sabin, Albert, 127
Sadi Carnot, 334
safety'bicycle, 270
safety match, 291–93
Sagebrush Foot Coverings, 88
Saint Catherine's Monastery, 43
Salk, Jonas, 127
salmonellosis, 134
saltpeter, 179
Salvator-Dormuspistol, 178
Samarkand, 40
SAMHSA, 256
Samuel, Herbert, 135
Samurai swords, 67
Sanaa, 79
sandals, 87
    bark, 87
Sandouping, 264
San Francisco Bay Area, 250
sanitation, 76, 134
Sapporo, 67
Sasson, Steven, 209
satellite communications, 137
satellites, 26, 235–39, 304–5, 314
Sati, 61
Saturn, 235
Saturn V, 234
Saudi Arabia, 79–80, 155, 191, 284, 341
Sauria, Charles, 292
Savannah, 62, 150
Savery, Thomas, 103
Saving Private Ryan, 13, 230
SBR (styrene butadiene rubber), 196
scalpels, 68
Scanning Electron Microscope, 187

scanning probe microscopes, 188
schistosomiasis, 134
schools, 10, 34, 128, 335–36, 340
science, 37–38, 44, 53–54, 183, 186
    social, 148
science history institute, 286
Science Museum, 66, 124, 229, 270, 301
scintillators, 251
Scottish trader, 225
scribes, 38, 43, 337
    human, 125
sculpture, 73
sea, 58, 61, 172, 268, 304
    mediterranean, 53
Sea-Me-We, 314
Sears, 283
seasonal flu, 194
seasons, 46, 49–50, 132
Seattle, 87, 227, 264, 323
Second Continental Congress, 86
security guards, 200
sedentary lifestyle, 32
Seiko, 47
Selden, Samuel, 71
selenium cell, 214
semaphore, 114
semi-automatic, 178
semiconductors, 166, 314
Semi-synthetic penicillins, 130
Senator Edward Kennedy, 168
Sendai, 272
Sergei Korolov, 235
Sergey Prokudin-Gorsky, 208
services, public, 86
Sevoflurane, 159–61
sewing machine, 122, 124–25, 149, 284
sextant, 135
shale frackingrevolution, 191
Shang dynasty, 66
shareholders, 336

Sharp, James, 91
sheeps, 31, 193
sheets, 40, 117
Sheffeld thwitel, 68
Sheffield, 68
Sherak, Tom, 230
Sherwin, Henry, 224
Sherwin-Williams, 224
shigellosis, 134
Shildon, 105
Shimantan, 266
shintai, 67
ships, 58–62, 65, 103, 105–8, 110, 135–36, 267–68
Shockley, William, 166
shoe company, 89
shoe maker, 68
shoemaker, 72–73, 88
shoes, 87–89, 197
    sports, 89
Shojiro Ishibashi, 197
Sibudu Cave, 22
siderealyear, 50
Sidney, 71
Siemens, 241
Siemens, Ernst, 233
sikkatu, 199
silica, 64, 156, 276
Silicon Valley, 9, 107
silicon wafer, 166
silk, 194
    natural, 193
silkroad, 40
Silliman'sAmerican JournalofScience, 204
silver, 33, 53, 156
silverhalide, 208
SIM, 241
Simpson, James, 159
Sinai Bible, 43
Singapore, 52, 150, 218, 284
Singer, Isaac, 334

skin, 160, 245, 275
skyscrapers, 63, 65, 76–77, 276, 278, 280–82, 284
SLAC, 169
ß-lactam antibiotics, 130–31
ß-lactamase, 131
slaves, 33, 62, 152, 335
SLS, 245
smallpox, 126, 128, 131
smallpox disease, 127
Smart locks, 200
smartphone features, 242
smartphones, 203, 233, 242, 304, 311
Smith, Adam, 32, 44, 52
Smith, Kirk, 90
Smiths Industries, 322
Smithsonian Institution, 216
Smithsonian Magazine, 200, 271
smoke, 90
smokers, 260, 302
SMW3, 314
Snellen, Herman, 185
Snellen Chart, 185
sniper rifle, 179
Snow, John, 133
Snuff, 260
soap, 16, 243–45
Social Responsibility, 335
Soda, 156
soda ash, 156
soda-lime glass, 156
Sodium Carbonate, 156
sodium hypochlorite, 134
Sodium laureth sulfate, 245
Sodium Stearate, 244
Sodium thiopental, 160
Sodium Thiosulphate, 208
solar cells, 101
soldiers, 325
solidfuels, 90
Solingen, 15, 67–68

Somerville, 71
Song dynasty, 53
Sony, 203, 215, 232
sounds, 34, 45, 141-42, 230-31, 233
sound waves, 233, 275
soup spoon, 172
South Africa, 22, 73, 323
South America, 55, 58, 95, 196, 199
South ForkDam, 266
South Kensington Museum, 45
South Korea, 118, 241, 284, 314
South Pittsburg, 70
south pole, 90, 205
South Sudan, 76
Southworth, James, 247
spacecraft, 56
space flights, 236
spacetime, 46
SpaceX, 238
Spain, 77, 79-80, 251
Spaniards, 259
spark plug, 109
sparks, 293
speaker, public, 233
specialrelativity, 137
speed oflight, 46, 55, 115, 137-38, 249
Spencer, Percy, 92, 299
Spherical Geometry, 304
spin, 149
spinal column, 273
spinal cord, 273
spindle, 149
spine, 273
spinning frame, 149
spinningjenny, 329
spinning Jenny, 13, 16, 125, 144, 147-49, 152
Spinning Mule, 125, 149
spinning wheel, 125, 147-49, 331
spiralgrooves, 232
sports equipment, 192

Sputnik, 237
Sputnikrocket, 235
square ofthe number, 164
Squawk, 307
Sri lanka, 130, 227
stabilizers, 179
Staedtler, 83
Standardbank, 323
Standard Oil, 190
Stanford University, 169
Starbucks, 227
starch, 24
Starley, John, 271
starlink, 238
stars, 50, 250
starvation, 32, 120
state department, 54, 181
statistics, 154, 338
Statue ofLiberty, 235
steam, 102, 104-5, 108-9, 271, 279
steamboat, 62, 103
steam condenser, 41
steam-engine, 280
steam engine condenser, 329
steam engines, 12-13, 41, 99, 101-4, 108, 285, 331-32
steam pump, 103
steam ships, 62
steam turbines, 103-4
stéar, 244
stearic acid, 244
steel, 22, 46, 63-65, 277, 283
    stainless, 64, 71
steel production, 65
steel rods, 276
stele, 33-34, 38, 88
Stephenson, George, 105
St.Louis, 261
Stockholm, 99, 292
stocks, 52
stocktickers, 139

Stockton andDarlington railway, 105
Stockton-on-Tees, 292
stomach cramping, 212
stone dam, 265
stores, 24, 139, 311
storms, 308
stove, 90–91, 299
    electric, 91
St. Petersburg, 22, 284
St. Petersburg-TampaAirboatLine, 220
Strand, 226
Strasbourg, 153
Straus, Nathan, 211
Strauss, Joseph, 263
Strauss, Levi, 328
stream turbines, 104
Strowgerswitch, 169
Sturgeon, William, 205
Stuttgart, 15, 111, 136, 153, 223
styrene, 196
styrene butadiene rubber. See SBR
Suez Canal, 61, 267–69, 277
sufentanil, 247
sulphide ofantimony, 292
Sulphur, 158
sulphuric acid, 293
Sumerians, 34, 49
sun, 23, 48–50, 59, 251
Sunnyvale, 16
Superbowl, 215
super-emitron, 215
supermarkets, 314
supernatural, 78
surfactant, 161, 244
Suspension bridges, 263
Swaminathan, 120
Sweden, 159–60, 202, 292, 295, 302
Swedish, 200
Swiss border, 286
Swiss companies, 67

Switzerland, 58, 68, 277–78, 289, 292
swords, 66–68
Sydney, 301
SydneyHarbourBridge, 262
symbiotic relationship, 155
synthetic detergents, 243–44
synthetic dyes, 83
synthetic rubber, 197
Syracuse, 184, 314
Syria, 78, 80, 199, 226
system
    base-10, 38
    base-20, 38
    base-60, 38
    human immune, 127, 194
    metric, 55
    solar, 239
    unary, 37
    universal healthcare, 325

T

Tadao Yoshida, 328
Taiwan, 60, 284
Taiwan Semiconductor Manufacturing Company, 167
Tallahassee, 289
tallow, 244
Talon Zipper, 328
Tamil Nadu, 130, 272
Tampere, 240
Tanakh, 80
Tandoori, 71
tanks, 324–26
Tarbela Dam, 265
Tarkhan dress, 22
TB, 127
Tchi, 40
TCP/IP, 169
tea, 225–27
    green, 227
    indian, 225

Technical University of Berlin, 187
Technicolor, 230
technology, 47, 169, 242, 318, 320
    digital, 313
technology companies, 335
Teflon, 71
telegraph, 12, 137–43, 154, 173, 203, 242, 331
telegraph devices, 139
telegraph wires, 137, 214
telegraphy, 143
Telemobiloscope, 307
telephone cables, 140, 168, 213, 313
telephone company, 143
telephone exchange, 143
telephone industry, 143
telephones, 140–43, 201, 203, 209, 241–42
    landline, 137
    long distance, 299
telephone traffic, 314
telescopes, 13, 158, 185, 236, 249–51
    optical, 187
television, 128, 131, 140, 155, 158, 213–15, 298–99
    mechanical, 214
Television Infrared Observation Satellite, 239
TeliaSonera, 241
Telstar, 238
temperature, 63–64, 99–100, 112
    thermodynamic, 100
Temple of Bel, 34
temples, 78, 80
Tennant, Charles, 134
terminals, 206
terrestrial meridian, 55
Terrible Valley, 35
Tesla, 201
tetraethyllead, 216
tetrafluoroethane, 217
Texas, 265, 289, 322–23
Texas Transport Institute, 106
text, 58, 139

textiles, 89, 104, 146, 152, 193
Thailand, 197
Thales, 314
Thames, 77, 263
theaflavins, 227
thearubigins, 227
theatre, ring, 119
theory of evolution, 10, 188, 249
thermionic emission, 297
thermionic valves, 298–99
thermodynamics, 100–101, 334
thermometer, 13, 38, 99, 101, 338
Thirukkural, 35, 340
Thomas Alva Edison. See Edison, Thomas
Thomas Bradford, 301
Thomas Twining, 226
Thomson, William, 100
Thor, 301
threads, 14, 16, 40, 144–46, 148
throne, 73
through-arch bridge, 262
thwitel, 68
Tiananmen Square, 325–26
Tibet, 226
Tide, 243–45
Tierra del Fuego, 95
Tilley, 296
Tim Berners Lee, 169
timekeeping, 65
Tinderbox, 293
tires, 192, 197
TIROS-1, 239
Titanic, 103, 135–36, 192
Titanium Dioxide, 224
titanium oxide, 83
Titusville, 190
tobacco, 255, 258–60
    chewable, 260
    fermented, 260
    smokeless, 260

uncured, 260
tobacco cultivation, 259
tobacco use, 256
toilet, 10, 75–77
    squat, 77
Tokyo, 272, 316, 328
Tomlinson, Raymond, 170
TomTom, 304
Tonga, 60
Toothpastes, 245
tornadoes, 239, 308
Toronto, 309, 314
Toronto Police Department, 309
Tower Bridge, 263
TowerBridge, 262–63
Townes, Charles, 313
Toyota, 201
toys, 29, 167, 192–93, 203
tracks, 146, 233, 310
trade guild, 184
traffic light, 315, 317
trains, 63, 107, 110, 137, 142
    electric, 205
transatlantic cable, 140
transformers, 116
transistors, 101, 165–67, 206, 209, 299
transmitter, 115, 142
transponder, 311
Transportation Statistics, 113
Trans-siberian railway, 105
treadles, 271
Treviso, 183
tri-axis magnetometer, 173
trinitroglycerin, 286
triode, 298
triphenylmethane, 83
tropical cyclone, 266
tropicalyear, 49–50
Tropical year, 50
trucks, 106, 263

refrigerated, 217–18
Truss bridges, 262
truth serum, 160
Tschunkur, Eduard, 196
tuberculosis, 127, 255
Tufts, Otis, 280
Tufts University, 275
tungsten, 64, 298
Tunneling Electron Microscopes, 187
Turing Complete, 164
Turkey, 31–32, 77, 310, 314, 324
Tutankhaman tomb, 67
Tutankhamun, 28, 73
two-stroke, 109
typhoid, 133–34
Typhoon Nina, 266

U

UAE, 218, 284
Uighurs, 226
Ukraine, 28, 334
Ultramarine, 223
ultrasound, 10, 275
UMTS, 241
uncertaintyprinciple, 100
Uniform Product Code Council, 320
Uniform Resource Identifier, 169
Unilever, 245
UnitedAirlines, 220
United Kingdom, 55
United Nations, 75
United States, 70–71, 141–43, 168–70, 181–82, 190–93, 255–59, 261–66, 310–11, 332–36
UniversalProduct Code, 319
Universal Turing Machine (UTM), 164
University College London, 22, 63
UniversityofBologna, 115
UniversityofCambridge, 163
UniversityofCopenhagen, 205
UniversityofKarlsruhe, 307
UniversityofManchester, 164

468

University of Oxford, 117, 130, 203
University of Pavia, 115, 201
University of Pennsylvania, 177
University of Pittsburgh, 127
UPC, 319–20
UPCC, 320
Uppsala university, 99
UPS, 86
Upton, Emory, 302
Upton Machine Company, 302
Uranium, 122
Uranium-235, 181
Uri Friedman, 72
US population, 126, 160, 338
US postal service, 86
USSR, 311, 313
Uzbekistan, 40

V

vaccines, 10, 13, 15, 126–28, 131, 161
    coronavirus, 128
vacuum tubes, 114, 165, 167, 297–99
Valley of the Kings, 46
valves, 293, 298–99
vapor-compression refrigeration, 217
vellum, 40–41, 44, 79
velocipede, 270
Venetian glasses, 157
Venetian island, 66
Versace, 123
Via Appia, 94
Vibrio Cholerae, 133
Victorinox, 68
video camera, 214
Vietnam war, 207, 209
Vincennes, 306–7
viruses, 131
visas, 310, 335
visible light spectrum, 157
Vision Council of America, 184

Vitamins, 226
Vitascope, 229
Vivaldi brothers, 172
Vogue, 185
volcanic eruptions, 239
Volta, Alessandro, 115
voltaic pile, 201
Volta laboratory, 231
volts, 201
Volvo, 201
Vostok, 235
vulcanization, 195–96

W

Wadi-al-Natuf, 31
Wadielhol, 35
Wagner, Bernard, 71
Wagner Manufacturing Company, 71
Wall street Journal, 221
Walmart, 123, 311
warp, 14, 16, 40, 144–46, 163
washing machines, 114, 167, 205–6, 301–2
Washington, 43, 54, 87
Washington DC, 169, 220, 229, 317
Washington Post, 83, 154–55
watches, 13, 45–46, 151
waterframe, 149
Waterman, Lewis, 82
Waterman Pen Company, 82
water mills, 28
water pumps, 77, 205
Watt, James, 41, 102, 149, 331
W-CDMA, 241
Wealth of Nations, 32
weaving, 40, 144–45
web browser, 170
web page, 163, 169
weft, 14, 16, 40, 144
welts, 88–89
Wenner-Gren, Axel, 302

Wernhervon Braun, 234
West Coast, 105, 118
Western Union, 140
Wetzel, Don, 322
whales, 275
wheat, 24, 269
Wheatstone, Charles, 139
Wheelan, Charles, 35, 51
wheels, 23, 26–29, 74, 79, 94–95, 148
    automobile, 28
    rotating, 28
Whirlpool, 302
Whitcomb Judson, 328
White House, 318
Whitewash, 223
Whitney, Eli, 150–51, 333
Whittingham, Stanley, 203
WHO (World Health Organization), 127–29, 133–34, 255, 257, 260
whooping cough, 127–28
Wikipedia, 12, 170
Wilhelm Röntgen, 274
William, Louis, 70
William Boeing, 220
William Colgate, 245
Williams, Edward, 224
wines, 212
wireless telegraph, 136, 140, 214, 249, 299
WizardofOz, 230
women's suffrage movement, 270
Wonderwerkcave, 25
Wooden footwear, 88
Wooden furniture, 73
wooden spokes, 271
wool, 194
    natural, 193
workers, 148, 150, 268, 335–36, 338–39
world
    developed, 92, 272, 311
    developing, 91–92, 140, 143, 311
World Health Organization. See WHO

World War I, 296
world war II, 62, 79, 130–31, 196, 324–26
world wars, 131, 177, 179, 272, 296
world wide web, 169
Wrangham, 24–25
Wrangham, Richard, 25
Wright brothers, 334
Wright Flyer, 220
wringers, 301
wroughtiron, 331
Wuhan, 226, 278
Wuhu, 278
Würzburg, 44
Wüsthof, 67
Wyeth, Nathaniel, 193

X

Xi'an, 226
Xianrendong Cave, 70
Xinjiang, 154–55
x-ray crystallography, 187–88
x-ray imaging, 79
x-rays, 10, 188, 251, 273–75

Y

Yafa, Stephen, 151
Yahoo, 169–70
Yale & Towne, 200
Yale University, 271
Yamaha, 272
Yamnaya, 28
Yangtze, 269
Yangtze River, 262, 264, 278
yarn, 144, 148–49
YBC, 37
Yerkes Observatory, 250
Yersinia enteritis, 134
YKK, 328
yttrium oxide, 296
Yuri Gagarin, 235

Z

Zacharias Janssen, 186

ZambeziRiver, 265

Zambia, 25, 265

zebra crossing, 316

Zimbabwe, 265

zipper, 327–29

Zippos, 293

Zurich, 184

www.ingramcontent.com/pod-product-compliance
Lightning Source LLC
Chambersburg PA
CBHW060821220526
45466CB00003B/927